原 版 影 印 说 明

1.《聚合物百科词典》（5册）是 Springer Reference *Encyclopedic Dictionary of Polymers*（2nd Edition）的影印版。为使用方便，由原版2卷改为5册：

第1册 收录 A–C 开头的词组；

第2册 收录 D–I 开头的词组；

第3册 收录 J–Q 开头的词组；

第4册 收录 R–Z 开头的词组；

第5册 为原书的附录部分及参考文献。

2. 缩写及符号、数学符号、字母对照表、元素符号等查阅说明各册均完整给出。

由 Jan W. Gooch 主编的《聚合物百科词典》是关于高分子科学与工程领域的参考书，2007年出版第一版，2011年再版。本书收录了7 500多个高分子材料方面的术语，涉及高分子材料的各个方面，如粘合剂、涂料、油墨、弹性体、塑料、纤维等，还包括生物化学和微生物学方面的术语，以及与新材料、新工艺相关的术语；并且不仅包括其物理、电子和磁学性能方面的术语，还增加了数据处理的统计和数值分析以及实验设计方面的术语。每个词条方便查找，并给出了简洁的定义，以及相互参照的相关术语。为了说明得更清晰，全书给出1 160个图、73个表。有的词条还给出方程式、化学结构等。

材料科学与工程图书工作室

联系电话 0451-86412421

　　　　　0451-86414559

邮　　箱 yh_bj@aliyun.com

　　　　　xuyaying81823@gmail.com

　　　　　zhxh6414559@aliyun.com

Springer 词典精选原版系列

聚合物百科词典

Jan W. Gooch

Encyclopedic Dictionary of Polymers

2nd Edition

VOLUME 1

A – C

哈尔滨工业大学出版社
HARBIN INSTITUTE OF TECHNOLOGY PRESS

黑版贸审字08-2014-010号

Reprint from English language edition:
Encyclopedic Dictionary of Polymers
by Jan W.Gooch
Copyright © 2011 Springer New York
Springer New York is a part of Springer Science+Business Media
All Rights Reserved

This reprint has been authorized by Springer Science & Business Media for distribution in China Mainland only and not for export therefrom.

图书在版编目（CIP）数据

聚合物百科词典.1,A~C:英文/（美）古驰（Gooch,J.W.）主编.—哈尔滨:哈尔滨工业大学出版社，2014.3

（Springer词典精选原版系列）

ISBN 978-7-5603-4442-3

Ⅰ.①聚… Ⅱ.①古… Ⅲ.①聚合物–词典–英文 Ⅳ.①O63-61

中国版本图书馆CIP数据核字（2013）第292196号

材料科学与工程
图书工作室

责任编辑 杨 桦 许雅莹 张秀华
出版发行 哈尔滨工业大学出版社
社　　址 哈尔滨市南岗区复华四道街10号 邮编150006
传　　真 0451-86414749
网　　址 http://hitpress.hit.edu.cn
印　　刷 哈尔滨市石桥印务有限公司
开　　本 787mm×1092mm 1/16 印张 13.75
版　　次 2014年3月第1版 2014年3月第1次印刷
书　　号 ISBN 978-7-5603-4442-3
定　　价 108.00元

（如因印刷质量问题影响阅读，我社负责调换）

Acknowledgements

The editor wishes to express his gratitude to all individuals who made available their time and resources for the preparation of this book: James W. Larsen (Georgia Institute of Technology), for his innovations, scientific knowledge and computer programming expertise that were invaluable for the preparation of the Interactive Polymer Technology Programs that accompany this book; Judith Wiesman (graphics artist), for the many graphical presentations that assist the reader for interpreting the many complex entries in this publication; Kenneth Howell (Springer, New York), for his continued support for polymer science and engineering publications; and Daniel Quinones and Lydia Mueller (Springer, Heidelberg) for supporting the printed book and making available the electronic version and accompanying electronic interactive programs that are important to the scientific and engineering readers.

Preface

The second edition of Encyclopedic Dictionary of Polymers provides 40% more entries and information for the reader. A Polymers Properties section has been added to provide quick reference for thermal properties, crystallinity, density, solubility parameters, infrared and nuclear magnetic spectra. Interactive Polymer Technology is available in the electronic version, and provides templates for the user to insert values and instantly calculate unknowns for equations and hundreds of other polymer science and engineering relationships. The editor offers scientists, engineers, academia and others interested in adhesives, coatings, elastomers, inks, plastics and textiles a valuable communication tool within this book. In addition, the more recent innovations and biocompatible polymers and adhesives products have necessitated inclusion into any lexicon that addresses polymeric materials. Communication among scientific and engineering personnel has always been of critical importance, and as in any technical field, the terms and descriptions of materials and processes lag the availability of a manual or handbook that would benefit individuals working and studying in scientific and engineering disciplines. There is often a challenge when conveying an idea from one individual to another due to its complexity, and sometimes even the pronunciation of a word is different not only in different countries, but in industries. Colloquialisms and trivial terms that find their way into technical language for materials and products tend to create a communications fog, thus unacceptable in today's global markets and technical communities.

The editor wishes to make a distinction between this book and traditional dictionaries, which provide a word and definition. The present book provides for each term a complete expression, chemical structures and mathematic expression where applicable, phonetic pronunciation, etymology, translations into German, French and Spanish, and related figures if appropriate. This is a complete book of terminology never before attempted or published.

The information for each chemical entry is given as it is relevant to polymeric materials. Individual chemical species (e.g., ethanol) were taken from he *CRC Handbook of Chemistry and Physics*, 2004 Version, the Merck Index and other reference materials. The reader may refer to these references for additional physical properties and written chemical formulae. Extensive use was made of ChemDraw®, CambridgeSoft Corporation, for naming and drawing chemical structures (conversion of structure to name and vice versa) which are included with each chemical entry where possible. Special attention was given to the IUPAC name that is often given with the common name for the convenience of the reader.

The editor assembled notes over a combined career in the chemical industries and academic institutions regarding technical communication among numerous colleagues and helpful acquaintances concerning expressions and associated anomalies. Presently, multiple methods of nomenclature are employed to describe identical chemical compounds by common and IUPAC names (eg. acetone and 2-propanone) because the old systems (19th century European and trivial) methods of nomenclature exists with the modern International Union of Pure and Applied Chemistry, and the conflicts between them are not likely to relent in the near future including the weights and measures systems because some nations are reluctant to convert from English to metric and, and more recently, the International Systems of Units (SI). Conversion tables for converting other systems to the SI units are included in this book for this purpose. In addition, there are always the differences in verbal pronunciation, but the reasons not acceptable to prevent cogent communication between people sharing common interests.

In consideration of the many challenges confronting the reader who must economize time investment, the structure of this book is optimized with regard the convenience of the reader as follows:

- Comprehensive table of contents
- Abbreviations and symbols
- Mathematics signs
- English, Greek, Latin and Russian alphabets
- Pronunciation/phonetic symbols
- Main body of terms with entry term in English, French German and Italian
- Conversion factors

- Microbiology nomenclature and terminology
- References

The editor acknowledges the utilization of many international sources of information including journals, books, dictionaries, communications, and conversations with people experienced in materials, polymer science and engineering. A comprehensive reference section contains all of the sources of information used in this publication. Pronunciation, etymological, cross-reference and related information is presented in the style of the 11th Edition of the Meriam-Webster Dictionary, where known, for each term. The spelling for each term is presented in German, French, and Spanish where translation is possible. Each term in this book includes the following useful information:

- Spelling (in **bold** face) of each term and alternative spellings where more than one derivation is commonly used
- Phonetic spelling \-\ using internationally published phonetic symbols, and this is the first book that includes phonetic pronunciation information missing in technical dictionaries that allows the reader to pronounce the term
- Parts of speech in English following each phonetic spelling, eg. *n.*, *adj.*
- Cross-references in CAPITALS letters
- Also called *example* in italics
- Etymological information [-] for old and new terms that provides the reader the national origins of terms including root words, prefixes and suffixes; historical information is critical to the appreciation of a term and its true meaning
- French, German, Italian and Spanish spellings of the term { - }
- A comprehensive explanation of the term
- Mathematical expressions where applicable
- Figures and tables where applicable
- A comprehensive reference section is included for further research

References are included for individual entries where a publication(s) is directly attributable to a definition or description. Not all of the references listed in the Reference section are directly attributable to entries, but they were reviewed for information and listed for the reader's information. Published dictionaries and glossaries of materials were very helpful for collecting information in the many diverse and smaller technologies of the huge field of polymers. The editor is grateful that so much work has been done by other people interested in polymers.

The editor has attempted to utilize all relevant methods to convey the meaning of terms to the reader, because a term often requires more information than a standard entry in a textbook dictionary, so this book is dedicated to a complete expression. Terminology and correct pronunciation of technical terms is continuously evolving in scientific and industrial fields and too often undocumented or published, and therefore, not shared with others sometimes leading to misunderstandings. Engineering and scientific terms describe a material, procedure, test, theory or process, and communication between technical people must involve similar jargon or much will be lost in the translation as often has been the editor's experience. The editor has made an attempt to provide the reader who has an interested in the industries that have evolved from adhesives, coatings, inks, elastomers, plastics and textiles with the proper terminology to communicate with other parties whether or not directly involved in the industries. This publication is a single volume in the form of a desk-handbook that is hoped will be an invaluable tool for communicating in the spoken and written media.

Physics, electronic and magnetic terms because they are related to materials and processes (e.g., *ampere*).

Biomolecular materials and processes have in the recent decade overlapped with polymer science and engineering. Advancements in polymeric materials research for biomolecular and medical applications are rapidly becoming commercialized, examples include biocompatible adhesives for sutureless tissue bonding, liquid dressings for wounds and many other materials used for *in vitro* and *in vivo* medical applications. To keep pace with these advancements, the editor has included useful terms in the main body that are commonly used in the material sciences for these new industries.

A microbiology section has been included to assist the reader in becoming familiar with the proper nomenclature of bacteria, fungi, mildew, and yeasts – organisms that affect materials and processes because they are ubiquitous in our environment. Corrosion of materials by microorganisms is commonplace, and identification of a specific organism is critical to prevent its occurrence. Engineers and materials scientists will appreciate the extensive sections on different types of microorganisms together with a section dedicated to microbiology terminology that is useful for communicating in the jargon of biologists instead of referring to all organisms as "bugs."

New materials and processes, and therefore new terms, are constantly evolving with research, development and global commercialization. The editor will periodically update this publication for the convenience of the reader.

Statistics, numerical analysis other data processing and experimental design terms are addressed as individual terms and as a separate section in the appendix, but only as probability and statistics relate to polymer technology and not the broad field of this mathematical science. The interactive equations are listed in the Statistics section of the Interactive Polymer Technology program.

Interactive Polymer Technology Programs

Along with this book we are happy to provide a collection of unique and useful tools and interactive programs along with this Springer Reference. You will find short descriptions of the different functions below. Please download the software at the following website: http://extras.springer.com/2011/978-1-4419-6247-8

Please note that the file is more than 200 MB. Download the ZIP file and unzip it. It is strongly recommended to read the **ReadMe.txt** before installing. The software is started by opening the file InPolyTech.pdf and following the instructions. Detailed instructions can be found under 'Help Instructions'.

The software consists of 15 programs and tools that are briefly described in the appendix.

Abbreviations and Symbols

Abbreviations	Symbols
An	absorption (formerly extinction) (= $\log t_i^{-1}$)
A	Area
A	surface
A	Helmholtz energy ($A = U - TS$)
A	preexponential constant [in $k = A \exp(-E^{\ddagger}/RT)$]
A_2	second virial coefficient
a	exponent in the property/molecular weight relationship ($E^{\ddagger} = KM^a$); always with an index, e.g., a_{η}, a_s, etc.
a	linear absorption coefficient, $a = l^{-1}$
absolute	abs
acre	spell out
acre-foot	acre-ft
air horsepower	air hp
alternating-current (as adjective)	a-c
A^m	molar Helmholtz energy
American Society for Testing and Materials	ASTM
amount of a substance (mole)	n
ampere	A or amp
ampere-hour	amp-hr
amplitude, an elliptic function	am.
angle	β
angle, especially angle of rotation in optical activity	α
Angstrom unit	Å
antilogarithm	antilog
a_o	constant in the Moffit–Yang equation
Area	A
Atactic	at
atomic weight	at. wt
Association	Assn.
atmosphere	atm

Abbreviations	Symbols
average	avg
Avogadro number	N_L
avoirdupois	avdp
azimuth	az or α
barometer	bar.
barrel	bbl
Baumé	Bé
b_o	constant in the Mofit–Yang equation
board fee (feet board measure)	fbm
boiler pressure	spell out
boiling point	bp
Boltzmann constant	k
brake horsepower	bhp
brake horsepower-hour	bhp-hr
Brinell hardness number	Bhn
British Standards Institute	BSI
British thermal unit[1]	Btu or B
bushel	bu
C	heat capacity
c	specific heat capacity (formerly; specific heat); c_p = specific isobaric heat capacity, c_v = specific isochore heat capacity
c	"weight" concentration (= weight of solute divided by volume of solvent); IUPAC suggests the symbol ρ for this quantity, which could lead to confusion with the same IUPAC symbol for density
c	speed of light in a vacuum
c	speed of sound
calorie	cal
candle	c
candle-hour	c-hr
candlepower	cp
ceiling temperature of polymerization, °C	T_c

Abbreviations	Symbols
cent	c or ¢
center to center	c to c
centigram	cg
centiliter	cl
centimeter or centimeter	cm
centimeter-gram-second (system)	cgs
centipoise	cP
centistokes	cSt
characteristic temperature	Θ
chemical	chem.
chemical potential	μ
chemical shift	δ
chemically pure	cp
circa, about, approximate	ca.
circular	cir
circular mils	cir mils
cis-tactic	ct
C^m	molar heat capacity
coefficient	coef
cologarithm	colog
compare	cf.
concentrate	conc
conductivity	cond, λ
constant	const
continental housepower	cont hp
cord	cd
cosecant	csc
cosine	cos
cosine of the amplitude, an elliptic function	cn
cost, insurance, and freight	cif
cotangent	cot
coulomb	spell out
counter electromotive force	cemf
C_{tr}	transfer constant ($C_{tr} = k_{tr}/k_p$)
cubic	cu
cubic centimeter (liquid, meaning milliliter. ml)	cu, cm, cm^3
cubic centimeter	cm^3 cubic expansion coefficient α
cubic foot	cu ft
cubic feet per minute	cfm
cubic feet per second	cfs
cubic inch	cu in.
cubic meter	cu m or m^3
cubic micron	cu μ or cu mu or μ^3
cubic millimeter	cu mm or mm^3
cubic yard	cu yd
current density	spell out
cycles per second	spell out or c
cylinder	cyl
D	diffusion coefficient
D_{rot}	rotational diffusion coefficient
day	spell out
decibel	db
decigram	d.g.
decomposition, °C	T_{dc}
degree	deg or °
degree Celsius	°C
degree centigrade	C
degree Fahrenheit	F or °
degree Kelvin	K or none
degree of crystallinity	α
degree of polymerization	X
degree Réaumur	R
delta amplitude, an elliptic function	dn
depolymerization temperature	T_{dp}
density	ρ
diameter	diam
Dictionary of Architecture and Construction	DAC
diffusion coefficient	D
dipole moment	p
direct-current (as adjective)	d-c
dollar	$
dozen	doz
dram	dr
dynamic viscosity	η
E	energy (E_k = kinetic energy, E_p = potential energy, $E^\ddagger$ = energy of activation)
E	electronegativity
E	modulus of elasticity, Young's modulus ($E = \sigma_{ii}/\varepsilon_{ii}$)
E	general property

Abbreviations	Symbols
E	electrical field strength
e	elementary charge
e	parameter in the Q-e copolymerize-tion theory
e	cohesive energy density (always with an index)
edition	Ed.
Editor, edited	ed.
efficiency	eff
electric	elec
electric polarizability of a molecule	α
electrical current strength	I
electrical potential	V
electrical resistance	R or X
electromotive force	emf
electronegativity	E
elevation	el
energy	E
enthalpy	H
entropy	S
equation	eq
equivalent weight	equiv wt
et alii (and others)	et al.
et cetera	etc.
excluded volume	u
excluded volume cluster integral	β
exempli gratia (for example)	e.g.
expansion coefficient	α
external	ext
F	force
f	fraction (excluding molar fraction, mass fraction, volume fraction)
f	molecular coefficient of friction (e.g., f_s, f_D, f_{rot})
f	functionality
farad	spell out or f
Federal	Fed.
feet board measure (board feet)	fbm
feet per minute	fpm
feet per second	fps
flash point	flp

Abbreviations	Symbols
fluid	fl
foot	ft
foot-candle	ft-c
foot-Lambert	ft-L
foot-pound	ft-lb
foot-pound-second (system)	fps
foot-second (see cubic feet per second)	
fraction	$\int$
franc	fr
free aboard ship	spell out
free alongside ship	spell out
free on board	fob
freezing point	fp
frequency	spell out
fusion point	fnp
G	Gibbs energy (formerly free energy or free enthalpy) ($G = H - TS$)
G	shear modulus ($G = \sigma_{ij}$/angle of shear)
G	statistical weight fraction ($G_i = g_i/\Sigma_i\, g_i$)
g	gravitational acceleration
g	statistical weight
g	*gauche* conformation
g	parameter for the dimensions of branched macromolecules
G^m	molar Gibbs energy
gallon	gal
gallons per minute	gpm
gallons per second	gps
gauche conformation	g
Gibbs energy	G
grain	spell out
gram	g
gram-calorie	g-cal
greatest common divisor	gcd
H	enthalpy
H^m	molar enthalpy
h	height
h	Plank constant
haversine	hav

Abbreviations and Symbols

Abbreviations	Symbols
heat	Q
heat capacity	C
hectare	ha
henry	H
high pressure (adjective)	h-p
hogshead	hhd
horsepower	hp
horsepower-hour	hp-hr
hour	h or hr
hundred	C
hundredweight (112 lb)	cwt
hydrogen ion concentration, negative logarithm of	pH
hyperbolic cosine	cosh
hyperbolic sine	sinh
hyperbolic tangent	tanh
I	electrical current strength
I	radiation intensity of a system
i	radiation intensity of a molecule
ibidem (in the same place)	ibid.
id est (that is)	i.e.
inch	in.
inch-pound	in-lb
inches per second	ips
indicated horsepower	ihp
indicated horsepower-hour	ihp-hr
infrared	IR
inside diameter	ID
intermediate-pressure (adjective)	i-p
internal	int
International Union of Pure and Applied Chemistry	IUPAC
isotactic	it
J	flow (of mass, volume, energy, etc.), always with a corresponding index
joule	J
K	general constant
K	equilibrium constant
K	compression modulus ($p = -K \Delta V/V_o$)
k	Boltzmann constant

Abbreviations	Symbols
k	rate constant for chemical reactions (always with an index)
Kelvin	K (Not °K)
kilocalorie	kcal
kilocycles per second	kc
kilogram	kg
kilogram-calorie	kg-al
kilogram-meter	kg-m
kilograms per cubic meter	kg per cu m or kg/m^3
kilograms per second	kgps
kiloliter	Kl
kilometer or kilometer	km
kilometers per second	kmps
kilovolt	kv
kilovolt-ampere	kva
kilowatt	kw
kilowatthour	kwhr
Knoop hardness number	KHN
L	chain end-to-end distance
L	phenomenological coefficient
l	length
lambert	L
latitude	lat or ϕ
least common multiple	lcm
length	l
line*a*r expansion coefficient	Y
linear foot	lin ft
liquid	liq
lira	spell out
liter	l
logarithm (common)	log
logarithm (natural)	log. or ln
kibgutyde	kibg. or λ
loss angle	δ
low-pressure (as adjective)	l-p
lumen	1*
lumen-hour	1-hr*
luments per watt	lpw
M	"molecular weight" (IUPAC molar mass)
m	mass
mass	spell out or m
mass fraction	w

Abbreviations	Symbols
mathematics (ical)	math
maximum	max
mean effective pressure	mep
mean horizontal candlepower	mhcp
meacycle	mHz
megohm	MΩ
melting point, -temperature	mp, T_m
meter	m
meter-kilogram	m-kg
metre	m
mho	spell out
microsmpere	µa or mu a
microfarad	µf
microinch	µin.
micrometer (formerly micron)	µm
micromicrofarad	µµf
micromicron	µµ
micron	µ
microvolt	µv
microwatt	µw or mu w
mile	spell out
miles per hour	mph
miles per hour per second	mphps
milli	m
milliampere	ma
milliequivalent	meq
milligram	mg
millihenry	mh
millilambert	mL
milliliter or milliliter	ml
millimeter	mm
millimeter or mercury (pressure)	mm Hg
millimicron	mµ or m mu
million	spell out
million gallons per day	mgd
millivolt	mv
minimum	min
minute	min
minute (angular measure)	′

Abbreviations	Symbols
minute (time) (in astronomical tables)	m
mile	spell out
modal	m
modulus of elasticity	E
molar	M
molar enthalpy	H_m
molar Gibbs Energy	G_m
molar heat capacity	C_m
mole	mol
mole fraction	x
molecular weight	mol wt or M
month	spell out
N	number of elementary particles (e.g., molecules, groups, atoms, electrons)
N_L	Avogadro number (Loschmidt's number)
n	amount of a substance (mole)
n	refractive index
nanometer (formerly millimicron)	nm
National Association of Corrosion Engineers	NACE
National Electrical Code	NEC
newton	N
normal	N
number of elementary particles	N
Occupational Safety and Health Administration	OSHA
ohm	Ω
ohm-centimeter	ohm-cm
oil absorption	O.A.
ounce	oz
once-foot	oz-ft
ounce-inch	oz-in.
outside diameter	OD
osomotic pressure	Π
P	permeability of membranes
p	probability
p	dipole moment
$\mathbf{p}_i$	induced dipolar moment
p	pressure

Abbreviations and Symbols

Abbreviations	Symbols
p	extent of reaction
Paint Testing Manual	PTM
parameter	Q
partition function (system)	Q
parts per billion	ppb
parts per million	ppm
pascal	Pa
peck	pk
penny (pency – new British)	p.
pennyweight	dwt
per	diagonal line in expressions with unit symbols or (see Fundamental Rules)
percent	%
permeability of membranes	P
peso	spell out
pint	pt.
Planck's constant (in $E = h\nu$) (6.62517 +/− 0.00023 × 10^{-27} erg sec)	h
polymolecularity index	Q
potential	spell out
potential difference	spell out
pound	lb
pound-foot	lb-ft
pound-inch	lb-in.
pound sterling	£
pounds-force per square inch	psi
pounds per brake horsepower-hour	lb per bhp-hr
pounds per cubi foot	lb per cut ft
pounds per square foot	psf
pounds per square inch	psi
pounds per square inch absolute	psia
power factor	spell out or pf
pressure	p
probability	p
Q	quantity of electricity, charge
Q	heat
Q	partition function (system)
Q	parameter in the Q–e copolymerize-tion equation

Abbreviations	Symbols
Q, Q	polydispersity, polymolecularity in-dex ($Q = \overline{M_w}/\overline{M_n}$)
q	partition function (particles)
quantity of electricity, charge	Q
quart	qt
quod vide (which see)	q.v.
R	molar gas constant
R	electrical resistance
R_G	radius of gyration
R_n	run number
R_ϑ	Rayleigh ratio
r	radius
r_o	initial molar ratio of reactive groups in polycondensations
radian	spell out
radius	r
radius of gyration	R_G
rate constant	k
Rayleigh ratio	R_ϑ
reactive kilovolt-ampere	kvar
reactive volt-ampere	var
reference(s)	ref
refractive index	n
relaxation time	τ
resistivity	ρ
revolutions per minute	rpm
revolutions per second	rps
rod	spell out
root mean square	rms
S	entropy
S^m	molar entropy
S	solubility coefficient
s	sedimentation coefficient
s	selectivity coefficient in osmotic measurements)
Saybolt Universal seconds	SUS
secant	sec
second	s or sec
second (angular measure)	″
second-foot (see cubic feet per second)	

Abbreviations and Symbols

Abbreviations	Symbols
second (time) (in astronomical tables)	s
Second virial coefficient	A_2
shaft horsepower	shp
shilling	s
sine	sin
sine of the amplitude, an elliptic function	sn
society	Soc.
Soluble	sol
solubility coefficient	S
solubility parameter	δ
solution	soln
specific gravity	sp gr
specific heat	sp ht
specific heat capacity (formerly: specific heat)	c
specific optical rotation	$[\alpha]$
specific volume	sp vol
spherical candle power	scp
square	sq
square centimeter	sq cm or cm^2
square foot	sq ft
square inch	sq in.
square kilometer	sq km or km^2
square meter	sq m or m^2
square micron	sq μ or μ^2
square root of mean square	rms
standard	std
Standard	Stnd.
Standard deviation	σ
Staudinger index	$[\eta]$
stere	s
syndiotactic	st
T	temperature
t	time
t	*trans* conformation
tangent	tan
temperature	T or temp
tensile strength	ts
threodiisotactic	tit
thousand	M
thousand foot-pounds	kip-ft
thousand pound	kip

Abbreviations	Symbols
ton	spell out
ton-mile	spell out
trans conformation	t
trans-tactic	tt
U	voltage
U	internal energy
U^m	molar internal energy
u	excluded volume
ultraviolet	UV
United States	U.S.
V	volume
V	electrical potential
v	rate, rate of reaction
v	specific volume always with an in-dex
vapor pressure	vp
versed sine	vers
versus	vs
volt	v or V
volt-ampere	va
volt-coulomb	spell out
voltage	U
volume	V or vol.
Volume (of a publication)	Vol
W	weight
W	work
w	mass function
watt	w or W
watthour	whr
watts per candle	wpc
week	spell out
weight	W or w
weight concentration*	c
work	y yield
X	degree of polymerization
X	electrical resistance
x	mole fractio y yield
yard	yd
year	yr
Young's	E
Z	collision number
Z	z fraction
z	ionic charge

Abbreviations	Symbols
z	coordination number
z	dissymmetry (light scattering)
z	parameter in excluded volume theory
α	angle, especially angle of rotation in optical activity
α	cubic expandion coefficient [$\alpha = V^{-1} (\partial V/\partial T)_p$]
α	expansion coefficient (as reduced length, e.g., α_L in the chain end-to-end distance or α_R for the radius of gyration)
α	degree of crystallinity (always with an index)
α	electric polarizability of a molecule
[α]	"specific" optical rotation
β	angle
β	coefficient of pressure
β	excluded volume cluster integral
Γ	preferential solvation
γ	angle
γ	surface tension
γ	linear expansion coefficient
δ	loss angle
δ	solubility parameter
δ	chemical shift
ε	linear expansion ($\varepsilon = \Delta l / l_o$)
ε	expectation
ε_r	relative permittivity (dielectric number)
η	dynamic viscosity
[η]	Staudinger index (called J_o in DIN 1342)
Θ	characteristic temperature, especial-ly theta temperature
θ	angle, especially angle of rotation
ϑ	angle, especially valence angle
κ	isothermal compressibility [$\kappa = V^{-1} (\partial V/\partial p)_T$]
κ	enthalpic interaction parameter in solution theory

Abbreviations	Symbols
λ	wavelength
λ	heat conductivity
λ	degree of coupling
μ	chemical potential
μ	moment
μ	permanent dipole moment
ν	mement, with respect to a reference value
ν	frequency
ν	kinetic chain length
ξ	shielding ratio in the theory of random coils
Ξ	partition function
Π	osmotic pressure
ρ	density
σ	mechanical stress (σ_{ii} = normal stress, σ_{ij} = shear stress)
σ	standard deviation
σ	hindrance parameter
τ	relaxation time
τ_i	internal transmittance (transmission factor) (represents the ratio of transmitted to absorbed light)
ϕ	volume fraction
$\varphi(r)$	potential between two segments separated by a distance r
Φ	constant in the viscosity-molecular-weight relationship
[Φ]	"molar" optical rotation
χ	interaction parameter in solution theory
ψ	entropic interaction parameter in solution theory
ω	angular frequency, angular velocity
Ω	angle
Ω	probability
Ω	skewness of a distribution

*(= weight of solute divided by volume of solvent); IUPAC suggests the symbol ρ for this quantity, which could lead to confusion with the same IUPAC symbol for density.

Notations

The abbreviations for chemicals and polymer were taken from the "Manual of Symbols and Terminology for Physicochemical Quantities and Units," *Pure and Applied Chemistry* **21***1) (1970), but some were added because of generally accepted use.

The ISO (International Standardization Organization) has suggested that all extensive quantities should be described by capital letters and all intensive quantities by lower-case letters. IUPAC doe not follow this recommendation, however, but uses lower-case letters for specific quantities.

The following symbols are used above or after a letter.

Symbols Above Letters

— signifies an average, e.g., $\overline{M}$ is the average molecular weight; more complicated averages are often indicated by $\langle \rangle$, e.g., $\langle R_G^2 \rangle$ is another way of writing $\overline{(R_G^2)}_z$

— stands for a partial quantity, e.g., $\tilde{v}_A$ is the partial specific volume of the compound A; V_A is the volume of A, wherea $\tilde{V}_A^m$xxx is the partial molar volume of A.

Superscripts

°	pure substance or standard state
∞	infinite dilution or infinitely high molecular weight
m	molar quantity (in cases where subscript letters are impractical)
(q)	the *q* order of a moment (always in parentheses)
‡	activated complex

Subscripts

Initial	State
1	solvent
2	solute
3	additional components (e.g., precipitant, salt, etc.)
am	amorphous
B	brittleness
bd	bond
cr	crystalline
crit	critical
cryst	crystallization
e	equilibrium

Initial	State
E	end group
G	glassy state
i	run number
i	initiation
i	isotactic diads
ii	isotactic triads
Is	heterotactic triads
j	run number
k	run number
m	molar
M	melting process
mon	monomer
n	number average
p	polymerization, especially propagation
pol	polymer
r	general for average
s	syndiotactic diads
ss	syndiotactic triads
st	start reaction
t	termination
tr	transfer
u	monomeric unit
w	weight average
z	z average
Prefixes	
at	atactic
ct	*cis*-tactic
eit	erythrodiisotactic
it	isotactic
st	syndiotactic
tit	threodiisotactic
tt	*trans*-tactic

Square brackets around a letter signify molar concentrations. (IUPAC prescribes the symbol c for molar councentrations, but to date this has consistently been used for the mass/volume unit.)

Angles are always given by °.

Apart from some exceptions, the meter is not used as a unit of length; the units cm and mm derived from it are used. Use of the meter in macromolecular science leads to very impractical units.

Mathematical Signs

Sign	Definition
Operations	
+	Addition
−	Subtraction
×	Multiplication
·	Multiplication
÷	Division
/	Division
∘	Composition
∪	Union
∩	Intersection
±	Plus or minus
∓	Minus or plus
Convolution	
⊕	Direct sum, variation
⊖	Various
⊗	Various
⊙	Various
:	Ratio
Ц	Amalgamation
Relations	
=	Equal to
≠	Not equal to
≈	Nearly equal to
≅	Equals approximately, isomorphic
<	Less than
<<	Much less than
>	Greater than
>>	Much greater than
≤	Less than or equal to
≦	Les than or equal to
≦	Less than or equal to
≥	Greater than or equal to
≥	Grean than or equalt o
≧	Greater than or equal to
≡	Equivalent to, congruent to
≢	Not equivalent to, not congruent to
|	Divides, divisible by
~	Similar to, asymptotically equal to
:=	Assignment

Sign	Definition
∈	A member of
⊂	Subset of
⊆	Subset of or equal to
⊃	Superset of
⊇	Superset of or equal to
∝	Varies as, proportional to
≐	Approaches a limit, definition
→	Tends to, maps to
←	Maps from
↦	Maps to
↪ or ↩	Maps into
□	d'Alembertian operator
Σ	Summation
Π	Product
∫	Integral
∮	Contour integral
Logic	
∧	And, conjunction
∨	Or, distunction
¬	Negation
⇒	Implies
→	Implies
⇔	If and only if
↔	If and only if
∃	Existential quantifier
∀	Universal quantifier
∈	A member o
∉	Not a member of
⊢	Assertion
∴	Hence, therefore
∵	Because
Radial units	
′	Minute
″	Second
°	Degree
Constants	
π	pi (≈ 3.14159265)
e	Base of natural logarithms (≈ 2.71828183)

Sign	Definition
Geometry	
⊥	Perpendicular
∥	Parallel
∦	Not parallel
∠	Angle
∢	Spherical angle
$\stackrel{v}{=}$	Equal angles
Miscellaneous	
i	Square root of -1
′	Prime
″	Double prime
‴	Triple prime
√	Square root, radical
$\sqrt[3]{}$	Cube root
$\sqrt[n]{}$	nth root
!	Factorial
!!	Double factorial
∅	Empty set, null set
∞	Infinity

Sign	Definition
∂	Partial differential
Δ	Delta
∇	Nabla, del
∇^2, Δ	Laplacian operator

English-Greek–Latin Numerical Prefixes

English	Greek	Latin
2	bis	di
3	tris	tri
4	tetrakis	tetra
5	pentakis	penta
6	hexakis	hexa
7	heptakis	hepta
8	octakis	octa
9	nonakis	nona
10	decakis	deca

Greek-Russian-English Alphabets

Greek letter		Greek name	English equivalent	Russian letter		English equivalent
A	α	Alpha	(ä)	А	а	(ä)
B	β	Beta	(b)	Б	б	(b)
				В	в	(v)
Γ	γ	Gamma	(g)	Г	г	(g)
Δ	δ	Delta	(d)	Д	д	(d)
E	ϵ	Epsilon	(e)	Е	е	(ye)
Z	ζ	Zeta	(z)	Ж	ж	(zh)
				З	з	(z)
H	η	Eta	(ā)	И	и	(i, ē)
Θ	θ	Theta	(th)	Й	й	(ē)
I	ι	Iota	(ē)	К	к	(k)
				Л	л	(l)
K	k	Kappa	(k)	М	м	(m)
Λ	λ	Lambda	(l)	Н	н	(n)
				О	о	(ô, o)
M	μ	Mu	(m)	О	о	(ô, o)
				П	п	(p)
N	ν	Nu	(n)	Р	р	(r)
Ξ	ξ	Xi	(ks)	С	с	(s)
				Т	т	(t)
O	o	Omicron	α	У	у	ōō
Π	π	Pi	(P)	Ф	ф	(f)
				Х	х	(kh)
P	ρ	Rho	(r)	Х	х	(kh)
				Ц	ц	(t$_s$)
Σ	σ	Sigma	(s)	Ч	ч	(ch)
T	τ	Tau	(t)	Ш	ш	(sh)
Υ	υ	Upsilon	(ü, ōō)	Щ	щ	(shch)
				Ъ	ъ	8
Φ	ø	Phi	(f)	Ы	ы	(ë)
X	χ	Chi	(H)	ь	ь	(ë)
Ψ	ψ	Psi	(ps)	Э	э	(e)
				Ю	ю	(ū)
Ω	ω	Omega	(ō)	Я	я	(yä)

English-Greek-Latin Numbers

English	Greek	Latin
1	mono	uni
2	bis	di
3	tris	tri
4	tetrakis	tetra
5	pentakis	penta
6	hexakis	hexa
7	heptakis	hepta
8	octakis	octa
9	nonakis	nona
10	decakis	deca

International Union of Pure and Applied Chemistry: Rules Concerning Numerical Terms Used in Organic Chemical Nomenclature (specifically as prefixes for hydrocarbons)

1	mono- or hen-	10 deca-	100 hecta-	1000 kilia-
2	di- or do-	20 icosa-	200 dicta-	2000 dilia-
3	tri-	30 triaconta-	300 tricta-	3000 trilia-
4	tetra-	40 tetraconta-	400 tetracta	4000 tetralia-
5	penta-	50 pentaconta-	500 pentactra	5000 pentalia-
6	hexa-	60 hexaconta-	600 hexacta	6000 hexalia-
7	hepta-	70 hepaconta-	700 heptacta-	7000 hepalia-
8	octa-	80 octaconta-	800 ocacta-	8000 ocatlia-
9	nona-	90 nonaconta-	900 nonactta-	9000 nonalia-

Source: IUPAC, Commission on Nomenclature of Organic Chemistry (N. Lorzac'h and published in *Pure and Appl. Chem* 58: 1693–1696 (1986))

Elemental Symbols and Atomic Weights

Source: International Union of Pure and Applied Chemistry (IUPAC) 2001Values from the 2001 table *Pure Appl. Chem.*, **75**, 1107–1122 (2003). The values of zinc, krypton, molybdenum and dysprosium have been modified. The *approved name* for element 110 is included, see *Pure Appl. Chem.*, **75**, 1613–1615 (2003). The *proposed name* for element 111 is also included.

A number in parentheses indicates the uncertainty in the last digit of the atomic weight.

List of Elements in Atomic Number Order

At No	Symbol	Name	Atomic Wt	Notes
1	H	Hydrogen	1.00794(7)	1, 2, 3
2	He	Helium	4.002602(2)	1, 2
3	Li	Lithium	[6.941(2)]	1, 2, 3, 4
4	Be	Beryllium	9.012182(3)	
5	B	Boron	10.811(7)	1, 2, 3
6	C	Carbon	12.0107(8)	1, 2
7	N	Nitrogen	14.0067(2)	1, 2
8	O	Oxygen	15.9994(3)	1, 2
9	F	Fluorine	18.9984032(5)	
10	Ne	Neon	20.1797(6)	1, 3
11	Na	Sodium	22.989770(2)	
12	Mg	Magnesium	24.3050(6)	
13	Al	Aluminium	26.981538(2)	
14	Si	Silicon	28.0855(3)	2
15	P	Phosphorus	30.973761(2)	
16	S	Sulfur	32.065(5)	1, 2
17	Cl	Chlorine	35.453(2)	3
18	Ar	Argon	39.948(1)	1, 2
19	K	Potassium	39.0983(1)	1
20	Ca	Calcium	40.078(4)	1
21	Sc	Scandium	44.955910(8)	
22	Ti	Titanium	47.867(1)	
23	V	Vanadium	50.9415(1)	
24	Cr	Chromium	51.9961(6)	
25	Mn	Manganese	54.938049(9)	
26	Fe	Iron	55.845(2)	
27	Co	Cobalt	58.933200(9)	
28	Ni	Nickel	58.6934(2)	
29	Cu	Copper	63.546(3)	2

At No	Symbol	Name	Atomic Wt	Notes
30	Zn	Zinc	65.409(4)	
31	Ga	Gallium	69.723(1)	
32	Ge	Germanium	72.64(1)	
33	As	Arsenic	74.92160(2)	
34	Se	Selenium	78.96(3)	
35	Br	Bromine	79.904(1)	
36	Kr	Krypton	83.798(2)	1, 3
37	Rb	Rubidium	85.4678(3)	1
38	Sr	Strontium	87.62(1)	1, 2
39	Y	Yttrium	88.90585(2)	
40	Zr	Zirconium	91.224(2)	1
41	Nb	Niobium	92.90638(2)	
42	Mo	Molybdenum	95.94(2)	1
43	Tc	Technetium	[98]	5
44	Ru	Ruthenium	101.07(2)	1
45	Rh	Rhodium	102.90550(2)	
46	Pd	Palladium	106.42(1)	1
47	Ag	Silver	107.8682(2)	1
48	Cd	Cadmium	112.411(8)	1
49	In	Indium	114.818(3)	
50	Sn	Tin	118.710(7)	1
51	Sb	Antimony	121.760(1)	1
52	Te	Tellurium	127.60(3)	1
53	I	Iodine	126.90447(3)	
54	Xe	Xenon	131.293(6)	1, 3
55	Cs	Caesium	132.90545(2)	
56	Ba	Barium	137.327(7)	
57	La	Lanthanum	138.9055(2)	1
58	Ce	Cerium	140.116(1)	1
59	Pr	Praseodymium	140.90765(2)	
60	Nd	Neodymium	144.24(3)	1
61	Pm	Promethium	[145]	5
62	Sm	Samarium	150.36(3)	1
63	Eu	Europium	151.964(1)	1
64	Gd	Gadolinium	157.25(3)	1
65	Tb	Terbium	158.92534(2)	
66	Dy	Dysprosium	162.500(1)	1
67	Ho	Holmium	164.93032(2)	
68	Er	Erbium	167.259(3)	1

At No	Symbol	Name	Atomic Wt	Notes
69	Tm	Thulium	168.93421(2)	
70	Yb	Ytterbium	173.04(3)	1
71	Lu	Lutetium	174.967(1)	1
72	Hf	Hafnium	178.49(2)	
73	Ta	Tantalum	180.9479(1)	
74	W	Tungsten	183.84(1)	
75	Re	Rhenium	186.207(1)	
76	Os	Osmium	190.23(3)	1
77	Ir	Iridium	192.217(3)	
78	Pt	Platinum	195.078(2)	
79	Au	Gold	196.96655(2)	
80	Hg	Mercury	200.59(2)	
81	Tl	Thallium	204.3833(2)	
82	Pb	Lead	207.2(1)	1, 2
83	Bi	Bismuth	208.98038(2)	
84	Po	Polonium	[209]	5
85	At	Astatine	[210]	5
86	Rn	Radon	[222]	5
87	Fr	Francium	[223]	5
88	Ra	Radium	[226]	5
89	Ac	Actinium	[227]	5
90	Th	Thorium	232.0381(1)	1, 5
91	Pa	Protactinium	231.03588(2)	5
92	U	Uranium	238.02891(3)	1, 3, 5
93	Np	Neptunium	[237]	5
94	Pu	Plutonium	[244]	5
95	Am	Americium	[243]	5
96	Cm	Curium	[247]	5
97	Bk	Berkelium	[247]	5
98	Cf	Californium	[251]	5
99	Es	Einsteinium	[252]	5
100	Fm	Fermium	[257]	5
101	Md	Mendelevium	[258]	5
102	No	Nobelium	[259]	5
103	Lr	Lawrencium	[262]	5
104	Rf	Rutherfordium	[261]	5, 6
105	Db	Dubnium	[262]	5, 6
106	Sg	Seaborgium	[266]	5, 6
107	Bh	Bohrium	[264]	5, 6
108	Hs	Hassium	[277]	5, 6
109	Mt	Meitnerium	[268]	5, 6
110	Ds	Darmstadtium	[281]	5, 6
111	Rg	Roentgenium	[272]	5, 6
112	Uub	Ununbium	[285]	5, 6
114	Uuq	Ununquadium	[289]	5, 6
116	Uuh	Ununhexium		see Note above
118	Uuo	Ununoctium		see Note above

1. Geological specimens are known in which the element has an isotopic composition outside the limits for normal material. The difference between the atomic weight of the element in such specimens and that given in the Table may exceed the stated uncertainty.
2. Range in isotopic composition of normal terrestrial material prevents a more precise value being given; the tabulated value should be applicable to any normal material.
3. Modified isotopic compositions may be found in commercially available material because it has been subject to an undisclosed or inadvertent isotopic fractionation. Substantial deviations in atomic weight of the element from that given in the Table can occur.
4. Commercially available Li materials have atomic weights that range between 6.939 and 6.996; if a more accurate value is required, it must be determined for the specific material [range quoted for 1995 table 6.94 and 6.99].
5. Element has no stable nuclides. The value enclosed in brackets, e.g. [209], indicates the mass number of the longest-lived isotope of the element. However three such elements (Th, Pa, and U) do have a characteristic terrestrial isotopic composition, and for these an atomic weight is tabulated.
6. The names and symbols for elements 112-118 are under review. The temporary system recommended by J Chatt, *Pure Appl. Chem.*, **51**, 381–384 (1979) is used above. The names of elements 101-109 were agreed in 1997 (See *Pure Appl. Chem.*, 1997, **69**, 2471–2473) and for element 110 in 2003 (see *Pure Appl. Chem.*, 2003, **75**, 1613–1615). The proposed name for element 111 is also included.

List of Elements in Name Order

At No	Symbol	Name	Atomic Wt	Notes
89	Ac	Actinium	[227]	5
13	Al	Aluminium	26.981538(2)	
95	Am	Americium	[243]	5
51	Sb	Antimony	121.760(1)	1

At No	Symbol	Name	Atomic Wt	Notes	At No	Symbol	Name	Atomic Wt	Notes
18	Ar	Argon	39.948(1)	1, 2	36	Kr	Krypton	83.798(2)	1, 3
33	As	Arsenic	74.92160(2)		57	La	Lanthanum	138.9055(2)	1
85	At	Astatine	[210]	5	103	Lr	Lawrencium	[262]	5
56	Ba	Barium	137.327(7)		82	Pb	Lead	207.2(1)	1, 2
97	Bk	Berkelium	[247]	5	3	Li	Lithium	[6.941(2)]	1, 2, 3, 4
4	Be	Beryllium	9.012182(3)		71	Lu	Lutetium	174.967(1)	1
83	Bi	Bismuth	208.98038(2)		12	Mg	Magnesium	24.3050(6)	
107	Bh	Bohrium	[264]	5, 6	25	Mn	Manganese	54.938049(9)	
5	B	Boron	10.811(7)	1, 2, 3	109	Mt	Meitnerium	[268]	5, 6
35	Br	Bromine	79.904(1)		101	Md	Mendelevium	[258]	5
48	Cd	Cadmium	112.411(8)	1	80	Hg	Mercury	200.59(2)	
55	Cs	Caesium	132.90545(2)		42	Mo	Molybdenum	95.94(2)	1
20	Ca	Calcium	40.078(4)	1	60	Nd	Neodymium	144.24(3)	1
98	Cf	Californium	[251]	5	10	Ne	Neon	20.1797(6)	1, 3
6	C	Carbon	12.0107(8)	1, 2	93	Np	Neptunium	[237]	5
58	Ce	Cerium	140.116(1)	1	28	Ni	Nickel	58.6934(2)	
17	Cl	Chlorine	35.453(2)	3	41	Nb	Niobium	92.90638(2)	
24	Cr	Chromium	51.9961(6)		7	N	Nitrogen	14.0067(2)	1, 2
27	Co	Cobalt	58.933200(9)		102	No	Nobelium	[259]	5
29	Cu	Copper	63.546(3)	2	76	Os	Osmium	190.23(3)	1
96	Cm	Curium	[247]	5	8	O	Oxygen	15.9994(3)	1, 2
110	Ds	Darmstadtium	[281]	5, 6	46	Pd	Palladium	106.42(1)	1
105	Db	Dubnium	[262]	5, 6	15	P	Phosphorus	30.973761(2)	
66	Dy	Dysprosium	162.500(1)	1	78	Pt	Platinum	195.078(2)	
99	Es	Einsteinium	[252]	5	94	Pu	Plutonium	[244]	5
68	Er	Erbium	167.259(3)	1	84	Po	Polonium	[209]	5
63	Eu	Europium	151.964(1)	1	19	K	Potassium	39.0983(1)	1
100	Fm	Fermium	[257]	5	59	Pr	Praseodymium	140.90765(2)	
9	F	Fluorine	18.9984032(5)		61	Pm	Promethium	[145]	5
87	Fr	Francium	[223]	5	91	Pa	Protactinium	231.03588(2)	5
64	Gd	Gadolinium	157.25(3)	1	88	Ra	Radium	[226]	5
31	Ga	Gallium	69.723(1)		86	Rn	Radon	[222]	5
32	Ge	Germanium	72.64(1)		75	Re	Rhenium	186.207(1)	
79	Au	Gold	196.96655(2)		45	Rh	Rhodium	102.90550(2)	
72	Hf	Hafnium	178.49(2)		111	Rg	Roentgenium	[272]	5, 6
108	Hs	Hassium	[277]	5, 6	37	Rb	Rubidium	85.4678(3)	1
2	He	Helium	4.002602(2)	1, 2	44	Ru	Ruthenium	101.07(2)	1
67	Ho	Holmium	164.93032(2)		104	Rf	Rutherfordium	[261]	5, 6
1	H	Hydrogen	1.00794(7)	1, 2, 3	62	Sm	Samarium	150.36(3)	1
49	In	Indium	114.818(3)		21	Sc	Scandium	44.955910(8)	
53	I	Iodine	126.90447(3)		106	Sg	Seaborgium	[266]	5, 6
77	Ir	Iridium	192.217(3)		34	Se	Selenium	78.96(3)	
26	Fe	Iron	55.845(2)		14	Si	Silicon	28.0855(3)	2

At No	Symbol	Name	Atomic Wt	Notes
47	Ag	Silver	107.8682(2)	1
11	Na	Sodium	22.989770(2)	
38	Sr	Strontium	87.62(1)	1, 2
16	S	Sulfur	32.065(5)	1, 2
73	Ta	Tantalum	180.9479(1)	
43	Tc	Technetium	[98]	5
52	Te	Tellurium	127.60(3)	1
65	Tb	Terbium	158.92534(2)	
81	Tl	Thallium	204.3833(2)	
90	Th	Thorium	232.0381(1)	1, 5
69	Tm	Thulium	168.93421(2)	
50	Sn	Tin	118.710(7)	1
22	Ti	Titanium	47.867(1)	
74	W	Tungsten	183.84(1)	

At No	Symbol	Name	Atomic Wt	Notes
112	Uub	Ununbium	[285]	5, 6
116	Uuh	Ununhexium		see Note above
118	Uuo	Ununoctium		see Note above
114	Uuq	Ununquadium	[289]	5, 6
92	U	Uranium	238.02891(3)	1, 3, 5
23	V	Vanadium	50.9415(1)	
54	Xe	Xenon	131.293(6)	1, 3
70	Yb	Ytterbium	173.04(3)	1
39	Y	Yttrium	88.90585(2)	
30	Zn	Zinc	65.409(4)	
40	Zr	Zirconium	91.224(2)	1

Pronounciation Symbols and Abbreviations

ə	Banana, collide, abut		ȯ	saw, all, gnaw, caught
ˈə, ˌə	Humdrum, abut		o͞o	fool
ᵊ	Immediately preceding \l\, \n\, \m\, \ŋ\, as in battle, mitten, eaten, and sometimes open \ˈō-pᵊm\, lock and key \-ᵊ ŋ-\; immediately following \l\, \m\, \r\, as often in French table, prisme, titre		o͝o	took
			œ	French coeuf, German Hölle
			œ̄	French feu, German Höhle
			ȯi	coin, destroy
ər	further, merger, bird		p	pepper, lip
ˈə-r, ˌə-r	As in two different pronunciations of hurry \ˈhər-ē, \ˈhə-rē\		r	red, car, rarity
			s	source, less
a	mat, map, mad, gag, snap, patch		sh	as in shy, mission, machine, special (actually, this is a single sound, not two); with a hyphen between, two sounds as in grasshopper \ˈgras-ˌhä-pər\
ā	day, fade, date, aorta, drape, cape			
ä	bother, cot, and, with most American speakers, father, cart			
á	father as pronounced by speakers who do not rhyme it with bother; French patte		t	tie, attack, late, later, latter
			th	as in thin, ether (actually, this is a single sound, not two); with a hyphen between, two sounds as in knighthood \ˈnīt-ˌh----d\
aú	now, loud, out			
b	baby, rib			
ch	chin, nature \ˈnā-chər\		th	then, either, this (actually, this is a single sound, not two)
d	did, adder			
e	bet, bed, peck		ü	rule, youth, union \ˈyün-yən\, few \ˈfyü\
ˈē, ˌē	beat, nosebleed, evenly, easy		ú	pull, wood, book, curable \ˈky ú r-ə-bəl\, fury \ˈfy----r-ē\
ē	easy, mealy			
f	fifty, cuff		ue	German füllen, hübsch
g	go, big, gift		ue̱	French rue, German fühlen
h	hat, ahead		v	vivid, give
hw	whale as pronounced by those who do not have the same pronunciation for both whale and wail		w	we, away
			y	yard, young, cue \ˈkyü\, mute \ˈmyüt\, union \ˈyün-yən\
i	tip, banish, active			
ī	site, side, buy, tripe		ʸ	indicates that during the articulation of the sound represented by the preceding character the front of the tongue has substantially the position it has for the articulation of the first sound of yard, as in French digne \dēnʸ\
j	job, gem, edge, join, judge			
k	kin, cook, ache			
k̲	German ich, Buch; one pronunciation of loch			
l	lily, pool		z	zone, raise
m	murmur, dim, nymph		zh	as in vision, azure \ˈa-zhər\ (actually this is a single sound, not two).
n	no, own			
ⁿ	Indicates that a preceeding vowel or diphthong is pronounced with the nasal passages open, as in French un bon vin blanc \œⁿ-bōⁿvaⁿ-bläⁿ\		\	reversed virgule used in pairs to mark the beginning and end of a transcription: \ˈpen\
			ˈ	mark preceding a syllable with primary (strongest) stress: \ˈpen-mən-ˌship\
ŋ	sing \ˈsiŋ\, singer \ˈsiŋ-ər\, finger \ˈfiŋ-gər\, ink \ˈiŋk\		ˌ	mark preceding a syllable with secondary (medium) stress: \ˈpen-mən-ˌship\
ō	bone, know, beau		-	mark of syllable division

( )	indicate that what is symbolized between is present in some utterances but not in others: *factory* \ **❙**fak-t(ə-)rē
÷	indicates that many regard as unacceptable the pronunciation variant immediately following: *cupola* \ **❙**kyü-pə-lə, ÷- **❙**lō\

Explanatory Notes and Abbreviations

(date)	date that word was first recorded as having been used
[...]	etomology and origin(s) of word
{...}	usage and/or languages, including French, German, Italian and Spanish
adj	adjective
adv	adverb
B.C.	before Christ
Brit.	Britain, British
C	centigrade, Celsius
c	century
E	English
Eng.	England
F	French, Fahrenheit
Fr.	France
fr.	from
G	German
Gr.	Germany
L	Latin
ME	middle English

n	noun
neut.	neuter
NL	new Latin
OE	old English
OL	old Latin
pl	plural
prp.	present participle
R	Russian
sing.	singular
S	Spanish
U.K.	United Kingdom
v	verb

Source: From *Merriam-Webster's Collegiate© Dictionary*, Eleventh Editioh, ©2004 by Merriam-Webster, incorporated, (www.Merriam-Webster.com). With permission.

Languages

French, German and Spanish translations are enclosed in {--} and preceded by F, G, I and S, respectively; and gender is designated by f-feminine, m-masculine, n-neuter. For example: **Polymer**--{F polymere m} represents the French translation "polymere" of the English word polymer and it is in the masculine case. These translations were obtain from multi-language dictionaries including: *A Glossary of Plastics Terminology in 5 Languages*, 5th Ed., Glenz, W., (ed) Hanser Gardner Publications, Inc., Cinicinnati, 2001. By permission).

A

A \ā\ *n* (1) SI abbreviation for prefix ATTO-, (2) Symbol for acceleration.

"a" or "α" *n* Redness–greenness coordinate in certain transformed color spaces, generally used as the Δa, or difference in "a" between a specimen and a standard reference color. If "a" or Δa is plus, there is more redness than greenness; if "a" or Δa is minus, there is more greenness than redness. It is normally used with b or β as part of the chromaticity or chromaticity difference (McDonald R (1997) Colour physics for industry, 2nd edn. Society of Dyers and Colourists, West Yorkshire, England; Billmeyer FW, Saltzman M (1966) Principles of color technology. Wiley, New York). See ▶ Uniform Chromaticity Coordinates.

"a" Kubelka-Munk Equation *n* Mathematical constant characteristic of a color at complete hiding; dependent on the optical constants K and S: $a = \frac{1}{2}[1/R\infty + R\infty] = 1 + K/S$. (McDonald R (1997) Colour physics for industry, 2nd edn. Society of Dyers and Colourists, West Yorkshire, England).

au *n* Abbreviation for ▶ Atomic Unit.

A *n* Abbreviation for ▶ Ampere.

Å \ˈaŋ-strəm\ *n* [Anders J. *Angstrom*] (1892) {*d* Angströmeinheit *f*, *f* unité f Angtröm, *s* unidad *f* Angström} A unit of length equal to 1×10^{-12} m. Abbreviation for deprecated Angstrom Unit. See ▶ Angstrom Unit (Weast RC (1971) Handbook of chemistry and physics, 52nd edn. CRC Press, Boca Raton, FL).

A-Acid \ā-ˈa-səd\ *n* [F. or L.; F. *acide*] (1626) $NH_2C_6H_4COOH$. Trade abbreviation for ▶ Anthranilic Acid, an intermediate used in the manufacture of the pigment, Lake Red D.

AATCC *n* Abbreviation for the American Association Of Textile Chemists And Colorists.

AB (= absolute) A prefix attached to the names of practical electrical units to indicate the corresponding unit in the old cgs system (emu), e.g., abampere, abvolt.

A-B-A Model Polymers *n* Two phase block copolymers, predictable molecular weights, narrow molecular weight distribution, convenient end-capping, thermoplastic, anionically polymerized, i.e., Kraton-G^R and Hytrel^R. The B block is usually styrene that forms hard and amorphous domains.

A-B-A *n* **Thermoplastic Elastomers** Three block thermoplastic polymer elastomer, high-strength rubber, no vulcanization, completely soluble, two glass and two glass transition temperatures, i.e., styrene and butadiene.

Abbe' Number \a-ˈbā, ˈa-ˌbā-\ *n* [F, fr. LL *abbat-, abass*] (1530) The refractive index varies with the wavelength of incident light, and the abbe' number *v* is given as a measure of this dispersion; and the capacity to separate the colors of white light increases as *v* decreases.

Abbe' Refractometer \-ˌrē-ˌfrak-ˈtä-mə-tər\ *n* Common form of refractometer used for determining the refractive index of oils and other liquids, or of greaselike products which are capable of liquefaction at moderate temperatures. Good accuracy is attainable in the range of 1.3–1.7, readings being given to the fourth decimal place. The prisms, which constitute the most important part of the instrument, and hence the liquid held between their faces, are capable of being maintained accurately at the temperature of the determination. With the use of special liquids to form an optical seal to the prisms and a special technique of viewing, it is also used for determining the refractive index of solids such as plastics cast in sheets with polished surfaces and edges. The refractometer measures the real part of the refractive index and thus helps to answer three different types of questions. First, and most simply, it is useful in the empirical identification of pure substances, it can act as a criterion of purity, and it serves in the quantitative analysis of solutions. These characterizations are made possible by the precision and accuracy of refractometers. Second, the evaluation of dipole moments of substances via measures of dielectric constant at a single temperature requires the knowledge of their refractive indexes. Third, refractive index measured as a function of wavelength, in concert with measurement of molar absorptivity characterize the optical properties of a given molecule. These measures in turn provide information on the electronic structures of molecules. As an example, refractometry can be useful in the determination of chain length and isomerism in organic molecules. The development of modern NMR and mass spectrometers have largely displaced the use of refractometry in such studies, giving less ambiguous answers regarding molecular structures, but at a great increase in instrumental complexity and cost. The modern Abbe refractometer invented at the Carl Zeiss Works was exclusively manufactured by Zeiss until the early twentieth century. The explosive growth of laboratory work after WW 1 led a number of other companies to begin

Jan W. Gooch, *Encyclopedic Dictionary of Polymers*, DOI 10.1007/978-1-4419-6247-8,
© Springer Science+Business Media LLC 2011

its manufacture as well, including Adam Hilgar and Stanley in Great Britain, and Spencer Lens Co., Bausch & Lomb, Gaertner, and Valentine in the U. S. Ernst Abbe constructed the first "Abbe"[3] refractometer in 1869. Five years later, in 1874, he published a comprehensive booklet. In it he discusses the theory and described.

Abbozzo *adj* Underpainting of an oil painting, either in monochrome or color. *Sometimes called Bozzo or Deadcoloring.*

Abcoulomb *n* The abcouloumb, the emu of charge, is defined as the charge which passes a given surface in 1 s if a steady current of 1 abampere flows across the surface. Its dimensions are, therefore, $cm^{0.5}$ $gm^{0.5}$ which differ from the dimensions of the statcoulomb by a factor which has the dimensions of speed. This relationship is connected with the fact that the ratio $2K_e/K_m$ must have the value of the square of the speed of light in any consistent system of units. It follows further that

$$1\ \text{abcouloumb} = 2.99793 \times 10^{10}\ \text{statcoulomb},$$

the speed of light in vacuo being $(2.99793 \pm 0.000003) \times 10^{10}$ cm/s (Weast RC (ed) Handbook of chemistry and physics, 52nd edn. CRC Press, Boca Raton, FL).

Abegg's Rule \ˈä-ˌbegz-\ *n* [Abegg, Richard Wilhelm Heinrich; Danish chemist, major work on chemical valence] (1869–1910) A Chemistry: For a given chemical element (as sulfur) the sum of the absolute value of its negative valence of maximum value (as −2 for sulfur in H_2S) and its positive value (as +6 in H_2SO_4) is often equal to 8. For use in regard to a helical periodic system. This tendency is exhibited especially by the elements of the fourth, fifth, sixth, and seventh groups and is known as Abegg's rule ((ed) (2002) General chemistry. Brookes/Cole, New York).

Aberration \ˌa-bə-ˈrā-shən\ *n* [L. *aberrare*] (1594) In optical systems, the failure of light rays from one object point to converge to a single focal point. See ▶ Chromatic Aberration and ▶ Spherical Aberration.

ABFA *n* See ▶ Azobisformamide.

Abherent \ab-ˈhir-ənt\ *n* (adhesive) A coating or film applied to one surface to prevent or reduce its adhesion to another surface brought into intimate contact with it. Abherents applied to plastic films are often called *antiblocking agents*. Those applied to molds, calendar rolls, etc., are sometimes called *release agents* or ▶ Parting Agents (Skeist I (ed) (1990, 1977, 1962) Handbook of adhesives. Van Nostrand Reinhold, New York).

Abhesive \-əb-ˈhē-siv, -ziv\ *n* (1670) Material that resists adhesion; applied to surfaces to prevent sticking, heat-sealing, etc. (Skeist I (ed) (1990, 1977, 1962) Handbook of adhesives. Van Nostrand Reinhold, New York).

Abietic Acid \a-bē-ə-tek, a-səd\ *n* $C_{19}H_{30}COOH$. A monocarboxylic acid derived from rosin. Plasticizers derived from it include ▶ Hydroabietyl Alcohol, ▶ Hydrogenated Methyl Abietate, and ▶ Methyl Abietate.

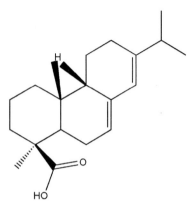

Abietic Acid, Commercial Grade *n* $C_{20}H_{30}O_2$. Product consisting chiefly of rosin acids in substantially pure form, separated from rosin or tall oil commercially for specific purposes and in which Abietic acid and its isomers are the principal components. Syn: ▶ Sylvic Acid.

Abietates *n* Esters or salts of abietic acid, a principal constituent of ordinary rosin from which the products of commerce are derived, no attempt being made to separate abietic acid from the other acids which rosin is likely to contain. Metallic abietates, as such, are rarely encountered under this name but generally as resonates (Langenheim JH (2003) Plant resins: chemistry, evolution ecology and ethnobotany. Timber Press, Portland, OR; (2001) Paint: pigment, drying oils, polymers, resins, naval stores, cellulosics esters, and ink vehicles, vol 3. American Society for Testing and Material). Esters of rosin, however, are commonly described as abietates and not as resonates. For example, methyl abietate (Trademark – Abalyn), a mixture of the methyl esters of the rosin acids. $C_{19}H_{29}COOCH_3$. The article of commerce is colorless to yellow, almost odorless, thick liquid. d_{20}^{20} 1.040. bp 360–365° with decompn. n_D^{20} 1.530. Flash pt 180–218°C. Insoluble in water, miscible with usual organic solvents, also with aliphatic hydrocarbons. Dissolves ester gums, rosin, many synthetic resins as well as ethyl cellulose, rubber, etc., bp 360–365°F with decomposition; use as a solvent for ester gums, rosin, many synthetic resins, ethyl cellulose, rubber, etc.; in the manufacture of varnish resins; as ingredient in adhesives (Merck index (2001) 13th edn. Merck and Company, Whitehouse Station, New Jersey). Esters of rosin are described as abietates and include the methyl, ethyl, and benzyl derivatives, usually used as plasticizers. The ester abietates which have enjoyed some popularity are the methyl, ethyl, and benzyl derivatives. They are soft, resinous materials and are used chiefly as plasticizers (Wypych G (ed) (2003) Plasticizer's data base. Noyes Publication, New York).

Ablation \a-ˈblā-shən\ *n* (15c) Derived from the Latin *ablatio*, meaning "a carrying away," this term has been used by astrophysicists to describe the erosion and disintegration of meteors entering the atmosphere, and more recently by space scientists and engineers for the layer-by-layer decomposition of a plastic surface when heated quickly to a very high temperature. Usually, the decomposition is highly endothermic and the absorption of energy at the surface slows penetration of high temperature to the interior. In other words, it is the ability of a material such as a polymer to form a protective thermal layer when carbonized by extreme heat (Kidder RC (1994) Handbook of fire retardant coatings and fire testing services. CRC Press, Boca Raton, Fl; Rosato DV (ed) (1992) Rosato's plastics encyclopedia and dictionary. Hanser-Gardner, New York) {G ablative, F ablative, S ablative, I ablative}.

Ablative Coatings *n* Thick, mastic-like materials which absorb heat; they are designed to char and sacrifice themselves while protecting the metal substrate underneath. This type of coating is similar with ▶ Intumescent coatings that produce a foam on exposure to high heat to protect the substrate, but do not char as ablative coatings. These coatings are used for missiles and re-entry rockets (Kidder RC (1994) Handbook of fire retardant coatings and fire testing services. CRC Press, Boca Raton, FL; Nelson G (1990) Fire and polymers: hazards identification and prevention. Oxford University Press, UK). See also ▶ Ablative Plastic.

Ablative Plastic *n* Material which absorbs heat while part of it is being consumed by heat through a decomposition process (pyrolysis) which takes place near the surface exposed to the heat (Nelson G (1990) Fire and polymers: hazards identification and prevention. Oxford University Press, UK; Pittance JC (ed) (1990) Engineering plastics and composites. SAM International, Materials Park, OH).

ABL Bottle *n* A filament-wound test vessel about 46 cm in diameter and 61 cm long, subjected to rising internal hydrostatic pressure to determine the quality and strength of the composition from which it was made.

Abnormal Crimp \(ˌ)ab-ˈnór-məl, əb-ˈkrimp\ *n* A relative term for crimp that is either too low or too high in frequency and/or amplitude or that has been put into the fiber with improper angular characteristics.

ABR *n* Copolymers from acrylic esters and butadiene.

Abraded Yarn \ə-ˈbrādəd\ *n* A filament yarn in which filaments have been cut or broken to create hairiness (fibrillation) to simulate the surface character of spun yarns. Abraded yarns are usually plied or twisted with other yarns before use (Kadolph SJJ, Langford AL (2001) Textiles. Pearson Education, New York).

Abraser *n* An instrument used for measuring resistance to abrasion using a sample on a turntable rotating under a pair of weighted abrading wheels that produce abrasion through side-slip (Koleske JV (ed) (1995) Paint and coating testing manual. American Society for Testing and Materials. West Conshohoka, Pennsylvania, UDA).

Abrasiometer *n* One of the many devices used to test abrasion of a coating by using an air blast to drive an abrasive against the test film, or by rotating a film submerged in an abrasive, or by simply dropping a stream of abrasive onto the film (Koleske JV (ed) (1995) Paint and coating testing manual. American Society for Testing and Materials; www.gardco.com).

Abrasion \ə-ˈbrā-zhən\ *n* [ML *atrasion-, abrasion,* fr. L *abradere*] (1656) The wearing away of a surface in service by action such as rubbing, scraping, or erosion {G Abrieb m, F abrasion f, S abrasión f, I abrasione f}.

Abrasion Coefficient \-ˌkō-ə-ˈfi-shənt\ *n* Method for reporting the result of an abrasion test using the falling sand abrasion tester, in which it is assumed that the abrasion resistance is proportional to the film thickness.

$$\text{Abrasion Coefficient} = \frac{(W_1 - W)_2}{T}$$

where:
W_1 = grams of abrasive and holder before tests,
W_2 = grams of abrasive and holder after test, and
T = thickness of coating in mils (0.001 in.) (0.025 mm)
(Koleske JV (ed) (1995) Paint and coating testing manual. American Society for Testing and Materials; Gardner-Sward handbook (1995) MNL 17, 14th edn. ASTM, Conshohocken, PA).

Abrasion Cycle *n* The number of abrading motions or cycles to which a test specimen is subjected in a test of abrasion resistance (Paint and coating testing manual (Gardner-Sward handbook) (1995) MNL 17, 14th edn. ASTM, Conshohocken, PA).

Abrasion Resistance *n* (1) This test method (See www.astm.org) covers the determination of the resistance or organic coatings produced by an air blast of abrasive material on coatings applied a plane rigid substrate such as a glass or metal. (2) The ability of a coating to resist being worn away and to maintain its original appearance and structure as when subjected to rubbing, scraping, or erosion such as measured by the ▶ Taber Abraser The resistance to shearing of material from a surface, i.e., rubber has abrasion resistance from sand. The ability of a fiber or fabric to sustain wearing of its surface. See ▶ Abrasion. (3) The ability of a material to withstand mechanical actions such as subbing, scraping, grinding, sanding, or erosion that tends progressively to remove material from its surface (Gardner-Sward handbook) (1995) MNL 17, 14th edn. ASTM, Conshohocken, PA).

Abrasion Test n Tests designed to determine the ability to withstand the effects of rubbing and scuffing.

Abrasive n (1853) Any material which, by a process of grinding down, tends to make a surface smooth or rough.

Abrasive Finishing n (1) A method of removing flash, gate marks, and rough edges from plastics articles by means of grit-containing belts or wheels. The process is usually employed on large rigid or semi-rigid products with intricate surfaces that cannot be treated by tumbling or other more efficient methods of finishing. (2) To finish, dress, or decorate a surface using an material such as polishing grit.

Abrasive Forming n Formation of a part or shape using abrasives to chip away unwanted materials.

Abrasiveness \ə-ˈbrā-siv, ziv ˈnes\ n (1875) That property of a substance which causes it to wear or scratch other surfaces with which it is in contact. (Merriam-Webster's Collegiate Dictionary (2004) 11th edn. Merriam-Webster, Springfield, MA).

Abrasive Wheels n An abrasion material in the shape of a disk which is often turned on a power tool, i.e., abrasive polishing of granite.

Abraum A red ocher used to stain mahogany.

Abridged Spectrophotometer \ə-ˈbrij ˌspek-trō-fə-ˈtä-mə-tər\ n An instrument which measures spectral transmittance or reflectance at a limited number of wavelengths, usually employing filters rather than a monochromator (Skoog DA, Holler FJ, Nieman TA (1997) Principles of instrumental analysis, Brooks/Coles, New York; Willard HH, Merritt LL, Dean JA (1974) Instrumental methods of analysis, D. Van Nostrand, New York). See ▶ Filter Spectrophotometer.

ABS \ˌā-(ˌ)bē-ˈes\ n [*acrylonitrile-butadiene-sytrene*] (1966) Copolymer of acrylonitrile–butadiene–styrene segments. Abbreviation for ▶ Acrylonitrile-Butadiene-Styrene. See ▶ ABS Resin.

ABS Nylon Alloy n See ▶ Acrylonitrile Butadiene Styrene Polymer Nylon Alloy.

ABS PC Alloy n See ▶ Acrylonitrile Butadiene Styrene Polymer Polycarbonate Alloy.

ABS Polymers n Generic term for copolymers of polyblends from acrylonitrile, butadiene, and styrene.

ABS Resin n Any of a family of thermoplastics based on acrylonitrile, butadiene, and styrene combined by a variety of methods involving polymerization, graft copolymerization, physical mixtures, and combinations thereof. Hundreds of standard grades of ABS resins are available, plus many special grades, alloyed or otherwise modified to yield unusual properties. The standard grades are rigid, hard, and tough, and possess good impact strength. ABS compounds in pellet form can be extruded, blow molded, calendered, and injection molded. ABS powders are used as modifiers for other resins, e.g., PVC (Wickson EJ (ed) (1993) Handbook of polyvinyl chloride formulating. Wiley, New York). Typical applications for ABS resins are household appliances, automotive parts, business-machine and telephone components, pipe and pipe fittings, packaging and shoe heels (Harper CA (ed) (2002) Handbook of plastics, elastomers and composites, 4th edn. McGraw-Hill, New York). See also ▶ Acrylonitrile Butadiene Styrene Polymer.

Absolute \ˈab-sə-ˌlüt\ *adj* [ME *absolut*, fr. L *absolutus*, fr. pp of *absolvere* to set free, absolve] (14c) Adjective used to describe measurements in terms of fundamentally defined units (Merriam-Webster's Collegiate Dictionary (2000) 10th edn. Springfield, MA).

Absolute Alcohol n Ethyl alcohol that has been refined by azeotropic distillation to 99.9% purity (200 proof). Other commercial ethanols contain about 5% water and may contain denaturants that make the alcohol undrinkable. Pure anhydrous ethyl alcohol (ethanol). The term is used to distinguish it from the several varieties of alcohol which are available, and which contain varying amounts of water and/or other impurities.

Absolute Humidity n (1867) The actual weight of water vapor contained in a unit weight of air. See ▶ Humidity, Absolute.

Absolute Pressure See ▶ Pressure.

Absolute Reflectance n Reflectance measured relative to the perfect diffuser.

Absolute Temperature n Temperature measured from the absolute zero, at which all molecular motions cease; 0.00 K (Kelvin) = $-273.15°C$ (Whitten KW, Davis RE, Davis E, Peck LM, Stanley GG (2003) General chemistry. Brookes/Cole, New York). See ▶ Kelvin Temperature Scale.

Absolute Units *n* A system of units based on the smallest possible number of independent units. Specifically, one unit of force, work, energy, and power not derived from or dependent on gravitation.

Absolute Viscosity *n* (1) Tangential force on unit area of either of two parallel planes at unit distance apart, when the space between the planes in filled with fluid (in question) and one of the planes moves with unit velocity in its own plane relative to the other. (2) Force required to move in opposite directions at a velocity of 1 m per second, two parallel plans of liquid, 1 m^2 in area and separated from each other by a distance of 1 m. The absolute viscosity is designated by the Greek letter η (Goodwin JW, Goodwin J, Hughes RW (2000) Rheology for chemists. Royal Society of Chemistry, London, England). See ▶ Poise, Viscosity.

Absolute Zero *n* (1848) The temperature at which all particles in a substance are in their lowest energy states: 0 K or −273.15°C, the temperature at which all chemical activity ceases. It is equal to −273.15°C or −459.67°F. Absolute 0 K has never been achieved and does not exist in nature or perhaps not anywhere in the known universe (Serway RA, Faugh JS, Bennett CV (2005) College physics. Thomas, New York; Whitten KW, Davis RE, Davis E, Peck LM, Stanley GG (2003) General chemistry. Brookes/Cole, New York).

Abson *n* ABS. Manufactured by Goodrich, U.S.

Absorbance \əb-ˈsȯr-bən(t)s, -ˈzȯr-\ *n* (1947) Logarithm of the reciprocal of spectral internal transmittance. The ability of a substance to transform radiant energy into a different of energy, usually with a resulting rise in temperature. Mathematically, absorbance is the negative logarithm to the base 10 of transmittance (Willard HH, Dean JA, Merritt LL (1995) Instrumental methods of analysis. Wadsworth, New York). See ▶ Beer-Bouguer Law and ▶ Light Absorbance.

Absorbency *n* (1859) That property of a porous material, such as paper, which causes it to take up liquids or vapors (e.g., moisture) with which it is in contact (Merriam-Webster's Collegiate Dictionary (2004) 11th edn. Merriam-Webster, Springfield, MA).

Absorption *n* [F & L; F, fr. L] (1741) (1) The penetration of a substance into the mass of another substance by chemical or physical action. See, e.g., ▶ Water Absorption. (2) The process by which energy is dissipated within a specimen placed in a field of radiant energy. Since some part of the impinging energy may be transmitted through the specimen and another part be reflected, the energy absorbed will nearly always be less than that impinging. (3) The adhesion of a substance to the surface of a solid or liquid. Pollutants are extracted by adsorption on activated carbon or silica gel {G Absorption f, F absorption f, S absorción f, I assorbimento m}.

Absorption Coefficient *n* Absorption of radiant energy for a unit concentration through a unit path-length for a specified wavelength and angle of incidence and viewing (Skoog DA, Holler FJ, Nieman TA (1997) Principles of instrumental analysis. Brooks/Coles, New York). See ▶ Absorption Factor, Beer-Bouguer Law, Kubelka-Munk Theory, and ▶ MIE Theory.

Absorption Factor *n* The ratio of the intensity loss by absorption to the total original intensity of radiation. If I_o represents the original intensity, I_r, the intensity of reflected radiation, I_t, the intensity of the transmitted radiation, the absorption factor is given by the expression

$$\frac{I_o - (I_r + I_t)}{I_o}$$

Also called Coefficient of Absorption (McDonald R (1997) Colour physics for industry, 2nd edn. Society of Dyers and Colouritst, West Yorkshire, England).

Absorption Hygrometer *n* Any one of several types of hygrometers containing a hygroscopic substance, the change in length, thickness, or mass of which is a measurable index of the humidity of the atmosphere.

Absorption, Lambert's Law *n* If I_o is the original intensity, I the intensity after passing through a thickness x of a material whose absorption coefficient is k,

$$I = I_o c^{-kx}$$

The index of absorption k' is given by the relation $k = (4\pi k' n)/\lambda$ where n is the index of refraction and λ the wavelength in vacuo. The mass absorption is given by k/d when d is the density. The transmission factor is given by I/I_o (Barton AFM (ed) (1983) Handbook of solubility parameters and other cohesion parameters. CRC Press, Boca Raton, FL).

Absorption, Oil *n* Oil absorption of a pigment or extender is recorded as the amount of vegetable drying oil required to convert a given mass or volume of the dry powder to a very stiff putty-like paste, which does not break or separate. It is more usually expressed as the pounds of refined linseed oil required for 100 lb of pigment or g·100 g.

Absorption Spectrophotometry *n* (spectrophotometry) An analytical technique utilizing the absorption of electromagnetic radiation by a specimen (or solution) as a property related to the composition and quantity of a given material in the specimen. The radiation is usually is the ultraviolet, the visible, or the near-infrared portions of the electromagnetic spectrum. When the absorbing medium is in the gaseous state, the absorption spectrum consists of dark lines or bands, being the reverse of the mission spectrum of the absorbing gas. The spectrum of the transmitted light shows broad dark regions are that are not resolvable into lines and have no sharp or distinct edges when the absorbing medium is in the solid or liquid state. In quantitative spectrophotometry, the intensity of the radiation passing through a specimen or solution is compared with the intensity of the incident radiation and with radiation passing through a nonabsorbing solvent (*blank*). The percent absorbed by the solution is exponentially related to the solute concentration (Beer's law) (Skoog DA, Holler FJ, Nieman TA (1997) Principles of instrumental analysis. Brooks/Coles, New York; Willard HH, Dean JA, Merritt LL (1995) Instrumental methods of analysis. Wadsworth, New York). Modern spectrophotometers are capable of generating nearly monochromatic radiation, so they can develop plots of percent absorption vs wavelength – *absorption spectra* – for the test compound. See ▶ Infrared Spectrophotometry.

Absorption Spectrum *n* (1879) The spectrum obtained by the examination of light from a source, itself giving a continuous spectrum, after this light has passed through an absorbing medium in the gaseous state. The absorption spectrum will consist of dark lines or bands, being the reverse of the emission spectrum of the absorbing substance. The spectrum of the transmitted light shows broad dark regions that are not resolvable into lines and have no sharp or distinct edges when the absorbing medium is in the solid or liquid state (Skoog DA, Holler FJ, Nieman TA (1997) Principles of instrumental analysis. Brooks/Coles, New York; Willard HH, Merritt LL, Dean JA (1974) Instrumental methods of analysis, D. Van Nostrand, New York).

Absorption Tinting Strength *n* See ▶ Tinting Strength, Absorption.

Absorptive Power or Absorptivity *n* For any body, the body is measured by the fraction of the radiant energy falling upon the body which is absorbed or transformed into heat. This ratio varies with the character of the surface and the wavelength of the incident energy. It is the ratio of the radiation absorbed by any substance to that absorbed under the same conditions by a black body (Fox AM (2001) Optical properties of solids, Oxford University Press, UK; Driggers RC, Edwards T, Co P (1998) Introduction to infrared and electro-optical systems. Artech House, Oxford, England).

Abut \ə-ˈbət\ *v* [ME *abutten*, partly fr. OF *aboter* to border on, fr. a- (fr. L *ad*-) + *bout* blow, end, fr. *boter* to strike; partly fr. OF *abuter* to come to an end, fr. *a*- + *but* end] (15c) To adjoin at an end; to be contiguous.

Abvolt *n* The cgs electromagnetic unit of potential difference and electromotive force. It is the potential difference that must exist between two points in order that 1 erg of work be done when one abcoulomb of charge is moved from one point to the other. One abvolt is 10^{-8} V (Weast RC (ed) Handbook of chemistry and physics, 52nd edn. CRC Press, Boca Roton, FL).

Acacia Gum \ə-ˈkā-shə ˈgəm\ *n* Water-soluble gum obtained from trees of the acacia species, as an exudation from incisions in the bark. It is water-soluble and is used as an adhesive, thickening agent, and for transparent paints (Whistler JN, BeMiller JN (eds) (1992) Industrial gums: polysaccharides and their derivatives. Elsevier Science and Technology Books, Amsterdam, Netherlands). *Also known as Gum Arabic.*

Academy Board \ə-ˈka-də-mē ˈbōrd\ *n* A board which is given a surface in preparation for painting, primarily oil painting. It is made of paper containing chalk and size and has a face of pale gray or white ground, usually of a white lead, oil, and chalk mixture.

Accelerant \ik-ˈse-lə-rənt, ak-\ *n* (1916) A chemical used to speed up chemical or other processes. For example, accelerants are used in dyeing triacetate and polyester fabrics (Goldberg DE (2003) Fundamentals of chemistry. McGraw-Hill Science/Engineering/Math, New York) {G Beschleuniger m, F accélérateur m, S acelerador m, I acceleratore m}.

Accelerated Aging *n* Any set of conditions designed to produce in a short time the results obtained under normal conditions of aging. In accelerated aging test, the usual factors considered are heat, light, or oxygen either separately or combined (Koleske JV (ed) (1995) Paint and coating testing manual. American Society for Testing and Materials; Paint and coating testing manual (Gardner-Sward handbook) (1995) MNL 17, 14th edn. ASTM, Conshohocken, PA).

Accelerated Life See ▶ Accelerated Aging.

Accelerated Test *n* A test procedure in which conditions such as temperature, humidity, and ultraviolet radiation are intensified to reduce the time required to obtain a deteriorating effect similar to one resulting from exposure to normal service conditions for much longer times.

Accelerated Weathering *n* Tests designed to simulate, but at the same time to intensify and accelerate, the destructive action of natural outdoor weathering on coatings films. The tests involve exposure to artificially produced components of natural weather, e.g., light, heat, cold, water vapor, rain, etc., which are arranged and repeated in a given cycle. There is no universally accepted test, and different investigators use different cycles (Paint and coating testingmanual (Gardner-Sward handbook) (1995) MNL 17, 14th edn. ASTM, Conshohocken, PA). See ▶ Artificial Weathering.

Accelerated Weathering Machine *n* Device intended to accelerate the deterioration of coatings by exposing them to controlled sources of radiant energy, heat, water, or other factors that may be introduced (Koleske JV (ed) (1995) Paint and coating testing manual. American Society for Testing and Materials; Paint and coating testing manual (Gardner-Sward handbook) (1995) MNL 17, 14th edn. ASTM, Conshohocken, PA). See ▶ Weatherometer and ▶ Accelerated Weathering.

Acceleration \ik- ˌse-lə- ˈrā-shən, (ˌ)ak-\ *n* (1531) The time rate of change of velocity in either speed or direction. Cgs unit – 1 cm per second per second. Dimensions, $[l\ t^{-2}]$. See ▶ Angular Acceleration.

Acceleration Due to Gravity *n* The acceleration of a body freely falling in a vacuum. The International Committee on Weights and Measures has adopted as a standard or accepted value, 980.665 cm/s^2 or 32.174 ft/s^2 (Hartland S (ed) (2004) Surface and interfacial tension. CRC Press, Boca Raton, FL).

Acceleration Due to Gravity at any Latitude and Elevation *n* If ϕ is the latitude and H the elevation in centimeters the acceleration in cgs units is, $g = 980.616 - 2.5928 \cos 2\phi + 0.0069 \cos^2 2\phi - 3.086 \times 10^{-6}$ H (Helmert's equation).

Accelerator *n* (1611) (1) Any substance used in small proportion which increases the speed of a chemical reaction. In the paint industry, the term usually indicates materials that hasten the curing or crosslinking of a resin system. In the polyester resin field, it covers more specifically an additive which accelerates the action of the catalyst (Odian GC (2004) Principles of polymerization. Wiley, New York. (2) An organic or inorganic chemical which hastens the vulcanization of rubber, natural, or synthetic, causing it to take place in a shorter time or at a lower temperature. Accelerators, particularly organic, are not mere catalysts of vulcanization, however, because they produce different and generally beneficial states of cure and different degrees of stability or resistance to chemical attack in the vulcanization (Carley JF (ed) (1993) Whittington's dictionary of plastics. Technomic Publishing, Lancaster, Pennsylvania, USA). See ▶ Catalyst and ▶ Crosslinking Agent.

Accomodation *n* The adjustment of the eye to obtain maximum sharpness of the retinal imge for n object at which an observer is viewing. One of the important changes involves the shape of the eye lens.

Accra \ə- ˈkrä\ *n* Natural copal resin of African origin (Langenheim JH (2003) Plant resins: chemistry, evolution ecology and ethnobotany. Timber Press, Portland, OR; (2001) Paint: pigment, drying oils, polymers, resins, naval stores, cellulosics esters, and ink vehicles, vol 3. American Society for Testing and Material, West Conshohoken, Pennsylvania).

Accroides *n* Resinous accumulation which occurs on the leaf and stem of the *Xanthorrhoea* species. It is native to Australia and Tasmania. It appears on the market in red and yellow forms, both of which are soluble in industrial alcohol, and are used in spirit varnishes. The resin is also described as "Black Boy Gum," "Botany Bay Resin," "grass tree gum," "gum acaroid," "acaroid resin," "red gum," and "yacca" or "yacka" gum (Langenheim JH (2003) Plant resins: chemistry, evolution ecology and ethnobotany. Timber Press, Portland, OR; Whistler JN, BeMiller JN (eds) (1992) Industrial gums: polysaccharides and their derivatives. Elsevier Science and Technology Books; Langenheim JH (2003) Plant resins: chemistry, evolution ecology and ethnobotany. Timber Press, Portland, OR).

Accumulator \ə- ˈkyü-m(y)ə- ˌlā-tər\ *n* (1748) Series of rolls which festoon strip metal on a continuous line both at the beginning and at the end. This allows the beginning or the end of the line to stop while the rest of the line is in operation. The accumulator actually accumulates a considerable length of strip, and gives a portion of its strip to whichever end is stopped.

Accumulator *n* (1) In blow molding and injection molding, an auxiliary ram extruder providing fast parison delivery or fast mold filling. The accumulator cylinder is filled with plasticated melt from the main extruder between parison deliveries shots, and stores this melt until the plunger is called upon to deliver the next parison or shot. (2) A pressurized gas reservoir that stores energy in hydraulic systems (Strong AB (2000) Plastics materials and processing. Prentice-Hall, Columbus, OH) {G Akkumulator m, F accumulateur m, S acumulador m, I accumulatore m}.

Acenaphthene *n* $C_{10}H_6(CH_2)_2$. Solid with an mp of 95°C, obtained from coal tar. It is a dyestuff intermediate.

Aceta *n* Cellulose acetate, manufactured by Bayer, Germany.

Acetal \ˈa-sə-ˌtal\ *n* [G *Azetal*, fr. *azet-*, *acet-* + *Al*kohol] (1853) CH$_3$CH(OC$_2$H$_5$)$_2$. (1) A colorless, flammable liquid used in cosmetics and as a solvent. (2) Any of a class of compounds formed from aldehydes combined with alcohol. (3) A group of materials including polyoxymethylene (Delrin^R).

Acetaldehyde \ˌa-sə-ˈtal-də-hīd\ *n* [ISV] (1877) (ethanal, ethyl aldehyde, acetic aldehyde) CH$_3$CHO. Low boiling liquid (21°C). A colorless, flammable liquid made by the hydration of acetylene, the oxidation or dehydrogenation of ethyl alcohol, or the oxidation of saturated hydrocarbons or ethylene.

Acetaldehyde Resin *n* Product of auto-condensation of acetaldehyde.

Acetal Formation, Mechanism of *n* The formation of the (–CHO–) repeat unit which after initiation (i.e., Lewis acids) forms –CHO+ and propagates to the polymer, (–CHO–)$_n$.

Acetal Resin *n* (polyformaldehyde, polyoxymethylene, polycarboxane) A thermoplastic produced by the addition polymerization of a aldehyde through the carbonyl function, yielding unbranched polyoxymethylene (–O–CH$_2$–)n chains of great length. Examples are DuPont's "Delrin" and Hoechst–Celanese's "Celcon" (acetal copolymer based on trioxane). The acetal resins are among the strongest and stiffest of all thermoplastics, and are characterized by good fatigue life, resilience, low moisture sensitivity high resistance to solvents and chemicals, and good electrical properties. They may be processed by conventional injection molding and extrusion techniques, and fabricated by welding methods used for other thermoplastics. Their main area of application is industrial and mechanical products, e.g., gears, rollers, and many automotive parts (Strong AB (2000) Plastics materials and processing. Prentice-Hall, Columbus, OH; Carley JF (ed) (1993) Whittington's dictionary of plastics. Technomic Publishing).

Acetal Resins *n* High-molecular weight, stable, linear polymers of formaldehyde; structurally, an oxygen atom joins the repeating units in an ether rather than ester-type link. These also include butyrals (Strong AB (2000) Plastics materials and processing. Prentice-Hall, Columbus, OH).

Acetals *n* See ▶ Acetal Resins.

Acetamide \ə-ˈse-tə-ˌmīd, ˌa-sə-ˈta-ˌmīd\ *n* [ISV] (1873) CH$_3$CONH$_2$. Amide of acetic acid with a melting point of 81°C and boiling point of 222°C. Generally used as a plasticizer for cellulose esters.

Acetanilide \ˌa-sə-ˈta-nə-ˌlīd, -ləd\ *n* [ISV] (1864) C$_6$H$_5$NHCOCH$_3$. Its physical properties include: mp of 115°C; bp of 304°C; sp gr of 1.21; and flp of 165°C (329°F). *Also known as Phenyl Acetamide*.

Acetate \ˈa-sə-ˌtāt\ *n* (1827) (1) Generic name for fibers from cellulose-2$\frac{1}{2}$ -acetate. (2) A salt or ester of acetic acid. (3) A generic name for cellulose acetate plastics, particularly for their fibers. Where at least 92% of the hydroxyl groups have been acetylated, the term *triacetate* may be used as the generic name of the fiber. (4) A compound containing the acetate group, CH$_3$COO– (e.g., polyvinyl acetate).

Acetate Chromes *n* Lead chromate pigments prepared from lead acetate or basic lead acetate. Available as the lemon, primrose, medium, and orange shades. ((1996) Kirk-Othmer encyclopedia of chemical technology: pigments-powders. Wiley, New York; Solomon DH, Hawthorne DG (1991) Chemistry of pigments and fillers. Krieger Publishing, New York).

Acetate Dope *n* Term applied to cellulose acetate lacquers used for coating aircraft fabrics.

Acetate Fiber *n* A manufactured fiber in which the fiber-forming substance is cellulose acetate (FTC definition). Acetate is manufactured by treating purified cellulose refined from cotton linters and/or wood pulp with acetic anhydride in the presence of a catalyst. The resultant product, cellulose acetate flake, is precipitated, purified, dried, and dissolved in acetone to prepare the spinning solution. After filtration, the highly viscous solution is extruded through spinnerets into a column of warm air in which the acetone is evaporated, leaving solid continuous filaments of cellulose acetate. The evaporated acetone is recovered using a solvent recovery system to prepare additional spinning solution. The cellulose acetate fibers are intermingled and wound onto a bobbin or shippable métier cheese package, ready for use without further chemical processing. In the manufacture of staple fiber, the filaments from numerous spinnerets are combined into tow form, crimped, cut to the required length, and packaged in bales. Acetate fibers are environmentally friendly. Characteristics: Acetate fabrics are breathable, luxurious in appearance, fast-drying, wrinkle and shrinkage resistant, crisp or soft in hand depending upon the end use. End Uses: The end uses of acetate include women's and men's sportswear, evening wear, lingerie, dresses, blouses, robes, coats, other apparel, linings, draperies, bedspreads, upholstery, ribbons, formed fabrics, and filtration products (Complete textile glossary, Celanese Corporation, Three Park Avenue, New York, NY; Vincenti R (ed) (1994) Elsevier's textile dictionary. Elsevier Science and Technology Books, New York).

Acetate Green *n* Range of Brunswick or chrome greens derived from mixtures of acetate chrome and Prussian blue. These can be made wet or dry by mixing. The type made by wet mixing is preferred to paint making, as these greens have less tendency to partial separation in the film, a phenomenon known as "floating."

Acetate Rayon[R] *n* Cellulose acetate made from preswelling cellulose pulp with acetic acid followed by esterification with sulfuric acid–acetic anhydride mixture, the diacetate dissolved in acetone, the triacetate in methylene chloride, and dry spun into fibers (Kadolph SJJ, Langford AL (2001) Textiles. Pearson Education, New York).

Acetates *n* (1) Metallic salts derived from acetic acid by interaction of the metallic oxide, hydroxide, carbonate with the acid, or by the esters derived by interaction of alcohols with acetic acid (Goldberg DE (2003) Fundamentals of chemistry. McGraw-Hill Science/Engineering/Math, New York). Typical metallic salts are lead, cobalt, and manganese acetates. Common esters are ethyl, propyl, isopropyl, butyl, and amyl acetates. Acetate salts have the formula CH_3COOMe, where Me is a monovalent metal. Divalent metals like lead, etc., obviously combine with two acid radicals. The formula given above for the acetate salts apply also to esters, except that the Me becomes an alkyl radical. (2) It is also colloquially for cellulose acetate plastics (Vincenti R (ed) (1994) Elsevier's textile dictionary. Elsevier Science and Technology Books, New York).

Acetic Acid \ə-ˈsē-tik, ˈa-səd\ *n* (F *acétique* fr. L *acetum* vinegar] (1808) CH_3COOH (ethanoic acid, methanecarboxylic acid, vinegar acid). A colorless liquid with the familiar taste and odor of vinegar, it is the chief constituent in dilute form. Acetic acid was originally derived by souring wine and beer, but is synthesized today by oxidation of acetaldehyde in the presence of a catalyst. Among the uses of acetic acid in the plastics industry is the manufacture of cellulosics plastics such as cellulose acetate (CA), CA butyrate and CA propionate, vinyl acetate, and acetate esters for plasticizing thermoplastics. It is a monobasic acid. Its mp is 16°C and bp, 118°C. In the paint industry, its chief applications are in the manufacture of metallic acetates used for the production of driers, and in the manufacture of acetate esters employed as solvents or plasticizers (Goldberg DE (2003) Fundamentals of chemistry. McGraw-Hill Science/Engineering/Math, New York).

Acetic Aldehyde See ▶ Acetaldehyde.

Acetic Anhydride *n* (1976) $(CH_3CO)_2O$. The pungent liquid may be thought of as the condensation product of two molecules of acetic acid by removal of one molecule of water, though in fact it is made by reaction of acetic acid with ketene, $CH_2=C=O$ (Morrison RT, Boyd RN (1992) Organic chemistry, 6th edn. Prentice-Hall, Englewood Cliffs, NJ). It is a strong acetylating agent, used for many of the same purposes as its parent acid. It is an important reagent which has wide application in the manufacture of many raw materials and intermediates for the paint trade, and also in analysis. It is used for the acetylation of hydroxyl groups as in the manufacture of acetyl ricinoleates, and in the manufacture of cellulose acetate. It has a bp of 137°C ((1978) Paint/coatings dictionary, Federation of Societies for Coatings Technology, Philadelphia, Blue Bell, PA).

Acetic Ester and Acetic Ether *n* See ▶ Ethyl Acetate.
Acetic Ether See ▶ Ethyl Acetate.
Acetocopal *n* Product obtained when Congo Copal is reacted with acetic anhydride.
Acetone \ˈa-sə-ˌtōn\ *n* [Gr *Azeton,* fr. L *acetum*] (1839) (dimethyl ketone, 2-propanone) CH_3COCH_3. The simplest and most important member of the ketone family of solvents. All the cellulosics plastics and polyvinyl chloride, polyvinyl acetate, polymethyl methacrylate, epoxies, and some thermosetting resins are soluble in acetone. It is also an intermediate in the production of bisphenol. It is a typical low-boiling ketone. It is a liquid which flashes at ordinary room temperature, has a bp of 57.5°C, fp of −15°C; and a sp gr of 0.788 at 25°C (Ash M, Ash I (1996) Handbook of paint and coating raw materials: trade name products - chemical products dictionary with trade name cross-references. Ashgate Publishing, New York; Weast RC (ed) Handbook of chemistry and physics, 52nd edn. CRC Press, Boca Roton, FL). Also known by its chemical names *Dimethyl Ketone or Propanone.*

Acetone Extraction *n* In molded phenolic products, the amount of acetone-soluble material that can be extracted from the material is an indication of the degree of cure.
Acetone Resin *n* A synthetic resin produced by the reaction of acetone with materials such as phenol or formaldehyde.
Acetonyl Acetone *n* $CH_3COCH_2CH_2COCH_3$. Solvent containing two keto groups in each molecule. It has a boiling range of 188–193°C; a sp gr of 0.973/15°C; a refractive index of 1.449; flp of 85°C (185°F); and vapor pressure of less than 2 mmHg at 30°C.

Acetophenone *n* $CH_3COC_6H_5$. A solvent with a bp of 202°C; mp of 20°C; sp gr of 1.023/25°C; flp of 83°C (180°F); and refractive index of 1.536. *Also known as Phenyl Methyl Ketone or Acetyl Benzene.*

Acetyl \ə-ˈsē-tᵊl, ˈa-sə-; ˈa-sə-ˌtēl\ *n* (1864) Monovalent radical $CH_3CO–$.

Acetylated Congo See ▶ Acetocopal.
Acetylated Damar *n* Product obtained when dammar is reacted with acetic anhydride.
Acetylation *vt* (1900) Reaction wherein the hydrogen atom of an hydroxyl group is replaced by an acetyl radical (CH_3CO). It can really be regarded as an ester formation, except that it is a specific ester, the acetate, which is formed. When the acetyl value of castor or other oils containing free hydroxyl groups, or of the reaction mixture of monoglycerides, or alkyds, is determined, acetic anhydride is commonly used as the acetylating agent. This reacts with any free hydroxyl groups, whether they occur in the fatty acid chains of the vegetable oil or in unreacted polyhydric alcohols. This can also be written as a reaction with acetylene groups. Acetylene is an alkyne or HC≡CH (Morrison RT, Boyd RN (1992) Organic chemistry, 6th edn. Prentice-Hall, Englewood Cliffs, NJ).
Acetyl Benzene \-ˈben-ˌzēn, ben-ˈ\ See ▶ Acetophenone.

Acetyl Chloride *n* CH_3OCl. Acetic acid in which the –OH group has been replaced by –Cl; a very active acetylating agent.

Acetyl Coenzyme A *n* (1952) A compound $C_{25}H_{38}N_7O_{17}P_3S$ formed as an intermediate in metabolism and active as a coenzyme in biological acetylations.

Acetyl Cyclohexane Sulfonyl Peroxide *n* A polymerization initiator, often used in conjunction with a dicarbonate such as di-sec-butyl peroxydicarbonate in vinyl chloride polymerization. The initiators have largely replaced enzoÿl and lauroÿl peroxides, the principal initiators in the early years of PVC production.

Acetylene \ə-ˈse-tᵊl-ən, -tᵊl-ˌēn\ *n* (1864) (ethyne) HC ≡ CH. A colorless (but not ordorless) gas obtained by reacting water with calcium carbide, CAC_2, or by cracking petroleum hydrocarbons. In the plastics industry, it is an important intermediate in the production of vinyl chloride, neoprene, acrylonitrile, and trichloroethylene. See also ▶ Polyacetylene.

Acetylene Black *n* Particularly pure form of carbon black pigment, made by the controlled combustion of acetylene in air under pressure. It is graphitic in nature and has high electrical conductivity (Donnet J-B, Wang M-J (1993) Carbon black. Marcel Dekker, New York). See also ▶ Carbon Black.

Acetylene Polymers See ▶ Polyacetylene.

Acetyl Number (Value) *n* Number of milligrams of potassium hydroxide required to neutralize the acetic acid set free from 1 g of an acetylated compound when the latter is subjected to hydrolysis.

Acetyl Peroxide *n* (diacetyl peroxide) $(CH_3CO)_2O_2$. A polymerization catalyst.

4-Acetyl Resorcinol *n* (2,4-dihydroxyacetophenone) $C_6H_3(OH)_2CO-CH_3$. A light stabilizer for plastics.

Acetyl Ricinoleates *n* Plasticizers, such as butyl and ethyl acetyl ricinoleates. They can be regarded as esters of acetylated ricinoleic acid. The hydroxy group in the ricinoleic chain has been acetylated or esterified, and the carboxylic group has also undergone esterification, but with an alcohol (Wypych G (ed) (2003) Plasticizer's data base. Noyes Publication, New York).

Acetyl Triallyl Citrate *n* $CH_3COOC_3H_4(COOCH_2CH=CH_2)_3$. A cross-linking agent for polyesters and a polymerizable monomer. Easily polymerized with peroxide catalysts, it forms a clean, hard thermosetting resin.

Acetyl Triethyl Citrate *n* $CH_3COOC_3H_4(COOC_2H_5)_3$. A plasticizer produced by esterifying and acetylating citric acid, used in cellulose nitrate, cellulose acetate, and certain vinyls, e.g., polyvinyl acetate. It has been FDA-approved for food-contact use (Wypych G (ed) (2003) Plasticizer's data base. Noyes Publication, New York).

Acetyl Tri-2-Ethylhexyl Citrate n $CH_3COOC_3H_4(COOC_8H_{17})_3$. A plasticizer for vinyls, with limited compatibility for cellulose nitrate and ethyl cellulose.

Acetyl Value n The number of milligrams of potassium hydroxide (KOH) necessary to neutralize the acetic acid liberated by hydrolysis of 1 g of an acetylated compound. It can also be written as a measure of the degree of esterification or combination of acetyl radicals with cellulose in acetate or triacetate products.

Achromatic \▪a-krə-▪ma-tik, (▪)ā-\ *adj* (1766) Having no distinguishable hue; neutral. See ▶ Gray Scale.

Achromatic *n* **Objective** An objective corrected spherically for one wavelength (usually green light) and chromatically for two wavelengths.

ACI *n* Abbreviation for American Concrete Institute.

Acicular \ə▪si-kyə-lər\ *adj* [L *acicula*] (1794) Having a needlelike-shape. It is a term applied chiefly to describe the shape of pigment particles which are long and slender. To be acicular, the length of the particle must be at least three times the width.

Acicular Pigments *n* Pigments whose particles are needle-shaped (Solomon DH, Hawthorne DG (1991) Chemistry of pigments and fillers. Krieger Publishing, New York).

Acid \▪a-səd\ *n* [F *acide*, L *acidus* fr. *acēre*] (1626) (1) Inorganic compound characterized by an ionizable hydrogen atom. With organic acids, however, the definition must be extended to emphasize the ionizable hydrogen atom in question is directly attached, through an oxygen atom, to a carbon atom, which is also attached to another distinct oxygen atom. Thus, e.g., acetic acid has the formula CH_3COOH, and it will be seen from this how the hydrogen atom is located. The inorganic acids, sometimes referred to as mineral acids, are written HCl, HNO_3, H_3PO_4, etc. All these hydrogen atoms, whether organic or inorganic, are capable of being substituted by a monovalent metallic radical, or by monovalent alkyl or aryl groupings. (2) Is a proton donor (Brønsted-Lowry). (3) Is an electron-pair acceptor (Lewis). (4) Increases the concentration of dissolved cations related to the solvent (solvent system) (Goldberg DE (2003) Fundamentals of chemistry. McGraw-Hill Science/Engineering/Math, New York).

Acid Acceptor *n* A compound that acts as a stabilizer by chemically combining with acid that may be initially present in minute quantities in a plastic, or that may be formed during the decomposition in the resin (Odian GC (2004) Principles of polymerization. Wiley, New York). See also ▶ Stabilizer.

Acid Catalysts *n* Acids which may be either organic or inorganic, or salts from these acids which exhibit acidic characteristics. They are used to promote or accelerate chemical reactions, and find special applications in the manufacture and subsequent hardening of synthetic resins. Acid catalysts have been employed in the manufacture of polymerized drying oils, coumarone, urea, phenol- and melamine-formaldehyde resins, and in the cold-setting of compositions containing the last three named resins (Odian GC (2004) Principles of polymerization. Wiley, New York). See ▶ Catalyst and ▶ Urea Formaldehyde Resins.

Acid Curing (Hardening) *n* A process of curing or hardening resins through the use of acid catalysts. These are frequently employed with urea and melamine-formaldehyde resins (OdianGC (2004) Principles of polymerization. Wiley, New York).

Acid-Dyeable Variants *n* Polymers modified chemically to make them receptive to acid dyes. (Vigo TL (1994) Textile processing, dyeing, finishing and performance. Elsevier Science, New York).

Acid Dyes *n* Term given to dyestuffs that possess acidic groupings, e.g., carboxy, sulfonic acid, or both. They form salts with heavy metals like barium and calcium, and this reaction is used in lake formation. It is usually an azo, triarylmethane, or anthraquinone dye with acid substituents such as nitro, carboxy, or sulfonic acid. Acid dyes Include such dyes as eosine, erioglaucine, fluorescein, naphthol yellow, ponceau, quinoline, tartrazine, etc. They precipitate with calcium, barium, or titanium glycerol chlorides or other metallic salts, to give suitable insoluble pigment dyestuffs, or they may be precipitated on bases to form lakes (Kadolph SJJ, Langford AL (2001) Textiles. Pearson Education, New York; Wells K, Beal S, Woodburn C, Durant J, Brandimane J (1997) Fabric dyeing. Interweave Press Incorporated, Amsterdam, Netherlands; Vigo TL (1994) Textile processing, dyeing, finishing and performance. Elsevier Science, New York). See ▶ Dyes.

Acid Fading *n* See ▶ Gas Fading.

Acid Groups *n* Functional groups having the properties of acids. In cellulose and its derivatives, these are usually carboxyl groups.

Acidic \ə-ˈsi-dik, a-\ *adj* (1880) A term describing a material having a pH of less than 7.0 in water (Whitten KW, Davis RE, Davis E, Peck LM, Stanley GG (2003) General chemistry. Brookes/Cole, New York).

Acidic Solution *n* An aqueous solution in which the concentration of hydrogen (hydronium) ions exceeds that of hydroxide ions (Whitten KW, Davis RE, Davis E, Peck LM, Stanley GG (2003) General chemistry. Brookes/Cole, New York).

Acidity \ə-ˈsi-də-tē, a-\ *n* (1620) (1) Measure of the free acid present. (2) In oils, acidity denotes the presence of acid-type constituents whose concentrations are usually defined in terms of the neutralization number, called acid number (Dainth J (2004) Dictionary of chemistry. Oxford University Press, UK; Whitten KW, Davis RE, Davis E, Peck LM, Stanley GG (2003) General chemistry. Brookes/Cole, New York) {G Säuregrad m, F taux d'acidité, taux m, S grado de acidez, grado m, I acidità f}.

Acid Number or Value *n* The number of milligrams of KOH required to neutralize the free acids in one gram of an oil, resin, varnish, or other substance; generally reported on the nonvolatile. See ▶ Acid Value.

Acidolysis *n* A chemical reaction analogous to hydrolysis in which an acid plays a role similar to that of water. See also ▶ Ester Interchange.

Acid Recovery *n* A reclamation process in chemical processing in which acid is extracted from a raw material, by-product, or waste product. In the manufacture of cellulose acetate, acetic acid is a major by-product. Acid recovery consists of combining all wash water containing appreciable acetic acid and concentrating it to obtain glacial acetic acid.

Acid Refined Linseed Oil *n* Linseed oil that has been treated with acid, usually sulfuric, to remove mucilaginous matter.

Acid Resistance *n* The ability of materials to withstand attack by acids, specifically strong mineral acids. The type of acid should be stated (i.e., organic or inorganic). Most plastics have excellent acid resistance. Tests for resistance of plastics to some acids are included in ASTM 543.

Acid Salt *n* A salt, the anion of which can serve as an acid by losing an H^+ (donating a proton) (Goldberg DE (2003) Fundamentals of chemistry. McGraw-Hill Science/Engineering/Math, New York).

Acid Sludge *n* Residue which separates from mineral and related oils when they are refined with sulfuric acid.

Acid Value *n* The measure of the free-acid content of a substance. It is expressed as the number of milligrams of potassium hydroxide (KOH) neutralized by the free acid present in 1 g of the substance. This value, also called acid number, is sometimes used in connection with the end-group method of determining molecular weights of polyesters (Deligny P, Oldring PKT, Tuck N (eds) (2001) Resins for surface coatings, alkyds and polyester, vol 22. Wiley, New York; Patton TC (1962) Alkyd resin technology. Wiley, New York; Martens CR (1961) Alkyd resins. Reinhold Publishing, New York). It is also used in evaluating plasticizers, in which acid values should be as low as possible (Wypych G (ed) (2003) Plasticizer's data base. Noyes Publication, New York). See ▶ Acid Number.

Aclar *n* Fluorinated polycarbonate film. Manufactured by Allied Chemical, U.S.

ACM *n* Copolymers from acrylic ester and 2-chlorovinyl ether.

Acoustical Board \ə-ˈkü-stik or -sti-kəl-\ *n* A low-density, sound-absorbing structural insulating board having a factory-applied finish and a fissured, felted-fiber, slotted or perforated surface pattern provided to reduce sound reflection (Harris CM (2005) Dictionary of architecture and construction. McGraw-Hill, New York).

Acoustical Material *n* Any material considered in terms of its acoustical properties. Commonly, and especially, a material designed to absorb sound.

Acoustical Paint See ▶ Anti-Noise Paint.

Acoustical Plaster *n* A special low-density, sound-absorptive plaster, applied in the form of a finish-coat, to provide a continuous finished surface (Harris CM (2005) Dictionary of architecture and construction. McGraw-Hill, New York).

Acoustical Tile *n* An acoustical material in board form, usually having unit dimensions of 24 by 24 in. (approx. 61 by 61 cm) or less. Usually used on ceilings but also may be applied to sidewalls (Harris CM (2005) Dictionary of architecture and construction. McGraw-Hill, New York).

Acoustic Coating *n* Coating which absorbs or deadens sound (Harris CM (2005) Dictionary of architecture and construction. McGraw-Hill, New York). See ▶ Antinoise Paints.

Acoustic Emission Testing *n* A nondestructive test for determining material or structural integrity by detecting and recording location, amplitude, and frequency of sound emissions as test loads are applied.

Acrilan \ˈa-krə-ˌlan, -lən\ *n* Poly(acrylonitrile). Manufactured by Chemstrand Corp. (Monsanto), U.S. (Tortora PG (ed) (2000) Fairchild's dictionary of textiles, 7th edn. Fairchild Publications, New York).

Acrolein \ə-ˈkrō-lē-ən\ *n* [ISV *acr-* (fr. L *acr-*, *acer*) + L *olēre*] (1857) (propenal, acrylic, or allyl aldehyde) CH$_2$=CHCHO . A liquid derived from the oxidation of allyl alcohol or propylene, used as an intermediate in the production of polyester resins and polyurethanes. It is an unsaturated liquid aldehyde with a bp of 52°C. It possesses a very pungent odor, and has strong lachrymatory properties.

Acrolein Polymers and Resins *n* Homopolymers or copolymers of acrolein.

Acronal Dispersions based on uni- and copolymers of acrylic esters. Manufactured by BASF, Germany.

Acrylamide \ˌa-krˈl- ˈa-ˌmīd\ *n* [*acrylic* + *amide*] (1946) CH$_2$=CHCONH$_2$. A crystalline solid produced by hydrolysis of acrylonitrile; the monomer of ▶ Polyacrylamide and a useful comonomer (Odian GC (2004) Principles of polymerization. Wiley, New York).

2-Acrylamido-2-Methylpropanesulfonic Acid *n* (AMPS) A solid aliphatic sulfonic-acid monomer produced by Lubrizol Corp. Its homopolymers are water-soluble and hydrolytically stable. It can be incorporated into other polymers by crosslinking.

2-acrylamido-2-methylpropanesulfonic acid

Acrylan Rubber *n* Butyl acrylate/5–10% acrylonitrile copolymer. Manufactured by Monomer Corp., U.S.

Acrylate \ˈa-krə-ˌlāt\ *n* (1873) Ester formed from acrylic acid. The term also applies to the metallic salts of this acid (Odian GC (2004) Principles of polymerization. Wiley, New York).

Acrylate Elastomers *n* Elastomeric formed from acrylate and elastomeric monomers such as butadiene (Harper CA (ed) (2002) Handbook of plastics, elastomers and composites, 4th edn. McGraw-Hill, New York).

Acrylate Resin See ▶ Acrylic Resin.

Acrylate Styrene Acrylonitrile Polymer *n* Acrylic rubber-modified thermoplastic with high weatherability. ASA has good heat and chemical resistance, toughness, rigidity, and antistatic properties. Processed by extrusion, thermoforming, and molding. Used in construction, leisure, and automotive applications such as siding, exterior auto trim, and outdoor furniture. Also called ASA.

Acrylic \ə-ˈkri-lik\ *adj* [ISV *acrokein* + *-yl* + ^{1}*-ic*]) (1855) (1) The generic class of polymers and monomers (appox. 1942) derived from acrylic acid including polymethyl methacrylate. (2) Generic name for fibers from at least 85% poly(acrylonitrile) (Odian GC (2004) Principles of polymerization, Wiley, New York; Morrison RT, Boyd RN (1992) Organic chemistry, 6th edn. Prentice-Hall, Englewood Cliffs, NJ) {D Acrylglas n, F verre acrylique, verre m, S vidrio crílico, vidrio m, I vetro acrilico, vetro m}.

Acrylic Acid *n* (ca. 1855) (propenoic acid, vinylformic acid) CH$_2$=CHCOOH. A colorless, unsaturated acid, that polymerizes readily. The homopolymer is used as a thickener and textile-sizing agent and, crosslinked, as a cation-exchange resin. Acrylic-acid esters are widely used as monomers for acrylic resins. Mol wt., 72.06; mp, 14°C; bp, 141.0°C; sp gr, 1.422 (Odian GC (2004) Principles of polymerization. Wiley, New York).

Acrylic Aldehyde See ▶ Acrolein.

Acrylic Coating Polymers *n* Acrylic resins or polymers such as polymethyl methacrylate used for formulating coating.

Acrylic Ester *n* (acryl ester) An ester of acrylic or methacrylic acid or of structural derivatives thereof. Polymers derived from these monomers range from soft, elastic, film-forming materials to hard plastics. They are readily polymerized as homopolymers or copolymers with many other monomers, contributing improved resistance to heat, light, and weathering. Some members of the acrylic-ester family (e.g., butylenes dimethacrylate and trimethylolpropane trimethacrylate) function as reactive plasticizers in PVC and elastomers. They serve as plasticizers during processing, then polymerize while curing to impart hardness to the finished article (Odian GC (2004) Principles of polymerization. Wiley, New York). See also ▶ Acrylic Resin.

Acrylic Esters *n* Elastomeric polymers formed from acrylate and elastomeric monomers such as butadiene (Harper CA (ed) (2002) Handbook of plastics, elastomers and composites, 4th edn. McGraw-Hill, New York).

Acrylic Fiber *n* (1951) A manufactured fiber in which the fiber-forming substance is any long chain synthetic polymer composed of at least 85% by weight of acrylonitrile units [–CH_2–CH(CN)–] (FTC definition). Acrylic fibers are produced by two basic methods of spinning (extrusion), dry and wet (Vincenti R (ed) (1994) Elsevier's textile dictionary. Elsevier Science and Technology Books, New York). In the dry spinning method, material to be spun is dissolved is a solvent. After extrusion through the spinneret, the solvent is evaporated, producing continuous filaments which later may be cut into staple, if desired. In wet spinning, the spinning solution is extruded into a liquid coagulating bath to form filaments, which are drawn, dried, and processed. Uses of acrylic fibers include floor coverings, blankets, and apparel uses such as suitings, pile fabrics, coats, collars, linings, dresses, and shirts (Kadolph SJJ, Langford AL (2001) Textiles. Pearson Education, New York; Complete textile glossary, Celanese Corporation, Three Park Avenue, New York, NY; Tortora PG, Merkel RS (2000) Fairchild's dictionary of textiles, 7th edn. Fairchild Publications, New York).

Acrylic Foam *n* A cellular polymer used for lining drapes and made by mixing an emulsified acrylic resin with compressed air in the ration of one part emulsion to four or five parts air, spreading the foam on a substrate, then drying in an oven. The emulsion may contain fillers and pigments to provide opacity, and a foaming aid such as ammonium stearate. When the coated fabric must have abrasion resistance for washing and cleaning, the acrylic foam can be crushed between rollers to partly collapse the cell structure (Carley JF (ed) (1993) Whittington's dictionary of plastics. Technomic Publishing).

Acrylic Latex *n* Aqueous dispersion, thermoplastic or thermosetting, of polymers or copolymers of acrylic acid, methacrylic acid, esters of these acids, or acrylonitrile.

Acrylic Plastics *n* Thermoplastic or thermosetting plastics of polymers including copolymers of acrylic acid, methacrylic acid, esters of these acids or acrylonitrile.

Acrylic Resin *n* (1936) (1) A polymer composed of acrylic or methacrylic esters, sometimes modified with nonacrylic monomers such as the ABS group. The acrylates may be methyl, ethyl, butyl, or 2-ethylhexyl. Usual methacrylates are the methyl, ethyl, butyl, lauryl, and stearyl. The resins may be in the form of molding powders or casting syrups, and are noted for their exceptional clarity and optical properties. Acrylics are widely used in lighting fixtures because they are slow-burning or even, with additives, self-extinguishing, and do not produce harmful smoke or gases in the presence of flame ((2001) Paint: pigment, drying oils, polymers, resins, naval stores, cellulosics esters, and ink vehicles, vol 3. American Society for Testing and Material; Deligny P, Oldring PKT, Tuck N (eds) (2001) Resins for surface coatings, alkyds and polyester, vol 22. Wiley, New York). (2) Thermoplastic polymers of alkyl acrylates such as methyl methacrylates. Acrylic resins have good optical clarity, weatherability, surface hardness, chemical resistance, rigidity, impact strength, and dimensional stability. They have poor solvent resistance, resistance, rigidity, impact strength, and dimensional stability. They have poor solvent resistance, resistance to stress cracking, flexibility, and thermal stability. Processed by casting, extrusion, injection molding, and thermoforming. Used in transparent parts, auto trim, household items, light fixtures, and medical devices. Also called polyacrylates (Sepe MP (1998) Dynamic mechanical analysis. Plastics Design Library, Norwich, New York).

Acrylic Rubber *n* (AR) A synthetic rubber made at least partly from acrylonitrile, or from ethyl acrylate copolymerized with many of the monomers or block polymers of the synthetic-rubber family (Harper CA (ed) (2002) Handbook of plastics, elastomers and composites, 4th edn. McGraw-Hill, New York).

Acrylide Maroon *n* This group includes the azo pigments based on acrylides of beta hydroxy naphthoic acid (e.g., toluidine maroon); they are characterized by their excellent soap, acid, and alkali resistance and good bake resistance. Poor bleed resistance in aromatic and alcohol solvents, poor lightfastness in other than masstone shades (including metallics), low hiding power and high cost discourage their use except where chemical resistance requirements demand; to this extent they may be considered as specialty pigments (Herbst W, Hunger K (2004) Industrial organic pigments. Wiley, New York).

Acrylonitrile \ ˈa-krə-lō-ˈnī-trəl, -ˌtrēl\ *n* (1893) (propenenitrile, vinyl cyanide) (1) A monomer with the structure CH_2=CH-CN. It is most useful in copolymers. Its copolymer with butadiene is nitrile rubber, and several copolymers with styrene exist that are tougher than polystyrene. It is also used as a synthetic fiber and as a chemical intermediate (Kadolph SJJ, Langford AL (2001) Textiles. Pearson Education, New York). (2) A raw material for the manufacture of synthetic resins and rubbers. It is a liquid at room temperatures, with a bp of 77°C and flp of 0°C (Odian G (2004) Principles of polymerization. Wiley, New York). *Also known as Vinyl Cyanide.*

Acrylonitrile-Butadiene Copolymer *n* (NBR) Any of a family of copolymers ranging from about 18–50% acrylonitrile, and sometimes including small amounts of a third monomer (Odain G (2004) Principles of polymerization. Wiley, New York; Lenz RW (1967) Organic chemistry of synthetic high polymers. Interscience Publishers, New York). The family includes the German materials Perbunan and Buna-N, and the nitrile rubbers. The outstanding property of this nitrile-rubber family is excellent resistance to oils, fats, and hydrocarbons such as motor fuels, making them useful for motor gaskets, abrasion linings, conveyor belts, and hoses for oils and fuels (Carley JF (ed) (1993) Whittington's dictionary of plastics. Technomic Publishing).

Acrylonitrile-Butadiene-Styrene (ABS) *n* Acrylonitrile and styrene liquids and butadiene gas polymerized together in a variety of ratios to produce the family of ABS resins.

Acrylonitrile-Butadiene-Styrene Copolymers *n* Terpolymer of three monomers, forming ABS.

Acrylonitrile Butadiene Styrene Polymer *n* ABS resins are thermoplastics comprised of a mixture of styrene–acrylonitrile copolymer (SAN) and SAN-grafted butadiene rubber. They have high impact resistance, toughness, rigidity, and processability, but low dielectric strength, continuous service termperature, and elongation. Outdoor use requires protective coatings in some cases. Plating grades provide excellent adhesion to metals. Processed by extrusion, blow molding, thermoforming, calendaring, and injection molding. Used in household appliances, tools, nonfood packaging, business machinery, interior automotive parts, extruded sheet, pipe, and pipe fittings. Also called ABS, ABS resin, acrylonitrile–butadiene–styrene polymer.

Acrylonitrile–Butadiene–Styrene Polymer See ▶ Acrylonitrile Butadiene Styrene Polymer.

Acrylonitrile Butadiene Styrene Polymer Alloy *n* A thermoplastic processed by injection molding, with properties similar to ABS but higher elongation at yield. Also called ABS nylon alloy.

Acrylonitrile Butadiene Styrene Polymer Polycarbonate Alloy *n* A thermoplastic processed by injection molding and extrusion, with properties similar to ABS. Used in automotive application. Also called ABS PC alloy.

Acrylonitrile–Butadiene–Styrene Resin *n* See ▶ ABS Resin and ▶ Nitrile Barrier Resin.

Acrylonitrile-Chlorinated PE-Styrene *n* A terpolymer of three monomers by the same names.

Acrylonitrile Copolymer *n* A thermoplastic prepared by copolymerization of acrylonitrile with small amounts of other ensaturated monomers. Has good gas barrier properties and chemical resistance. Processed by extrusion, injection molding, and thermoforming. Used in food packaging.

Acrylonitrile–Styrene Copolymer *n* Any of a group of copolymers that have the transparency of polystyrene, but with improved resistance to solvents and stress cracking.

ACS *n* Abbreviation for the American Chemical Society, headquartered at 1155 Sixteenth St, NW, Washington, DC 20036. The Society's Polymer Chemistry Division holds national meetings and publishes several journals.

Actinic Degradation *n* See ▶ Ultraviolet Degradation.

Actinic Resistance *n* See ▶ Ultraviolet Resistance.

Actinide Series *n* Elements of atomic numbers 89–103 analogous to the lanthanide series of the so-called rare earths.

Actinoid *n* A member of the series of 14 elements following actinium in the periodic table. It is also called *actinide*.

Action \ˈak-shən\ *n* Action is measured by the product of work by time. Cgs units of action are the erg–second and the joule–second. Dimensions $[m\ l^2 t^{-1}]$. Plank's quantum or constant of action is $(6.62517 - 0.00023) \times 10^{-27}$ erg-s (Giambattista A, Richardson R, Richardson RC, Richardson B (2003) College physics. McGraw-Hill Science/Engineering/Math, New York).

Action Stretch *n* A term applied to fabrics and garments that give and recover in both the lengthwise and the widthwise directions. Action stretch is ideal for tight-fitting garments such as ski pants (Vincenti R (ed) (1994) Elsevier's textile dictionary. Elsevier Science and Technology Books, New York).

Activate \ˈak-tə-ˌvāt\ *v* (1626) To put into a state of increased chemical activity.

Activated *adj* Materials, specially treated to confer absorptive, adsorptive, or catalytic properties on them. Such substances include activated alumina, activated earths, and activated carbon.

Activated Carbon *n* (1921) (1) Any form of carbon characterized by high-adsorptive capacity of gases, vapors, and colloidal solids. (2) A highly adsorbent powdered or granular carbon made usually by carbonization and chemical activation and used chiefly for purifying by adsorption (Whitten KW, Davis RE, Davis E, Peck LM, Stanley GG (2003) General chemistry. Brookes/Cole, New York). Also known as *Activated Charcoal*.

Activated Complex *n* A short-lived combination formed by collision of reactant particles in an elementary process; also called *transition state*.

Activation *n* (1626) The process of making more active: to make (as moleclules) reactive or more reactive; inducing radioactivity in a specimen by bombardment with neutron or other types of radiation; rendering a thermoplastic surface more receptive to printing inks, paints, and adhesives by chemical treatment such as carbon and alumina; corona discharge or flame treatment; energetic elevation of a molecule to a state in which it becomes ready to react with another molecule (Merriam-Webster's Collegiate Dictionary (2004) 11th edn. Merriam-Webster, Springfield, MA; Carley JF (ed) (1993) Whittington's dictionary of plastics. Technomic Publishing) {G Aktivierung f, F activation f, S activación f, I attivazione f}.

Activation Energy *n* (1940) (E, E_A) The energy required to facilitate reaction between two molecules or, in the context of the Eyring theory of low, the energy required to cause a molecule of liquid or chain segment of a polymer to "jump" from its present position to a nearby hold (i.e., an empty volume of molecular or chain-segment size) in the liquid. Activation energies are usually expressed per mole of substance (SI: J/mol) and are evaluated by fitting reaction-rate or flow data at several temperatures to an equation of the Arrhenius form, $k = Ae^{-E_a/RT}$ (Watson P, Watson P (1997) Physical chemistry. Wiley, New York). See ▶ Arrhenius Equation.

Activator *n* (1626) (1) An agent added to the accelerator in natural or synthetic resins to enhance the action of the accelerator in the vulcanizing process. (2) A chemical additive used to initiate the chemical reaction in a specific mixture. See ▶ Accelerator and ▶ Catalyst.

Activator, Initiator *n* It is usually necessary to add an activator (e.g., $FeSO_4 \cdot 7H_2$) to attain reasonable rates of polymerization when redox initiators (lauroÿl peroxide and fructose) are used at 0° temperature or below.

Active Mass *n* The active mass of a substance is the number of gram molecular weights per liter in solution, or in gaseous form (Odian GC (2004) Principles of polymerization. Wiley, New York).

Active Metal *n* Metal in a condition of high-chemical activity that is more susceptible to corrosion.

Active Site *n* A location on the surface of a heterogeneous catalyst or an enzyme at which reactant molecules can combine and react with a low required activation energy (Smith MB, March J (2001) Advanced organic chemistry, 5th edn. Wiley, New York; Odian GC (2004) Principles of polymerization. Wiley, New York).

Activity \ak-ˈti-və-tē\ *n* (1530) A quantity which measures the parent or effective concentration (or, for a gas, partial pressure) of a species and which takes into account interparticle interactions which produce nonideal behavior. At low concentrations (or pressures) activity is essentially equal to concentration (or pressure).

Activity Coefficient *n* A factor which, when multiplied by the molecular concentration yields the active mass. The activity coefficient is evaluated by thermodynamic calculations, usually from data on the emf of certain cells, or the lowering of the freezing point of certain solutions. It is a correction factor which makes the thermodynamic calculations correct (Watson P (1997) Physical chemistry. Wiley, New York).

Acute Bisectrix (Bxa) The bisector of the acute optic axial angle for biaxial crystals (Rhodes G (1999) Crystallography made crystal clear: a guide for users of macromolecular models. Elsevier Science and Technology Books, New York).

ACV *n* Abbreviation for ▶ Adams Chromatic Value.

Acyclic \(ˌ)ā-ˈsī-klik, -ˈsi-\ *adj* (1878) Open chain, not ring formation.

Acyl, Acyl Groups *n* When the OH group is removed from a fatty acid molecule, the monovalent residue is described as an acyl radical. Examples of acyl radicals are CH_3CO-acetyl; C_2H_5CO=propionyl; and C_3H_7CO-butyrl. Radicals derived from carboxylic acids by removal of the hydroxyl group (Morrison RT, Boyd RN (1992) Organic chemistry, 6th edn. Prentice-Hall, Englewood Cliffs, NJ).

Acylation *vt* (1907) Formation or introduction of an acyl radical in or into an organic compound. An acyl group (RCO-) becomes attached to the aromatic ring, thus forming a ketone; the process is called acylation; includes Friedel-Craft acylation (Morrison RT, Boyd RN (1992) Organic chemistry, 6th edn. Prentice-Hall, Englewood Cliffs, NJ; Billmeyer FW, Saltzman M (1966) Principles of color technology. Wiley, New York).

Adams Chromatic Value Color Difference Equation
n One of the transformations of CIE color space into another space which is perceived as visually more uniform. It is used to describe the difference between two similar colors: it is based on the Munsell value function (Billmeyer FW, Saltzman M (1966) Principles of color technology. Wiley, New York).

$$Y_c/100 = 1.2219V_Y - 0.23111V_Y^2 + 0.23951V_Y^3 - 0.021009V_Y^4 + 0.000840V_Y^5$$

For determining the V_x, substitute $X_c/0.9804$ for Y_c; for determining V_z, substitute $Z_c/1.1810$ for Y_c in the above equation. The transformed space is then calculated according to the following equations:
a = $V_x - V_Y$
b = 0.4 $(V_z - V_Y)$ or 0.4 $(V_Y - V_Z)$
L = 0.23V_Y The total color difference, ΔE, is calculated as follows

$$\Delta E = \sqrt{(0.23\Delta V_Y + [\Delta(V_x - V_Y)]^2 + [0.4\Delta(V_z - V_Y)]^2}$$

Note that if b is calculated as 0.4 $(V_z - V_Y)$, a plus number indicates blueness; if the b is calculated as 0.4 $(V_Y - V_z)$, a plus number indicates yellowness. The equation is generally used with a normalizing constant placed outside the brackets. If the components L, a, and b are used individually they are also multiplied by the normalizing constant, f. The normalizing constant or factor, f, is used to convert the numbers obtained to the same magnitude as those obtained with the NBD equation. The factor used is generally 40, although it may vary between 40 and 50 and must therefore be specified. Unless specified otherwise, the value 40 is assumed to have been used (McDonald R (1997) Colour physics for industry, 2nd edn. Society of Dyers and Colourists, West Yorkshire, England).

Adaptation \ˈa-dəp-ˈtā-shən, –dəp-\ *n* (1610) Changes in the sensitivity of the eye resulting from changes in the viewing light sources, thus enabling the eye to meet the needs of vision in a very wide range of conditions.

Adapter *n* (1801) (die adapter) In an extrusion setup, the portion of the die assembly that attaches the die to the extruder and provides, inside a flow channel for the molten plastic between the extruder and the die.

Adapter Plate *n* In injection molding, the plate holding the mold to the press frame or platen.

Adapter Ring An annular retaining part of extrusion and injection apparatus.

ADC See ▶ Allyl Diglycol Carbonate.

Addition Polymer *n* Polymer made by addition polymerization (IUPAC).

Addition Polymerization *n* A reaction in which unsaturated monomer molecules join together to provide a polymer in which the molecular formula of the repeating unit is identical with that of the monomer. The molecular weight (i.e., M_w, M_n) of the polymer so formed is thus the total of the molecular weights of all of the combined monomer units. Example: n CH_2=CH_2 → (–CH_2CH_2)$_n$, with molecular weight = n × 28.03, and where n = 100, then 2,803 g/mole of polymer. Polymers formed from monomers without the lose of a small molecule unlike condensation polymerization, the composition of polymer is the same as the monomer, e.g., polyethylene (Odian GC (2004) Principles of polymerization. Wiley, New York; Lenz RW (1967) Organic chemistry of synthetic high polymers. Interscience Publishers, New York).

Additive \ˈa-də-tiv\ *adj* (1699) A supplementary material combined with a base material to provide special properties. For example, pigments are used as dope additives to give color in mass dyeing. Examples are slip additives, pigments, stabilizers, and flame retardants (Bart J (2005) Additives in polymers: industrial analysis and applications. Wiley, New York). Also sometimes called *Modifier* {G Additiv n, F additif m, S aditivo m, I additivo m}.

Additive Color Mixture *n* Color which results when the same area of the retina of the eye is illuminated by lights of different spectral distribution, such as by two or more colored lights. Pigments such as paints obey substractive color mixtures whereas light (sources of illumination) obeys Additive color mixture may result from addition of lights from two or more sources, by visual averaging of s colored dots (dot matrix) as on colored television screens or paper, or by additive color mixture rules. For example, yellow paint mixed with blue paint yields green paint, and yellow light mixed with blue light yields white light (McDonald R (1997) Colour physics for industry, 2nd edn. Society of Deyes and Colourists, West Yorkshire, England; Billmeyer FW, Saltzman M (1966) Principles of color technology. Wiley, New York).

Additive Reaction *n* Chemical reaction in which two components join together to form a single reaction product. In a pure additive reaction, neither of the reactants undergoes molecular fission or splitting, but attaches itself to the other reactant intact. In other additive reactions, one of the reactants may split into two separate parts, each of which attaches itself to the appropriate places of the other intact reactant. There is still, however, a single reaction product (Odian GC (2004) Principles of polymerization. Wiley, New York).

Adduct ə-ˈdəkt, a-\ *vt* [L *adductus*, pp of *adducere*] (ca. 1839) (1) The cyclic product of an addition reaction between one unsaturated compound, such as a diene and another. (2) A crystalline mixture, not a true compound, in which molecules of one of the components are contained within the crystal-lattice framework of the other component. Such complexes are stable at room temperature but the entrapped component can escape when the mixture is melted or dissolved (Odian GC (2004) Principles of polymerization. Wiley, New York).

Adduct Curing Agent See ▶ Crosslinking Agent.

Adeps Lanae, Anhydrous *n* Pharmaceutical name for lanolin or purified wood grease.

Adhere \ad-ˈhir, əd\ *v* [MF *adhérer*, L *adhaerēre* fr. *ad-* + *haerēre*] (1536) To cause two surfaces to be held together by adhesion.

Adherend *n* A body which is held to another body by an adhesive. See also ▶ Substrate.

Adherend Preparation *n* See ▶ Surface Preparation.

Adherometer An instrument that measures the force required to strip a coating from a metal surface ((1978) Paint/coatings dictionary. Federation of Societies for Coatings Technology, Philadelphia).

Adhesion \ad-ˈhē-zhən, əd-\ *n* [F or L; F *adhésion*, fr. L *adhaesion-*, *adhaesio*, fr. *adhaerēre*] (1624) The state in which two surfaces are held together by interfacial forces which may consist of valence forces or interlocking action, or both. One method for testing the strength of adhesive bonds (Skeist I (ed) (1990) Handbook of adhesives. Reinhold Publishing, New York). See also ▶ Adhesion, Mechanical, and ▶ Adhesion, Specific {G Adhäsion f, F adhésion f, S adhesión f, I adesione f}.

Adhesion, Mechanical *n* Adhesion between surfaces in which the adhesive holds the parts together by interlocking action such as "contact cement." See ▶ Adhesion, Specific.

Adhesion Promoter *n* A chemical coating that is applied to a substrate before it is coated with a plastic, to improve the adhesion of the plastic to the substrate. Adhesion promoters include materials such as silanes and silicones with hydrolysable groups on one end of their molecules that react with moisture to yield silanol groups, which in turn react with or adsorb to inorganic surfaces to enable strong bonds to be made. At the other ends of the molecules are reactive, but nonhydrolyzable groups that are compatible with resins or elastomers in adhesive formulations. Adhesion promoters are added to the adhesive as water or ethanol solutions (Skeist I (ed) (1990) Handbook of adhesives. Van Nostrand Reinhold, New York).

Adhesion, Specific *n* Adhesion between surfaces that are held together by valence forces of the same type as those that give rise to cohesion. See ▶ Adhesion, Mechanical.

Adhesive *n, adj, adv* (1670) *n* – Adhesives are used in to bond two or more surfaces. *Adj* – An adhesive material binds two or more surfaces. *Adv* – Two or surfaces or bonded adhesively. The above usages of the different forms of the term were derived from Skeist I (ed) (1990) Handbook of adhesives. Van Nostrand Reinhold, New York. The following information was derived from Skeist I (ed) (1990) Handbook of adhesives. Van Nostrand Reinhold, New York. Adhesives used in all of these applications can be classified into five types. A *monomeric cement* contains a monomer of at least one of the polymers to be joined and is catalyzed so that a bond is produced by polymerization. A *solvent cement* is a product that dissolves the plastics being joined, forming strong intermolecular bonds, then evaporates. *Bonded adhesives* are solvent solutions of resins, sometimes containing plasticizers that dry at room temperature. *Elastomeric adhesives* contain natural or synthetic rubbers either dissolved in solvents or suspended in water or other liquid, and are dried at room or elevated temperatures. *Reactive adhesives* are those containing partly polymerized resins, e.g., epoxies, polyesters, or phenolics that cure with the aid of catalytic hardeners to form a bond (usually thermoset). Further, various descriptive adjectives are applied to the term, adhesive, to indicate certain characteristics as follows: *Physical form* – liquid adhesive, tape adhesive; *Chemical type* – silicate adhesive, resin adhesive; *Material bonded* – paper adhesive, metal–plastic adhesive, can–label adhesive; and *Conditions of use* – hot-setting adhesive. In addition, contact-adhesives, moisture cure adhesives (cyanoacrylate and silicone) are among the many types of specialty adhesives in the ever growing number of products from the developing adhesives industry.

Adhesive Activated Yarns *n* Yarns treated by the fiber manufacturer to promote better adhesion to another material such as rubber and/or allowing easier processing.

Adhesive Assembly *n* An adhesive that can be used for bonding parts, such as in the manufacture of a boat, airplane, furniture, etc., NOTE - The term, "assembly adhesive," is commonly used in the wood industry to distinguish such adhesives (formerly called "joint glues"). It is applied to adhesives used in fabricating finished structures or goods, or subassemblies thereof, as differentiated from adhesives used in the production of sheet materials for sale as such, e.g., plywood, or laminates.

Adhesive Cellular See ▶ Adhesive, Foamed.

Adhesive, Cold-Setting *n* An adhesive that sets at temperatures below 20°C (68°F) (See also Adhesive, Hot-Setting; Adhesive, InterMediate Temperature Setting; and Adhesive, Room Temperature Setting).

Adhesive, Contact *n* An adhesive that is apparently dry to the touch and which will adhere to itself instantaneously upon contact; also called *Contact Bond Adhesive* or *Dry Bond Adhesive*.

Adhesive Dispersion *n* A two-phase system in which one phase is suspended in a liquid.

Adhesive, Edge Jointing *n* Adhesive used to bond strips of veneer together by their edges in the formation of larger sheets.

Adhesive Film *n* A thin film of dry resin, usually a thermoset, used as an interleaf in the production of laminates such as plywood. Heat and pressure applied in the laminating process cause the film to bond the layers together.

Adhesive, Foamed *n* An adhesive, the apparent density of which has been decreased substantially by the presence of numerous gaseous cells dispersed throughout its mass.

Adhesive, Heat Activated *n* A dry adhesive film that is rendered tacky or fluid by application of heat or heat and pressure to the assembly.

Adhesive, Hot Melt *n* An adhesive that is applied in a molten state and forms a bond on cooling to a solid state.

Adhesive, Hot-Setting *n* An adhesive that requires a temperature at or above 100° C (212°F) to set the adhesive.

Adhesive, Intermediate Temperature *n* An adhesive that sets in the temperature range 31–99°C (87–211°F).

Adhesive Migration *n* In nonwovens, the movement of adhesive together with its carrier solvent in a fabric during drying, giving it a nonuniform distribution within the web, usually increasing to the outer layers.

Adhesive, Multiple Layer *n* A film adhesive, usually supported, with a different adhesive composition on each side; designed to bond dissimilar materials such as the core to face bond of a sandwich composite.

Adhesive, Pressure-Sensitive *n* A viscoelastic material that in solvent-free form remains permanently tacky. The material will adhere instantaneously to most solid surfaces and can be removable (e.g., adhesive labels for garments).

Adhesive, Room Temperature *n* An adhesive that sets in the temperature range of 20–30°C.

Adhesives *n* In textiles, materials which cause fibers, yarns, or fabrics to stick together or to other materials.

Adhesive, Separate Application *n* A term used to describe an adhesive consisting of two parts, one part being applied to one adherend and the other part to the other adherend and the two brought together to form a joint.

Adhesive, Solvent *n* An adhesive having a volatile organic liquid as a vehicle. Note – This term excludes water-based adhesives.

Adhesive, Solvent Activated *n* A dry adhesive film that is rendered tacky just prior to use by application of a solvent.

Adhesive Tape Test See ▶ Tape Test.

Adhesive, Warm-Setting *n* A term that is sometimes used as a Syn: ▶ Intermediate Temperature Setting Adhesive. See ▶ Adhesive, Intermediate Temperature Setting.

Adiabatic \ ˌa-dē-ə-ˈba-tik, ˌā-ˌdī-ə-\ *adj.* [Gk *adiabatos* impassable, fr. *a-* + *diabatos* passable, fr. *diabainein* to go across, fr. *dia-* + *bainein* to go] (1890) (Merriam-Webster's Collegiate Dictionary (2004) 11th edn. Merriam-Webster, Springfield, MA). Denoting a process or system in heat is not added or removed, but while work may be delivered to or by the system (Ready RG (1996) Thermodynamics. Pleum Publishing, New York).

Adiabatic Change *n* A change which takes place with no gain or loss of heat (Ready RG (1996) Thermodynamics. Pleum Publishing, New York).

Adiabatic Extrusion See ▶ Autothermal Extrusion.

Adipate Plasticizer *n* For plasticizers derived from adipic acid (Wypych G (2003) Plasticizer's data base. Noyes Publication, New York), see

▶ Benzyloctyl Adipate
▶ Bis(2,2,4-trimethyl-1,3-pentanediol)diisodecyl–
▶ Monoisobutyrate–diisooctyl–
▶ Dibutoxyethoxy ethyl–dimethoxyethyl–
▶ Dibuto-xyethyl–di(methyl-cylo-
▶ Dicapryl–dinonyl–
▶ Dienthyl–furfuryl–
▶ Di(2-ethylhexyl) –*n*-octyl-*n*-decyl–
▶ Di-*n*-hexyl–
▶ Ditetrahydrofurfuryl

All of the above are ▶ Polypropylene Adipates (Carley JF (ed) (1993) Whittington's dictionary of plastics. Technomic Publishing).

Adipates *n* Esters of adipic acid.

Adipic Acid *n* [ISV] (1877) (hexanedioic acid, 1.4-butanedicarboxylic acid) A dicarboxylic acid used in the production of polyamides, alkyd resins, and urethane foams. Esters of adipic acid are used as plasticizers and lubricants. It is used in the polymerization reaction to form nylon 66 polymers and in the manufacture of polyurethane foams Acid value, 767; molecular weight, 146.1; mp, 151°C; and bp, 216°C/15 mmHg.

Adiponitrile Carbonate *n* (ADNC) See ▶ 5,5′-Tetramethylene Di-(1,3,4-Dioxazol−2−One).

Adiprene *n* Polyurethane elastomer. Manufactured by DuPont, U.S.

ADNC Abbreviation for ▶ Adiponitrile Carbonate.

Adobe Brick *n* Large, roughly molded, sundried clay brick of varying sizes.

Adronal See ▶ Cyclohexyl Acetate.

Adsorbed Water *n* Water which is held on the surface of a material by secondary bonding forces (e.g., hydrogen bonding), and its physical properties are substantially different from those of absorbed water or chemically combined water at the same temperature and pressure. This type of water is often called a ▶ Hygroscopic film (Complete Textile Glossary (2000) Celanese Acetate LLC, Three Park Avenue, New York; Goldberg DE (2003) Fundamentals of chemistry. McGraw-Hill Science/Engineering/Math, New York).

Adsorbent *n* \- bənt\ *n* [(1917) Substance offering a suitable active surface, upon which other substances may be adsorbed.

Adsorption *n* \ad-ˈsórp shən, -ˈzórp\ *n* [*ad* + ab*sorption*] (1882) (1) The concentration of molecules of a particular kind at the interface between two phases such as the pigment and vehicle in printing inks. Adsorption can effectively remove a component such as the drier from an ink vehicle (Leach RH, Pierce RJ, Hickman EP, Mackenzie MJ, Smith HG (eds) (1993) Printing ink manual, 5th edn. Blueprint, New York). (2) The attraction of gases, liquids, or solids to surface areas of textile fibers, yarns, fabrics, or any material (Tortora PG (ed) (1997) Fairchild's dictionary of textiles. Fairchild Books, New York).

Adulteration \ə-ˌdəl-tə-ˈrā-shən\ *n* (1506) Presence of inferior materials which reduce the quality of the standard.

Advanced Composite *n* Polymer, resin, or other matrix-material system in which reinforcement is accomplished via high-strength, high-modulus materials in continuous filament form or is discontinuous form such as staple fibers, filberts, and in-situ dispersions (Harper CA (ed) (2002) Handbook of plastics, elastomers and composites, 4th edn. McGraw-Hill, New York).

Advanced Fiber *n* Any reinforcing fibers characterized by either very high strength, modulus, or high-operating temperature, beyond those of more familiar fibers such as glass, nylon, and polyester. Examples are aramid, fibers of some metals, carbon, boron, silicon carbide and silicon nitride, and whiskers of metals and inorganics (Chung DD (1994) Carbon fiber composites. Elsevier Science and Technology Books, New York; Pittance JC (ed) (1990) Engineering plastics and composites. SAM International, Materials Park, OH).

Advanced Resin *n* Any of a new multiclass of thermoplastics of carious chemical natures, distinguished from the established and more common plastic materials by higher strength, modulus, or serviceability at 200+°C. These materials also command premium prices and some require special processing. Examples are polyimides, polyetheretherketone, liquid–crystal polymers, polytetrafluoroethylene, and polybenzimidazole (Odian GC (2004) Principles of polymerization. Wiley, New York; Modern plastics encyclopedia, McGraw-Hill/Modern Plastics, New York).

Advancing Colors *n* Colors that give the illusion of being closer to the eye than their complementary colors when both are adjacent and in the same plane; colors ranging from yellow-green to scarlet are examples.

Aeration Cell *n* Electrolytic cell in which the driving force to cause corrosion results from a difference in the amount of oxygen in solution at one point as compared to another. Corrosion is accelerated where the oxygen concentration is least or most.

Aerobic \ˌa(-ə)r-ˈō-bik, ˌe(-ə)r-\ *adj* (1884) Term referring to processes that can occur only in the presence of oxygen.

Aerogel *n* A porous, foam-like network with very small "cells," whose substance may be silica, a polymer or a carbonized polymer. Densities can range from half of solid values to as low as 3 mg/cm^3.

Aerograph *n* A spray gun for coatings, using air.

Aerosol *n* \ˈar-ə-ˌsäl, ˈer-, -ˌsól\ *n* (1923) A suspension of liquid or solid particles in a gas (e.g., smoke, fog mist). In the packaging industry, the term means a self-contained sprayable product in which the propellant force is supplied by a compressed or liquidified gas (e.g., isopentane) for an "aerosol spray paint."

Aerosol Coating *n* A conveniently packaged spray coating in a sealed can. Pressure is supplied by compressed liquefied gas.

Aerugo See ▶ Verdigris.

Aesthetics \es-▮the-tik, is-, *British usually* ēs-\ *adj* [Gr *ästhetisch*, NL *aestheticus*, Gk *aisthētikos* of sense of perception, fr. *aisthanesthai* to perceive] (1798) In textiles, properties perceived by touch and sight, such as the hand, color, luster, drape, and texture of fabrics or garments.

Affine Deformation *n* A deformation in which each element in the volume distorts in the same way as does the volume as a whole.

Affinity \ə-▮fi-nə -tē\ *n* [ME *affinite*, MF *afinité*, L *affinitas*, fr. bordering on, related by marriage, fr. *ad* + *finis* end, border] (14c) With respect to an adhesive, affinity is attraction or polar similarity between the adhesive and an ▶ Adherend.

African Ochre See Pigment Yellow (42/77492).

After-Bake *n* A technique used with phenolic and amino resins to increase the output of a molding press by ejecting moldings before they are fully cured, subsequently completing the cure by baking them. After-baking may also be done with fully cured parts to improve their electrical properties and heat resistance.

Afterburner \▮bər-nər\ *n* (1947) An air pollution abatement device that removes undesirable organic gases through incineration.

Aftercure *n* A continuation of the process of curing or vulcanization after the cure has been carried to the desired degree and the source of heat removed, generally resulting in over-cure and a product less resistant to aging than properly cured products.

After-Flame *n* In an ignition test, persistence of flame after removal of the ignition source.

After-Flame Time *n* The duration of ▶ After-Flame.

Afterglow \▮af-tər ▮glō\ *n* (1871) The flameless, glowing combustion of certain solid materials that occurs after the removal of an external source of ignition or after the cessation of combustion of the material.

After-Tack *n* Film defect in which the coated surface, having once reached a tack-free stage, subsequently develops a sticky condition. The effect may be due to syneresis (i.e., expulsion of liquid from a gel). Also applies to printing inks.

Aftertreatment *n* Any treatment done after fabric production. In dyeing, it refers to treating dyed material in ways to improve properties; in nonwovens, it refers to finishing processes carried out after a web has been formed and bonded. Examples are embossing, creping, softening, printing, and dyeing.

Ag Chemical symbol for the element silver (Tortora PG (ed) (1997) Fairchild's dictionary of textiles. Fairchild Books, New York).

Agar (or Agar-Agar) \▮a-gər\ *n* [Malay *agar-agar*] (1889) Gelling hydrocolloid from sea algae plants which are polymers of galactose.

Ager A steam chamber used for ageing printed or padded material.

Age Resistance Resistance to deterioration with time.

Agglomerate \ə-▮glä-mə- ▮rāt\ *vt* [L *agglomerates*, pp of *agglomerare* to heap up, join fr. *ad-* + *glomer-*, *glomus* ball] (1684) (1) Cluster of individual; particles. (2) Cluster of dry pigment particles held together by surface forces. The spaces between the particles are filled with air; may also be present in liquid paints of the pigment has not been properly dispersed. See ▶ Aggregate.

Agglomeration *n* (1774) Condition in which particles become united into clusters of individual particles. May be loosely used to refer to undispersed material.

Aggregate \▮a-gri-gət\ *adj* [ME *aggregate*, fr. L *aggregatus*, pp of *aggregare* to add to, fr. *ad-* + *greg-*, *grex* flock] (15c) (1) A group of dry pigment particles held together by their surface forces; the spaces between the particles are filled with air. See ▶ Agglomerate. (2) An inert granular material, such natural sand, manufactured sand, gravel, crushed gravel, crushed stone, vermiculite, perlite, and air-cooled blast furnace slag, which when bound together into a conglomerate mass by a matrix, forms concrete, or mortar. (3) In the reinforced-plastics industry, a mixture of a hard, fragmented material with an epoxy binder, used as a flooring or surfacing medium, or in epoxy tooling {*aggregation* G Aggregatzustand m, F état d'aggrégation, état m, S estado de agregación, estado m, I stato di aggregazione, stato m}.

Aggressive Tack See ▶ Tack, Dry.

Aging *n* (Ageing) (1) The deterioration of textile or other materials caused by gradual oxidation during storage and/or exposure to light. (2) The oxidation stage of alkali–cellulose in the manufacture of viscose rayon from bleached wood pulp. (3) Originally, a process in which printed fabric was exposed to a hot, moist atmosphere. Presently, the term is applied to the treatment of printed fabric in moist steam in the absence of air. Aging is also used for the development of certain colors in dyeing, e.g., aniline black. (4) The process, or the results of, exposure of plastics to natural or artificial environmental conditions for a prolonged period of time. See also ▶ Artificial Aging, Artificial Weathering. Storage of paints, varnishes, etc., under defined conditions of temperature, relative humidity, etc., in suitable containers, or of dry films of these materials, for the purpose of subsequent tests. See also ▶ Maturing {G Alterung f, F vieillissement m, S envejecimiento m, I invecchiamento m} (Complete textile glossary (2000) Celanese Acetate LLC, Three Park Avenue, New York; Glenz W (ed) (2001) A glossary of plastics terminology in 5 Languages, 5th edn. Hanser-Gardner, Cinicinnati).

Aging, Rubbers *n* The process of oxidation and other degradations of rubber or elastomeric materials.

Aging Time *n* See ▶ Time, Joint Conditioning.

Agitate \ˈa-jə-ˌtāt\ *v* [L *agitatus*, pp of *agitare*, frequentative of *agere* to drive] (15c) To stir or to mix, as in the case of a dyebath or solution.

Agitation *n* Process of mixing or stirring to achieve homogeneity, but not necessarily dispersion.

Agitator *n* Mechanical device used for mixing or stirring.

AI *n* Abbreviation of AMIDE-IMIDE (polymer). See ▶ Polyamide-Imide Resin.

AIA *n* Abbreviation for American Institute of Architects.

AICE Abbreviation for American Institute of Chemical Engineers.

AIMA *n* Abbreviation for Acoustical and Insulating Materials Association.

Air Assist Thermoforming *n* Method of thermoforming in which air is employed to perform the sheet immediately prior to the final pulldown onto the mold, using vacuum.

Air-Assist Vacuum Forming *n* A modification of the process of Sheet Thermoforming in which partial performing of the sheet is effected by air flow or air pressure before vacuum pull-down.

Air Bag *n* An automatically inflating bag in front of riders in an automobile to protect them from pitching forward in an accident. End use for manufactured textile fibers.

Air Blowing *n* Process employed for the production of low temperature blown oils or of high temperature isomerized oils. A large stream of air, sent through vegetable oils under suitable conditions, produces chemical changes in them. These changes are recognized by solubility characteristics, increased viscosity, and improved drying and polymerizing properties; but the changes which occur with a particular oil are governed by its chemical composition ((2001) Paint: pigment, drying oils, polymers, resins, naval stores, cellulosics esters, and ink vehicles, vol 3. American Society for Testing and Material; Shahidi F, Bailey AE (eds) (2000) Bailey's industrial oil and fat products. Wiley). See ▶ Blown Bitumen, Blown Castor Oil, Blown Oil, and ▶ Blown Stand Oil.

Air Brush *n* Very small spray gun, not much larger than a fountain pen, designed as an artist's tool.

Air Brushing *n* Blowing color on a fabric or paper with a mechanized pneumatic brush.

Air Bubble *n* Dry bubble in coating film caused by entrapped air.

Air-Bubble Viscometer *n* An instrument used to measure the viscosities of oils, varnishes, and resin solutions by matching the rate of rise of an air bubble in the sample liquid with the rate of rise in one of a series of standard liquids, whose viscosities are known. The Gardner-Holt bubble viscometer is such an instrument (Paul N. Gardner, Company, Inc., 316 N. E. Fifth Street, Pompano Beach, Fl, www.gardco.com).

Air Cap *n* Perforated housing for atomizing air at head of spray gun.

Air Columns Frequency of vibration.

Air Conditioning *n* (1) A chemical process for sealing short, fuzzy fibers into a yarn. Fabrics made from air-conditioned yarns are porous. Because they allow more air circulation, these fabrics are also cooler. (2) Control of temperature and/or humidity in work or living space.

Air Contamination *n* Foreign substances introduced into the air which make the air impure.

Air Cure *n* Vulcanization which takes place at room temperature with the use of ultrafast or fast-acting accelerator.

Air Drying See ▶ Drying.

Air Ducts *n* Pipes that carry warn air or cold air to rooms and back to furnace or air conditioning system.

Air Entangled Yarns See ▶ Compacted Yarns.

Air Entraining Agents *n* Natural wood resins, fats, inorganic materials, sulfonated compounds, and oils for air entrapment in concrete.

Air Entrapment *n* Inclusion of air bubbles in coating film or in other solids such as concrete. See ▶ Microvoids.

Air Flotation *n* Process used to separate light from heavy pigment particles by a strong current of air. The air stream is so arranged that the pigment particles are carried vertically from the grinding mill. By this means only the finest particles are carried away, the larger and heavier ones falling back to be re-ground.

Air Forming *n* A process in which air is used to separate and move fibers to fashion a web such as the Kroyer® process for short fibers, usually of wood pulp; or the Rando-Webber® process for staple-length fibers.

Air Gap *n* (1) In extrusion of film, sheet or a coating, the distance from the die opening to the nip formed by the pressure roll and the chill roll. (2) In the radio-frequency heating of plastics and corona treatment of films, the space between the electrode and the surface of the material.

Air Jet A type of sandblasting gun in which the abrasive is conveyed to the gun by partial vacuum.

Air Jet Spinning *n* A spinning system in which yarn is made by wrapping fibers around a core stream of fibers with compressed air. In this process, the fibers are drafted to appropriate sliver size, then fed to the air jet chambers where they are twisted, first in one direction, then in the reverse direction in a second chamber. They are stabilized after each twisting operation.

Air Jet Texturing See ▶ Texturing.

Air-Knife Coating *n* A coating technique especially suitable for thin coatings such as adhesives, wherein a high-pressure jet of air is forced through orifices in a knife to meter and control the thickness of the coating. See also ▶ Spread Coating.

Air-Laid Nonwovens *n* Fabrics made by an air-forming process. The fibers are distributed by air currents to give a random orientation within the web and a fabric with isotropic properties.

Airless Blast Deflashing *n* The process of removing flash from molded parts by bombarding them with tiny nonabrading pellets that break off flash by impact. See also ▶ Blast Finishing.

Airless Spray *n* A system of applying paint under high pressure in which the paint is broken up into droplets when it enters to lower-pressure region outside of the gun tip. Very little air is used as compared to conventional air spray.

Airless Spraying *n* Process of atomization of paint by forcing it through an orifice at high pressure. This effect is often aided by the vaporization of the solvents, especially if the paint has been previously heated. The term is not generally applied to those electrostatic spraying processes which do not use air for atomization. See also ▶ Hydraulic Spraying.

Air-Lock *n* Surface depression on a molded part, caused by trapped air between the mold surface and the plastic material.

Air Loss *n* Loss in mass by a plastic or coating on exposure to air at room temperature.

Air Permeability *n* The porosity or the ease with which air passes through material. Air permeability determines such factors as the wind resistance of sailcloth, the air resistance of parachute cloth, and the efficacy of various types of air filters. It also influences the warmth or coolness of a fabric.

Airplane Fabric *n* A plain, tightly woven, water-repellent fabric traditionally made of mercerized cotton. During World War I, the fabric was treated with a cellulose acetate dope and used to cover the wings, tail, and fuselage of airplanes. Today, similar fabrics made from nylon or polyester/cotton blends are used in rainwear and sportswear.

Air Pollutants, Hazardous *n* Materials discharged into the atmosphere that have a proven relationship to increase human death rates.

Air Pollution The presence of contaminants in the air in concentrations that interfere directly or indirectly with man's health, safety, or comfort.

Air Quality Control Regions *n* Geographical units of the country, as required by U.S. Law, reflecting common air pollution problems, for purposes of reaching national standards.

Air-Quality Regulations *n* Federal, state and/or local regulations constructed for the purpose of protecting air quality, e.g., low volatile organic compounds regulations.

Air Quality Standards *n* The prescribed level of pollutants in air that cannot be exceeded during a specified time in a specified geographical area.

Air Ring *n* (air-cooling ring) In the process of blowing tubular film, a circular manifold with one or more annular openings concentric with and just above the die lip that blows a uniform stream of air on or along the plastic tube. The air may be refrigerated.

Air Sampling *n* Determining quantities and types of atmospheric contaminants by measuring and evaluating a representative sample of air. The most numerous environmental hazards are chemical, and can be conveniently divided into (a) the particulates and (b) the gases or vapors. Particulates are mixtures or dispersions of solid or liquid particles in air and included dust, smoke, mist, and similar materials.

Air Separation See ▶ Air Flotation.

Air Shot *n* (air purge) In injection molding, a shot made with the nozzle withdrawn from the mold, so that the expelled melt may fall freely and be caught for measurement of ▶ Cup Temperature or for weighing.

Airslip Forming *n* (airslip vacuum forming) A variation of ▶ Snap-back Forming in which a male mold is enclosed in a box so that, as the mold advances toward the softened sheet, air is trapped between mold and sheet. The ballooning sheet is thus kept from touching the mold during most of the latter's advance. At the end of the stroke, vacuum is applied, destroying the air cushion, and the sheet is sucked against the plug. See also ▶ Sheet Thermoforming.

Air-Supported Roof *n* A fabric-based roofing system that is supported and held in place by air pressure.

Air Vent *n* A passageway, typically a fine groove or scratch, between a mold cavity and the outside edge of the mold face, that allows air to escape as melt is injected into the cavity. See ▶ Burn Mark.

Al Chemical symbol for the element aluminum.

Alabaster \\▮a-lə-▮bas-tər\\ *n* [ME *alabaster*, fr. MF, fr. L *alabaster* vase of alabaster, fr. Gk *alabastros*] (14c) Fine-grained, translucent variety of very pure gypsum, generally white or delicately shaded.

Alathon Ethylene/vinyl acetate copolymer. Manufactured by DuPont, U.S.

Albatross \\▮al-bə-▮trós, -▮träs\\ *n* [prob. alter. of obs. *alcatrace* frigate bird, fr. Sp or Pg *alcatraz* pelican, fr. Arabic *al-ghaṭṭās*, a kind of sea eagle] (1672) A soft, lightweight wool or wool blend fabric in a plain weave with a napped, fleecy surface that resembles in texture, the breast of the albatross. It is usually light-colored and is used in negligees, infants' wear, etc.

Albedo \\al-▮bē-(▮)dō\\ *n* [fr. L *albus*] (ca. 1859) The fraction of incident light or other electromagnetic radiation that is reflected by a surface.

Albertol, Alberlat *n* Modified phenolic resins. Manufactured by Chem. Werke Albert, Germany.

Albino Bitumen *n* Pale bitumen of petroleum origin, similar in general physical characteristics to the black types, but distinguished by its deep golden brown color.

Albite \\▮al-▮bīt\\ *n* [Sw *albit*, fr. L *albus*] (ca. 1843) $NaAlSi_3O_8$. A widely distributed white feldspar. One of the common rock-forming plagioclase groups.

Albumin \\al-▮byü-mən; ▮al-▮byü-, -byə-\\ *n* [ISV *albumen* + *-in*] (1869) Water-soluble protein derived from egg whites or animal blood, used in coatings and adhesives.

Alburnum Softer part of the wood between the inner part and the heart wood. Syn: ▶ Sapwood.

Alchemy \\▮al-kə-mē\\ *n* [ME *alkamie, alquemie*, fr. MF or ML; MF *alquemie*, fr. ML *alchymia*, fr. Arabic *al-kīmiyā'*, fr. *al* the + *kīmiyā'* alchemy, from LGk *chēmeia*] (14c) The chemical period from about 300 B.C. to A.D. 1500 during which a central goal was the transmutation of base metals into gold.

Alcohol \\▮al-kə-▮hól\\ *n* [Fr. L, fr. ML, powdered antimony, fr. OSp, from Arabic *al-kuÖul* the powdered antimony, from *kuÖl* kohl] (1672) (1) A generic term for organic compounds having the general structure ROH. In aliphatic alcohols, R has the formula C_nH_{2n+1}, as in methyl alcohol, CH_3OH, or n-butanol, C_4H_9OH. In more complex alcohols R may be other alkyl, acyclic, or alkaryl groups. Alcohols are classified according to the number of –OH groups they contain – monohydric, dihydric, trihydric, or polyhydric. Dihydric alcohols are called glycols; trihydric alcohols are also known as *glycerols*; and the term polyol is used for any polyhydric alcohol. Alcohols have many important applications in the plastics industry. They are used directly as solvent applications in the plastics industry. They are used directly as solvent and diluents. Many esters of alcohols with organic acids are plasticizers. As intermediates, alcohols are used in the production of resins such as acrylics, alkyds, aminos, polyurethanes, and epoxies. (2) Specifically, ethanol, C_2H_5OH.

Alcohol, Denatured *n* Ethyl alcohol that has been adulterated with a toxic material such as acetaldehyde or benzene so as to render it unfit for human consumption but still useful as an industrial solvent or, occasionally, as a reactant.

Alcohol Resistance *n* Ability of a dried or cured coating to withstand the damaging effects of (ethyl) alcohol.

Alcohols *n* A family of organic solvents containing the grouping C–OH, used in flexographic and gravure inks. The most common members of this group are: Methyl (wood) alcohol, Ethyl (grain) alcohol, Propyl and isopropyl alcohol.

Alcoholysis *n* By analogy to *hydrolysis*, any chemical reaction in which an alcohol acts in a way similar to that of water. See also ▶ Ester Interchange. Its general chemical reaction involves an ester exchange. The cleavage of a C–C bond by the addition of an alcohol.

Alcove \\▮al-▮kōv\\ *n* [F *alcôve*, fr. Sp *alcoba*, fr. Arabic *al-qubbah* the arch] (1676) A small recessed space, opening directly into a larger room.

Aldehyde \\▮al-də-▮hīd\\ *n* [Gr *Aldehyde*, fr. NL *al. dehyde.*, abb. *alcohol dehydrogenatum* dehydrogenated alcohol] (ca. 1846) (1) A generic term for organic compounds containing a double-bonded oxygen and hydrogen bonded to the same terminal carbon atom of the molecule, i.e., the –CHO group. Thus they may be represented by the general formula RCHO. The simplest one is formaldehyde, HCHO, in which R is hydrogen. For all other, R represents a hydrocarbon radical. (2) Any of a class of highly reactive organic chemical compounds obtained by oxidation of primary alcohols, characterized by the common group –CHO and used in the manufacture of resins, dyes, and organic acids.

Aldehyde Resin *n* Synthetic resin made by treating various aldehydes with condensation agents. Phenol, urea, aniline, and melamine react readily with aldehydes, such as formaldehyde, which are also called

Aldehyde Resins {G Aldehydharz n, F résine aldéhyde, résine f, S resina aldehídica, resina f I resina aldeidica, resina f}.

Aleurites *n* Botanical name for a species of shrub or tree which provides vegetable oils used in varnish manufacture. The various *Aleurites* include A. *fordii*, A. *Montana*, A. *cordate*, A. *trisperma*, and A. *moluccana*. The first three yield tung oils, the fourth, bagillumbang oil, and the last, candlenut oil.

Aleuritic Acid *n* $C_{15}H_{28}(OH)_3COOH$, DL-erythro -9, 10,6 – trihydroxy-hexadecanoic acid. Acid obtained from the major constituent of the hard lac portion of shellac. It is present as an interester acid, of monocarboxylic type, which on hydrolysis with caustic soda yields a mixture of two sodium salts, sodium aleuritate, and sodium lacollate.

Alfin Catalyst *n* A catalyst obtained from alkali alcoholates derived from a secondary alcohol, used for polymerizing olefins.

Alfin Polymerization *n* The "Alfin" catalyst (Morton, 1964; Reich, 1966), consists of a suspension in an inert solvent like pentane of a mixture of an alklenylsodium compound (such as allyl sodium), an alkoxide of a secondary alcohol (such as isoproxide), and an alkali halide (such as sodium chloride); the catalyst is highly specific for the polymerization of dienes into the 1,4 forms.

Alfrey-Price Equation *n* This copolymer equation enables you to predict the composition of the polymer as a function of the current ratio of monomers and the relative reactivities for monomers reacting with themselves or with the comonomer. *Also called the Copolymerization Equation*:

$$\frac{d[M_1]}{d[M_2]} = \frac{[M_1]}{[M_2]} \cdot \frac{r_1[M_1] + [M_2]}{[M_1] + r_2[M_2]}$$

The ratio of rate constant, r, for homo-addition, k_{11} or k_{22}, over cross-addition, k_{12} or k_{21}, has been termed the reactivity ratio, r_1 or r_2, for the monomer (Odian GC (2004) Principles of polymerization. Wiley, New York).

Algae \ˈal-jē\ *n* [L, seaweed] (1551) Unicellular or polycellular plants (*thallophytae*), a class of cryptograms, which live in fresh or salt water and are distinguished from fungi by the presence of chlorophyll and response to photosynthesis (e.g., seaweeds, kelps) (Black JG (2002) Microbiology, 5th edn. Wiley, New York). See ▶ Anti-Fouling Composition.

Algicide (Algaecide) \ˈal-jə-ˌsīd\ *n* (1904) Chemical agent used to destroy algae.

Algin \ˈal-jən\ *n* (1883) Any of various colloidal substances (as an alginate or alginic acid) derived from marine brown algae and used esp. as emulsifiers or thickeners. Soluble salts (sodium, ammonium, and potassium) of a polyuronic acid (mannuronic acid), a polymer obtained from brown sea plants.

Alginate \ˈal-jə-ˌnāt\ *n* (ca. 1909) Any derivative of alginic acid; alginates are used as emulsifying agents, as thickeners, and in films.

Alginate Fiber *n* Fiber formed from a metallic salt (normally calcium) of alginic acid, which is a natural polymer occurring in seaweed. Alginate fiber is soluble in water.

Alginates See ▶ Algin.

Alginic Acid *n* Polysaccharide composed of beta D mannuronic acid residues found in certain seaweeds.

Algoflon Poly(tetrafluoroethylene). Manufactured by Montedison, Italy.

Aliphatic \ˌa-lə-ˈfat-ik\ *adj* [ISV, fr. Gk *aleiphat-*, *aleiphar* oil, fr. *aleiphein* to smear, perhaps akin to Gk *lipos* fat] (1889) Designating a large class of organic compounds (and their radicals) having open-chain structures and consisting of the paraffin, olefin, and acetylene hydrocarbons and their derivatives. Examples are butane, isopropyl alcohol, many fats and oils, adipic acid, amyl acetate, ethyl amines. The name also applies to petroleum products which are straight-chain hydrocarbon derived from a paraffin-base crude oil {G aliphatisch, F aliphatique, S alifático, I alifatico}.

Aliphatic Amino Group *n* $-NH_2$ radical when attached to a chain.

Aliphatic Compounds *n* A class of organic compounds which are composed of open chains of carbon atoms. These include paraffins, olefins, etc.

Aliphatic Isocyanates *n* Open chained structures containing the isocyanate group, $–N=C=O$. Aliphatic isocyanates are used extensively in paints and coatings, providing excellent UV stability, chemical resistance and hardness with flexibility to a variety of coatings and elastomers.

Aliphatic Solvent *n* Hydrocarbon solvents comprised primarily of paraffinic and cycloparaffinic (naphthenic) hydrocarbon compounds. Aromatic hydrocarbon content may range from less than 1% to about 35%.

Aliphatic Solvents *n* Organic liquids having an open chain hydrocarbon structure and KB values below 40. They are relatively poor solvents for printing ink resins. Examples are VM&P naphtha, textile spirits and mineral oils (Cf., Aromatic Solvents).

Alizarin \ə-ˈli-zə-rən\ *n* [prob. fr. F *alizarine*] (ca. 1835) 1,2-dihydroxyanthraquinone, the raw material in making many pigments. It is obtained from madder roots, or made synthetically, and is also used in the manufacture of lake pigments.

Alizarin Madder Lake See ▶ Alizarin Red.

Alizarin Red *n* Pigment Red 83(58000). Light maroon pigment produced by treating alizarin with calcium and aluminum salts in the presence of alumina. It is identical with madder lake. Syn: ▶ Alizarin Madder Lake, Alizarin Red B, and ▶ Dihydroxy Anthraquinone Lake.

Alkali \ˈal-kə-ˌlī\ *n* [ME, fr. ML, from Arabic *al-qili*] (14c) Any of the hydroxides and carbonates of the alkali metals (lithium, sodium, potassium), and the radical ammonium. The term is also used more generally for any strong base in aqueous solution capable of forming salts. See ▶ Base.

Alkali Blue *n* Complex organic blue toner prepared by the phenylation of para rosaniline or fuchsine.

Alkali Cellulose *n* Also called regenerated cellulose, pure cellulose obtained from wood pulp, etc., reacted with strong alkali solutions at low temperatures, and is a further step in the reaction of mercerization; as a fiber is called Rayon, viscose. See ▶ Regenerated Cellulose.

Alkali Metal *n* (ca. 1885) A member of group AI in the periodic table.

Alkaline \ˈal-kə-lən, -ˌlīn\ *adj* (1677) A term used to describe a material having a pH greater than 7.0 in water.

Alkaline Catalysts *n* Hydroxides of sodium, potassium, lithium, and ammonium, or salts derived from these metallic radicals, which exhibit alkaline characteristics. Gaseous ammonia can also be used, as well as a number of basic organic compounds. The most important reactions in which alkaline catalysts are involved are in the condensation of phenols with formaldehyde, the condensation of urea with formaldehyde, and in the isomerization of drying oils.

Alkaline-Earth Metal *n* (1903) Any of the bivalent strongly basic metals of group II of the periodic table comprising beryllium, magnesium, calcium, strontium, barium, and radium. *Known also as Alkaline Earth.*

Alkali Refined Linseed Oil *n* Raw linseed oil treated with alkali to reduce the free acidity by formation of eater-soluble salts, which are subsequently removed by washing.

Alkali Resistance *n* (1) The ability of a plastic material to withstand the action of an alkali. ASTM D 543 lists several alkalis as reagents for testing the chemical resistance of plastics. (2) The degree to which a coating resists reaction with alkaline materials such as lime, cement, plaster, soap, etc.

Alkali-Resistant Paint See ▶ Alkali Resistance.

Alkali-Resistant Red *n* (12315,12355) Azo pigments based on acrylides of beta hydroxy naphthoic acid and sometimes called naphthol reds; the distinction between them and the acrylide maroons is strictly tinctorial. For the most part, they have identical properties except the reds are somewhat easier grinding but poorer in bake resistance than the maroons. These reds find some application in latex paints, particularly for application over alkaline substrates; they are not sufficiently light-fast for exterior exposure.

Alkali-Soluble Resins *n* These are generally lower molecular weight (than conventional lattices) polymers containing about 5–15% carboxyl groups which require amine and/or cosolvent to solubilizer them. These systems are generally dispersions of micelles rather than true solutions. Abbreviation for ASR.

Alkane \ˈal-ˌkān\ *n* [*alk*yl + -*ane*] (1899) The generic term for any saturated, aliphatic hydrocarbon, i.e., a compound consisting of carbon and hydrogen only and containing no double or triple bonds. Linear alkanes are representable as C_nH_{2n+2} while cyclic alkanes have the general formula C_nH_{2n}. Examples are propane, C_3H_8, and cyclohexane, C_6H_{12}.

Alkane-Imide Resin *n* A thermoplastic introduced by Raychem Corp under the trade name Polyimidal. The polymer retains high strength up to 200°C and melts at 302°C. It has good electrical properties, low water absorption, and high solvent resistance.

Alkathene Poly(ethylene) (high pressure) manufactured by ICI, Great Britain. Synonyms: Ethylene polymer, Ambythene, Etherin, Hizex, Grez, Biocolene C, Epolene C and E.

Alkene \ˈal-ˌkēn\ *n* [ISV *alk*yl + *ene*] (1899) An unsaturated hydrocarbon with the general formula C_nH_{2n}. Contains a C=C double bond. Same as ▶ Olefin

(Morrison RT, Boyd RN (1992) Organic chemistry, 6th edn. Prentice-Hall, Englewood Cliffs, NJ).

Alkyd \ˈal-kəd\ *n* [blend of *alkyl* and *acid*] (1929) Originally, alcohol and acid nouns formed the term "alkyd" (Wicks ZW, Jones FN, Pappas SP (1999) Organic coating. Wiley-Interscience, New York). Synthetic resins formed by the condensation of polyhydric alcohols with polybasic acids, including anhydrides. They may be regarded as complex esters. The most common polyhydric alcohol used is glycerol, and the most common polybasic acid is phthalic anhydride. Modified alkyds are those in which the polybasic acid is substituted in part by a monobasic acid, of which the vegetable oil fatty acids are typical (Gooch JW (2002) Emulsification and polymerization of alkyd resins. Kluwer Academic/Plenum Publishers, New York; Patton (1962) Alkyd resin technology. Wiley, New York) {G Alkydharz n, F résine alkyde, résine f, S resina alquídica, resina f, I resina alchidica, resina f}.

Alkydal *n* Trade name for a polyester resin, manufactured by Bayer, Germany.

Alkyd Molding Compound *n* A compound based on an ▶ Alkyd Resin containing fillers, pigments, lubricants, and other additives. Alkyd molding compounds are chemically similar to polyesters, but the term alkyd is usually applied to those polyester formulations that use lesser quantities of monomers of the high-viscosity or dry types, resulting in free-flowing granular and nodular types. The compounds are used for applications requiring good electrical properties and long-term dimensional stability such as automotive distributor caps, rotors, and coil caps. Alkyds can be compression molded at low pressure, they cure rapidly, and they present no venting problems because no volatiles are liberated during cure.

Alkyd Resin *n* A polyester resin resulting from the condensation of a polyfunctional alcohol and acid, typically glycerine, and phthalic anhydride (See ▶ Glyptal). Today the term is mostly used for (1) resins modified with drying oils and used as vehicles for varnishes and paints and (2) for crosslinking resins in ▶ Alkyd Molding Compounds. The word *alkyd* is an acronym, from *al-* for alcohol, and *–cid* (changed to *kyd*) for acid {G Alkydharz n, F résine alkyde, résine f, S resina alquídica, resina f, I resina alchidica, resina f}.

Alkyl \ˈal-kəl\ *adj* (1882) A general term for a monovalent aliphatic hydrocarbon radical, which may be represented as having been derived from an alkane by dropping one hydrogen from the formula, C_nH_{2n+2}. Examples of alkyl groups are C_2H_5- (ethyl) and $(CH_3)_2CH_2CH-$ (isobutyl).

Alkyl Aluminum Compound *n* Any of a family of organo-aluminum compounds widely used as catalysts in the Ziegler-process polymerization of olefins. Members include trialkyl compounds such as triethyl-, tripropyl-, and triisobutyl aluminums; alkyl aluminum hydrides such as diisobutyl aluminum hydride and diethyl aluminum hydride; and alkyl aluminum halides such as diethyl aluminum chloride.

Alkylaryl Phosphate *n* (octylphenyl phosphate) $(C_8H_{17}O)(C_6H_5O)PO$. A phosphate diester, a plasticizer for cellulose acetate butyrate, ethyl cellulose, polystyrene, and vinyl resins.

Alkylaryl Phthalate *n* Any of a family of diesters of phthalic acid containing two alkoxy groups, one aliphatic and one aromatic, used as plasticizers for cellulosic plastics, polymethyl methacrylate, polystyrene, and vinyl resins.

Alkylation \ˌal-kə-ˈlā-shən\ *n* (1900) The introduction of an alkyl radical into an organic molecule.

Alkyl Group *n* Monovalent aliphatic radicals derived from aliphatic hydrocarbons by removal of a hydrogen.

Alkyl Phenolic Resin *n* Phenol-formaldehyde resin in which the phenol used has an alkyl group in the para position. In resins used in coatings, the most common are the tertiary butyl and tertiary amyl phenols.

Alkylthio Cadmium *n* A stabilizer for PVC.

Alkyne \ˈal-ˌkīn\ *n* [*alkyl* + *-yne*, alter. of *–ine*] (ca. 1909) C_nH_{2n-2}. A hydrocarbon containing at least one pair of carbon atoms linked by a triple bond (–C≡C–). The simplest alkyne is acetylene, HC≡CH. *adj* Signifying the presence in a compound of the triple bond.

Alligatoring \ˈa-lə-ˌgā-tər-iŋ, ēŋ\ *n* [S *el lagarto* the lizard, fr. *el* the (fr. L *ille* that + *lagarto* lizard, fr. (assumed) VL *lacartus*, fr. L *lacertus, lacerta*] (1579) A form of paint failure in which cracks form on the surface layer only. It is caused by the application of thick films where the underlying surface remains relatively soft. The effect is often caused during weather aging.

Allobar *n* A form of an element differing in isotopic composition from the naturally occurring form.

Allomerism *n* A similarity of crystalline form with a difference in chemical composition. See ▶ Polyallomer.

Alloprene *n* Chlorinated rubber. Manufactured by ICI, Great Britain.

Allotonic Additive which changes the surface tension of water.

Allotropes \ˈa-lə-ˌtrōp\ *n* [ISV] (ca. 1889) Different forms of an uncombined element.

Allotropy \ə-ˈlä-trə-pē\ *n* (1850) (1) The existence of a substance in two or more solid, liquid, or gaseous forms due to differences in the arrangement of atoms of molecules. Examples are amorphous, graphite, and diamond forms of carbon; NO_2 and N_2O_4. See ▶ Monotropic and ▶ Enantiotropic. (2) Property which an element or compound possesses, of existing in different forms which in themselves have different characteristics. These various forms are described as allotropic modifications. Carbon, e.g., is found in an amorphous form as carbon black, and in the crystalline form as graphite, and as a diamond.

All-Over Pattern *n* The typical effect produced by a wallcovering. A pattern in which the units of design are evenly distributed over a surface, without undue emphasis.

Allowable Stress *n* In engineering design, the maximum stress to which a structure or structural element may be subjected under the expected operating conditions. The allowable stress is normally less, by a sizeable ▶ Factor of Safety than the stress of the same type that would cause the member to fail under the same conditions.

Alloy \ˈa-ˌlȯi *also* ə-ˈlȯi\ *n* [F *aloi*, fr. OF *alei*, fr. *aleir* to combine, fr. L *alligare* to bind] (1604) A blend of a polymer or copolymer with other polymers or elastomers. An important example is a blend of styrene–acrylonitrile copolymer with butadiene–acrylonitrile rubber. The term *polyblend* is sometimes used for such mixtures. Some writers restrict the term allow to mixtures of polymers that form a single phase, reserving the term *blend* for nonhomogeneous mixture. The sale of plastics blends and alloys worldwide was 1.3 billion pounds (0.95 Tg) in 1987 and was predicted to have increased by more than 50% by 1992.

Allyd Aldehyde See ▶ Acrolein.

Allyl \ˈal-əl\ *adj* [ISV, fr. L *allium* garlic] (1854) The group $CH_2=CH-$, e.g., allyl alcohol ($CH_2=CH-CH_2OH$). The unsaturated radical, $CH_2=CHCH_2-$, which upon liberation forms biallyl (1,5-hexadiene), a pungent, volatile liquid.

Allyl Alcohol *n* (propenyl alcohol, AA, 2-propene-1-ol) $CH_2=CH_CCH_2-OH$. A colorless liquid with a characteristic pungent odor synthesized by hydrolysis of allyl chloride (from propylene) with dilute caustic, or by the dehydration of propylene glycol. It is a basic material for all allyl resins, and its esters are used as plasticizers.

Allyl Chloride *n* (3-chloropropene, α-chloropropylene, AC) $CH_2=CH-CH_2Cl$. Used in the preparation of allyl alcohol and various thermosetting resins.

Allyl Cyanide *n* (3-butenoic acid nitrile, allyl carbylamine, vinylacetonitrile) $CH_2=CHCH_2CN$. Used as a crosslinking agent.

Allyl Diglycol Carbonate *n* (ADC) A colorless, water-clear monomer that can be polymerized and cast into a variety of transparent, optical-grade products. It can be copolymerized with other unsaturated monomers such as vinyl acetate, maleic anhydride, and methyl methacrylate to produce polymers with a wide spectrum of properties.

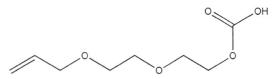

Allyl Diglycol Carbonate Resin *n* A thermosetting-resin group with outstanding optical clarity, good mechanical properties, and the highest scratch resistance of all transparent plastics. The resins are made by polymerizing the monomer of the same name with catalysts such as benzoÿl peroxide or, preferable, diisopropyl peroxy dicarbonate.

Allyl Esters *n* Esters of allyl alcohol, used in the production of plasticizers and resins.

Allyl Resins *n* Resins formed by the addition polymerization of compounds containing the group $CH_2=CHCH_2$, such as esters of allyl alcohol with dibasic acids. They are commercially available as monomers, as partly polymerized prepolymers, and as molding compounds. The dominant compound in the family is diallyl phthalate (DAP). Others are diallyl isophthalate (DAIP), diallyl maleate (DAM), and diallyl chlorendate (DAC). The monomers and partial polymers may be cured with peroxide catalysts to thermosetting resins that are stable at high temperatures and have good solvent and chemical resistance. The molding compounds may be reinforced with glass fibers or other reinforcements, and are easily molded by compression- and transfer-molding methods.

Allyl Starch *n* Soft, gummy mass prepared by the reaction of starch with allyl chloride in the presence of strong alkali.

Almaciga *n* Native Philippine name for manila copal.

Almond Oil *n* Vegetable nondrying oil, with an iodine value of about 99, and sp gr of 0.918/15°C.

Alpaca \al-ˈpa-kə\ *n* [Sp, fr. Aymara *allpaqa*] (1811) (1) Long, fine hair from Alpaca sheep. (2) A fabric from alpaca fibers or blends (originally a cotton cloth with alpaca filling), that is used for dresses, coats, suits, and sweaters. It is also used as a pile lining for jackets and coats (The term has been incorrectly used to describe a rayon fabric).

Alpaca Stitch *n* A 1 × 1 purl-links stitch that is knit so that the courses run vertically instead of horizontally as the fabric comes off the knitting machine. A garment made with an alpaca stitch is not always 100% alpaca; it can be made of other natural or manufactured fibers.

Alpha- \ˈal-fə\ *n* [ME, fr. L, fr. Gk, of Semitic origin; akin to Hebrew *āleph* aleph] (13c) A prefix, usually ignored in alphabetizing compound names, and usually abbreviated by the Greek letter α, signifying that the substitution is on the carbon atom immediately adjacent to the main functional group of the compound. An example is α-aminobutanol, CH_3–$CH_2CH(NH_2)CH_2OH$. Similarly, substituents on the second and third carbon atoms distant from the main functional group are designated β and γ, respectively.

Alpha Cellulose *n* One of three forms of cellulose. Alpha cellulose has the highest degree of polymerization and is the chief constituent of paper pulp and chemical dissolving-grade pulp. It is a colorless filler obtained by treating wood pulp with alkali, used in light-colored thermosetting resins such as urea formaldehyde and melamine formaldehyde. The material is sometimes treated with resinous agents to coat the individual particles and reduce water absorption of the finished articles. Also see ▶ Beta Cellulose and ▶ Gamma Cellulose.

Alpha-Hydroxypropionic Acid See ▶ Lactic Acid.

Alpha-Methylstyrene *n* $C_6H_5C(CH_3)$=CH_2. A colorless liquid which can be polymerized by acids to resinous products and will copolymerize with other monomeric vinyl compounds. It has a bp of 165°C; sp gr of 0.914 at 20°/4°.

Alpha Olefins *n* Olefins having 5–20 carbon atoms. See ▶ Olefin.

Alpha Paper *n* Paper made from purified wood cellulose, often beautifully preprinted when used as surfacing sheets of decorative laminates.

Alpha Particle *n* (1903) (alpha ray) A charged particle, essentially a helium nucleus, emitted during the radioactive decay of certain elements. Alpha particles have little penetrating power, typically dissipating their energy in passing through a few centimeters of air, but they can do harm if released within the human body. In simple form it is a radioactive emission consisting of two protons and two neutrons as in a ^4_2He nucleus.

Alpha-Pinene See ▶ Pinene.

Alpha-Protein *n* Protein obtained from soya beans which is chiefly glycinin. The process of extraction of the protein from the beans involves dissolution of the protein and separation from cellulose, removal of carbohydrates, and hydrolysis of the protein to reduce its chemical complexity. Its uses are similar to those of casein, with which it has much in common, being employed in water paints, emulsions, and adhesives of all types.

ALSIMAG® *n* Registered trademark of American Lava Corporation for ceramic materials. These materials are used in guides and discs on textile processing machines and fiber manufacturing equipment.

Alternating Copolymer *n* A polymer in which two different mer units alternate along the chain in a regular pattern of –A–B–A–B-... See also ▶ Graft and ▶ Block Polymer.

Alternating Current *n* (AC) Current in which the charge-flow periodically reverses, as opposed to direct current, and whose average value is zero. Alternating current usually implies a sinusoidal variation of current and voltage. This behavior is represented mathematically in various ways:

$$I = I_o \cos(2\pi ft + \phi)$$
$$I = I_o \angle \phi$$
$$I = I_I e^{jwt}$$

where f is the frequency; $w = 2\pi f$, the pulsatance, or radian frequency; ϕ the phase angle; I_o the amplitude; and I_1 the complex amplitude. In the complex rotation, it is understood that the actual current is the real part of I. For circuits involving also a capacitance C in farads and L in henrys, the impedance becomes,

$$\sqrt{R^2 + \left(2\pi fL - \frac{1}{2\pi fC}\right)^2}$$

Also known as A.C.

Alternating Polymers n Copolymers composed of monomers in uniform alternating succession along the chain which require highly specific copolymerization reactivity ratios.

Alternating Strain Amplitude n Related through the complex modulus to the ▶ Alternating Stress Amplitude.

Alternating Stress n A stress mode typical of fatigue tests in which the specimen is subjected to stress that varies sinusoidally between tension and compression, the two maximum stresses being equal in magnitude. In some tests, the stress cycles between zero and a tensile maximum, or other unsymmetrical limits. The term applies also to other modes of loading, such as bending and torsion.

Alternating-Stress Amplitude n A test parameter of a dynamic fatigue test, others being frequency and environment. It is one half the algebraic difference between the highest and lowest stress in one full cycle.

Alternating Twist n A texturing procedure in which S and Z twist are alternately inserted in the yarn by means of a special heating arrangement.

Altitudes with the Barometer n If b_1 and b_2 denote the corrected barometer readings at two stations, t the mean of the temperatures, t_1 and t_2 of the air at the two stations, e_1 and e_2 the tension of water vapor at the two stations, h the mean height above sea level, Φ the latitude; then the difference in elevation in centimeters is H = 1,843,000 (log b_i − log b_2) (1 + 0.00367t) (1 + 0.0026 cos 2ϕ + 0.00002h + $\frac{3}{8}k$) where

$$k = \frac{1}{2}\left(\frac{e_1}{b_1} + \frac{c_2}{b_2}\right)$$

An approximate formula, sufficient for differences not over 1,000 m is

$$H = 1,600,000 \frac{b_1 - b_2}{b_1 + b_2}(1 + 0.004t).$$

Alumina n \ə-ˈlü-mə-nə\ n [NL, fr. L *alumin-*. *alumen* alum] (1801) (corundum) The oxide of aluminum, Al_2O_3, very refractory and next to diamond and boron nitride in hardness, obtained by the calcinations of bauxite. Alumina powder is used as a fire-retardant filler in plastics and, over the past 2 decades, ▶ Alumina Fibers have enjoyed increasing use as reinforcements for plastics, metals, and even ceramics. Its density is 3.965 g/cm³.

Alumina-Blanc Fixe n Composite color base used in the manufacture of certain types of lakes. It is made by first reacting aluminum sulfate with soda ash, and then adding barium chloride solution to form barium sulfate from the sodium sulfate derived from the first reaction.

Alumina Fiber n A class of reinforcing fibers available as whiskers or continuous filaments, with quite different properties. Whiskers are almost pure Al_2O_3 (corundum) and are grown by passing a stream of moist hydrogen over aluminum powder heated to 1,300–1,500°C. Their strength ranges from 4 to 24 GPa, modulus ranges from 400 to 1,000 GPa, and they cost about \$15/g. Continuous filaments are lower in crystallinity and/or alumina content (densities range from 2.7 to 3.7 g/cm³), tensile strengths range from 1.3 to 2.1 GPa, and moduli from 105 to 380 GPa, depending on the manufacturer.

Alumina Hydrate n A white inorganic pigment used as an extender in inks and noted for its transparency. See ▶ Aluminum Oxide, Hydrated.

Alumina Trihydrate n (aluminum hydroxide, aluminum hydrate, hydrated aluminum oxide) $Al_2O_3 \cdot 3H_2O$ for $Al(OH)_3$. A white crystalline powder, alumina trihydrate accounts for about half of all flame retardants used in plastics. When heated above about 220°C, it releases water endothermically.

Aluminum n (in UK, Aluminium) \ə-ˈlü-mən-nəm\ n [NL, fr. *alumina*] (1812) See ▶ Aluminum Powder and ▶ Aluminum Paste.

Aluminum Alkyl n (aluminum trialkyl) See ▶ Alkyl Aluminum Compound.

Aluminum Arsenite n Admixture with alumina used as a base for lakes, its function being to improve the brilliance of the color. See ▶ Aluminum Phosphate.

Aluminum Chelate n Chemically modified aluminum secondary butoxide, used as a curing agent for epoxy, phenolic, and alkyd resins.

Aluminum Dihydroxy Stearate n $Al(OH)[OOC(CH_2)_{10}CHOH(CH_2)_5-CH_3]_2$. A white material used as a plastics lubricant.

Aluminum Distearate Al(OH)[OOC(CH$_2$)$_{16}$CH$_3$]$_2$. A white powder used as a lubricant for plastics.

Aluminum Hydrate See ▶ Aluminum Oxide, Hydrated.
Aluminum Ink See ▶ Silver Ink.
Aluminum Isopropylate (Aluminum isopropoxide) Al[OCH(CH$_3$)$_2$]$_3$. A white solid, crosslinking agent.

Aluminum Mixing Varnish *n* Any of the vehicles for aluminum paste or powder which have the property of causing the metallic flakes of the pigment to "leaf" or float, and that retains this property for an indefinite period of time. They are characterized by low acid number, usually less than 15; but in some specifications, the acid number is required to be less than 7. Ready mixed aluminum vehicles are those that are: made without lead drier, used in factory prepared paints, and capable of retention of excellent leafing properties for very long periods of time. Ready-to-mix aluminum vehicles are mixed with aluminum paste or powder, and used within a few hours at the job site. Vehicles for nonleafing aluminum paints are rarely referred to as aluminum mixing varnishes.

Aluminum Monostearate *n* Al(OH)$_2$[OOC(CH$_2$)$_{16}$CH$_3$]. A white or yellowish-white powder, a stabilizer.

Aluminum Naphthenate *n* Pale brown, jelly-like product, containing approximately 4.0% of aluminum, although the precise metal content varies with the type of naphthenic acid used in its manufacture.

Aluminum Oleate *n* Al[OOC(CH$_2$)$_7$CH=CH(CH$_2$)$_7$CH$_3$]$_3$. A plastics lubricant.

Aluminum Oxide See ▶ Aluminum Oxide, Hydrated.

Aluminum Oxide Abrasive *n* Produced in electric furnaces by purifying bauxite of a crystalline form and adding various amounts of titanium to impart extra toughness. The grain is a tough, durable abrasive characterized by the long life of its cutting edges. It is stable chemically and will not react with hydrochloric or nitric acids or with metals other than the alkaline metals. See ▶ Emery.

Aluminum Oxide Cloth *n* Extremely hard, sharp, and enduring coated abrasive cloth particularly suited for finishing metals, capable of the hardest service.

Aluminum Oxide, Hydrated *n* Al(OH)$_3$. Crystalline powder, balls, or granules. Used as a pigment or a base for organic lakes and as an extender for inks. Density, 2.4 g/cm^3 (19.8–20.2 lb/gal); O.A., 41–53 lb/100 lb. Particle size, 0.38–8.5 μm. Syn: ▶ Alumina Hydroxide, ▶ Alumina Trihydrate, ▶ Alumina Hydrate, ▶ Hydrated Alumina, and ▶ Gibbsite.

Aluminum Oxide Paper *n* Extremely hard, sharp and enduring coated abrasive paper, particularly suited for hardwoods, lacquer, and metals.

Aluminum Paint *n* A coating consisting of a mixture of metallic aluminum pigment in powder or paste form, dispersed in a suitable vehicle.

Aluminum Palmitate *n* Al$_2$(OH)$_2$[OOC (CH$_2$)$_{14}$CH$_3$]. A plastics lubricant.

Aluminum Paste *n* Metallic aluminum flake pigment in paste form, consisting of aluminum, solvent, and various additives. The metallic aluminum pigment can be in the form of very small, coated leaves or amorphous powder, known under the respective designations of "leafing" and "nonleafing." See ▶ Leafing.

Aluminum Phosphate *n* Admixture with alumina used as a base for lakes, its function being to improve the brilliance of the color. See ▶ Aluminum Arsenite.

Aluminum Pigment See ▶ Aluminum Powder.

Aluminum Potassium Silicate *n* 3Al$_2$O$_3$·K$_2$O·SiO$_2$·2H$_2$O. Pigment White 20 (77019). Generically known as mica, a complex hydrous aluminum silicate, based on several mineralogically related groups. The most common is Muscovite. Mica crystals have well developed basic cleavage that permit slitting into thin flaky pigment particles. Widely used in paints, rubber, and sealants. Density, 2.82 g/cm^3 (23.5 lb/gal); O.A., 56.74. Particle size 5–20 μm. Hardness (moh), 2.5. Syn: ▶ Mica, ▶ Ground Muscovite, ▶ Graphitic Mica, and ▶ Sericite.

Aluminum Powder *n* Minimum metal used as a pigment. Available in leafing or nonleafing forms. Density, 2.73 g/cm^3 (22.5 lb/gal).

Aluminum Primer *n* (1) Primer specifically formulated for aluminum metal. (2) Primer containing a proportion of aluminum pigment, but distinguished from aluminum paints in which the aluminum is designed to float to the top of the film giving metallic brilliance, a feature undesirable in a primer. Aluminum primers are used on resinous timber or timber which has been treated with oil-soluble wood preservatives.

Aluminum Resinate *n* Brown soft mass, insoluble in water; used for water proofing and as a drier in varnish.

Aluminum Silicate *n* Any of a large group of minerals with various proportions of Al$_2$O$_3$ and SiO$_2$, occurring naturally in clays. They are used as pigments and fillers in plastics.

Aluminum Silicate (Clay) *n* Al$_2$O$_3$·2SiO$_2$·2H$_2$O. Pigment White 19 (77005) White inert pigment of little color and opacity, obtained from certain natural deposits of china clay, kaolin, feldspar, and similar materials. Hydrated aluminum silicate, fine grain crystallized clay. Laminar structure with repeating alumina-silica configurations. Density, 2.58 g/cm^3 (21.5 lb/gal); O.A., 32.1–55 lb/100 gal. Particle size, 05.3.5 μm. Syn:

▶ Kaolin, ▶ Hydrated Aluminum Silicate, ▶ China Clay, ▶ Pipe Clay, and ▶ White Bole.

Aluminum Soaps *n* Aluminum compounds of fatty acids and naphthenic acid, etc. These materials are soluble generally in paint vehicles and thinners, and are employed as thickening, flatting, and water proofing agents.

Aluminum Stearate *n* Complex salt or soap of aluminum and stearic acid. It is a white powder, which forms colloidal solutions or gels with drying oils and certain solvents. It is used as a flatting agent, anti-setting agent for pigments, and to help prevent penetration.

Amalgam \ə-ˈmal-gəm\ *n* [ME *amalgame*, fr. MF, fr. ML *amalgama*] (15c) Solutions of certain metals in mercury, which is of considerable value in the preparation of unsupported films of surface coatings. If, e.g., a coated tinned-iron sheet or tinfoil is rubbed with or immersed in mercury, an amalgam is formed between the tin coating and the mercury. The initially hard tin surface is replaced by a soft, semiliquid coating of tin–amalgam, from which it is a relatively easy matter to detach dried coatings for further examination. The tin surface is said to be amalgamated.

Amber \ˈam-bər\ *n* [ME *ambre*, fr. MF, fr. ML *ambra* fr. Arabic *anbar* ambergris] (14c) A natural fossil resin formed during the Oligocene age by exudation from a species of pine now extinct. Its empirical formula is $C_{10}H_{16}O$, it softens at about 150°C, can be fabricated and polished. It is a fossil resin found chiefly in the blue earth of East Prussia. It is primary an exudation of *Pinus* succinifera. It has been used in jewelry, cigarette holders, and pipe mouthpieces.

Amberlite Synthetic ion-exchange resins. Manufactured by Roehm & Haas, U.S.

Ambient \ˈam-bē-ənt\ *adj* [L *ambient-*, *ambiens*, pp of *ambire* to go around, fr. *ambi-* + *ire* to go] (1596) Completely surrounding; indicative of the surrounding environmental conditions such as temperature, pressure, atmosphere, etc. When no values are given, the temperature is presumed to be room temperature (18–23°C) and the atmosphere to be air at standard pressure (101.3 kPa).

Ambient Air Quality *n* Average atmospheric purity, as distinguished from discharge measurement taken at the source of pollution. The general amount of pollution present in a broad area.

Ambient Conditions *n* A term used to denote the temperature, pressure, etc. of the surrounding air. See ▶ Atmospheric Conditions.

Ambient Cure See ▶ Self-Curing.

Ambient Temperature *n* (1) The temperature of the medium immersing an object. (2) The prevailing room temperature.

American Gallon See ▶ Gallon, U.S.

American National Standards Institute *n* (ANSI) address: 1430 Broadway, New York, NY 10018. Clearinghouse and national coordinator for voluntary standards of engineering, equipment, industrial processing, and safety. *Formerly known as American Standards Association (ASA)*. In the plastics field, ANSI works closely with the ▶ Society Of The Plastics Industry (SPI) and the ▶ Society Of Automotive Engineers (SAE) to develop and publish standards for plastics materials, processing equipment, operations, and operating safety.

American Process Zinc Oxide *n* Zinc oxide pigment made directly from zinc ores. Sometimes called "direct" process. See ▶ Zinc Oxide.

American Society for Testing and Materials *n* ASTM. A nonprofit corporation formed for the development of standards on characteristics and performance of materials, products, systems, and services, and the promotion of related knowledge. In ASTM terminology, standards include test methods, definitions, recommended practices, classifications, and specifications.

American Turpentine *n* Light-colored, volatile, essential oil obtained from resinous exudates or resinous wood associated with living or dead coniferous trees, particularly of the genus, *Pinus*, or more commonly pines. See ▶ Turpentine.

American Vermillion *n* A pigment usually consisting of a lead molybdate or as a basic lead chromate (as chrome red).

Ameripol *n* Poly(isoprene). Manufactured by Firestone, U.S.

Ameripol SM *n Cis*-1,4-poly(isoprene). Manufactured by Firestone, U.S.

Amide \ˈa-ˌmīd, -məd\ *n* [ISV, from New Latin *ammonia*] (ca. 1847) A compound containing the –$CONH_2$ group, formed by the reaction of an organic acid or an ester with ammonia. Except for formamide, all amides are crystalline solids at room temperature. Examples are acetamide, CH_3CONH_2, and urea, H_2NCONH_2.

Amide-Imide Resin See ▶ Amino Resin.

Amides \ˈa-ˌmīd, -məd\ *n* [ISV, fr. L *ammonia*] (ca. 1847) Carboxylic acids in which the hydroxyl group of the acid is replaced by an amino or amido group (NH_2). Thus, acetamide (CH_3CONH_2) is obtained by substituting the hydroxyl (OH) of acetic acid with the NH_2 group. They may also be regarded as derivates of ammonia (NH_3), in which one of the hydrogen atoms is replaced by an acyl group.

Amido \ə-ˈmē-(ˌ)dō, ˈa-mə-ˌdō\ *adj* [ISV *amide* + *-o-*] (1877) Terms "amido" and "amino" apply to the same grouping, NH_2. The former term is usually applied to the NH_2 group when it occurs in an acid amide.

Amine \ə-ˈmēn, ˈa-ˌmēn\ *n* [ISR, fr. NL *ammonia*] (1863) Organic bases derived from the parent compound, ammonia (NH_3) The hydrogens of the ammonia may be substituted by alkyl groups, in which case the series of aliphatic bases is produced. Similarly, aromatic bases are formed when the hydrogens are substituted with aryl groups. Primary, secondary, tertiary, and quaternary amines are formed as one, two, three, or four of the hydrogen atoms are substituted. Substitution of a fourth hydrogen atom is possible because it is considered available in the hypothetical compound, ammonium hydroxide (NH_4OH). A compound derived (in concept) from ammonia by substitution of one or more hydrogen atoms by a hydrocarbon radical. Amines in which one, two, or all three of the ammonia hydrogens have been substituted are termed *primary, secondary,* and *tertiary* amines.

Amine End Group *n* The terminating (–NH_2) group of a nylon polymer chain. Amine end groups provide dye sites for polyamides.

Amine Equivalent *n* See ▶ Amine Value.

Amine Equivalent Weight *n* Molecular weight of amine divided by the number of active hydrogens in the molecule.

Amine–Furfural Resin See ▶ Aniline–Furfural Resin.

Amine Nitrogen Content *n* Refer to Federal Test Method 141a for test procedure.

Amine Resin *n* Synthetic resin derived from the reaction of urea, thiourea, melamine, or allied compounds with aldehydes, particularly formaldehyde. See ▶ Amino Resin.

Amines *n* Organic bases derived from the parent compound, ammonia (NH_3) The hydrogens of the ammonia may be substituted by alkyl groups, in which case the series of aliphatic bases is produced. Similarly, aromatic bases are formed when the hydrogens are substituted with aryl groups. Primary, secondary, tertiary, and quaternary amines are formed as one, two, three, or four of the hydrogen atoms are substituted. Substitution of a fourth hydrogen atom is possible because it is considered available in the hypothetical compound, ammonium hydroxide (NH_4OH).

Amine Value *n* The number of milligrams of potassium hydroxide equivalent to the fatty amine basicity in 1 g of sample. Syn: ▶ Amine Equivalent.

Amino- \ə-ˈmē-(ˌ)nō\ *adj* [ISV *amine* + *-o-*] (1904) A prefix signifying the presence in a compound or resin of an –NH_2 or =NH group.

Amino Acid *n* An organic acid containing an amino group attached to the carbon atom adjacent to the –COOH group, obtained by the hydrolysis of a protein or by synthesis. Examples are glycine, $CH_2(NH_2)$–COOH (e.g., glycine +NH_3–CH_2–COO^+) and cysteine, $HSCH_2CH(NH_2)COOH$.

Amino Acids *n* (1898) Amino group and a carboxyl group, $H_2NCHRCOOH$, e.g., glycine +NH_3–CH_2–COO^+.

Amino Coatings *n* Made from amino resins which include resins from reacting urea, thiourea, melamine, or allied compounds, usually with formaldehyde.

2-Amino-2-Methyl-1-Propanol See ▶ AMP.

Amino Plastics Plastics based on resins made by the condensation of amines, such as urea and melamine, with aldehydes. See ▶ Amino Resin.

Aminoplasts *n* Thermosetting resins made by the polycondensation of formaldehyde with a nitrogen compound and a higher aliphatic alcohol. The two general types are *urea-formaldehyde* and *triazine-formaldehyde*. Melamine is the triazine most often used. See also ▶ Amino Resin {G Aminoplaste mpl, F aminoplastes mpl, S aminoplásticos mpl, I amminoplasti mpl}.

γ-Aminopropyltriethoxy Silane *n* $NH_2(CH_2)_3Si(OC_2H_5)_3$. A silane coupling agent used in reinforced epoxy, phenolic, melamine, and many thermoplastic resins.

Amino Resin *n* (polyalkene amide, aminoplast) A generic term for a group of nitrogen-rich polymers containing amino nitrogen or its derivatives. The starting amino-bearing material is usually reacted with formaldehyde to form a reactive monomer that is condensation-polymerized to a thermosetting resin. Included amino compounds are urea, melamine, copolymers of both with formaldehyde, and, of limited use, thiourea, aniline, dicyandiamide, toluenesulfonamide, benzoguanidine, ethylene urea, and acrylamide. Not included, because properties warrant separate classification, are polyamides of the nylon type, polyurethanes, polyacrylamide, and acrylamide copolymers. The most important members of the amino-resin family are melamine-formaldehyde and urea-formaldehyde resins. The basic resins are clear, water-white syrups or white powdered materials that can be dispersed in water to form colorless syrups. They cure to high temperatures with appropriate catalysts. Molding powders are made by adding fillers to the uncured syrups, forming a consistency suitable for compression and transfer molding. Amino resins are usually cured by baking and are blended with other resins (e.g., alkyds or epoxies). Amino resins are also cured by chemical means at normal air temperature, e.g., wood finishes.

AMMA *n* Abbreviation for ▶ Copolymers Of Acrylonitrile and ▶ Methyl Methacrylate. In Europe, written A/MMA.

Ammeter \ˈa-ˌmē-tər\ *n* [*amp*ere + -*meter*] (1882) Instrument for measuring the strength (amperage) of electric currents.

Ammine \ˈa-ˌmēn, a-ˈmēn\ *n* [ISV *ammonia* + ^{2}-*ine*] (1897) The name given to ammonia, NH_3, when it serves as a ligand.

Ammonia Cure *n* A modification of a hot air pressure cure for rubber, often used for curing, in which ammonia gas is used to accelerate vulcanization and to prevent the deteriorating effect of air.

Ammonium Caseinate *n* Casein solubilized by ammonium hydroxide generally employing a hot water presoak.

Ammonium Soaps *n* Soaps formed by reaction of ammonia with the higher molecular weight fatty acids, such as ammonium oleate, linoleate, stearate, etc., used as wetting or emulsifying agents.

Amoora Oil *n* Semidrying vegetable oil with an iodine value of 135 and sp gr of 0.939/15°C.

Amorphous \ə-ˈmȯr-fəs\ *adj* [Gk *amorphous*, fr. *a-* + *morphē* form] (ca. 1731) Devoid of crystallinity or stratification. Most plastics are amorphous at processing temperatures, many retaining this state under all normal conditions. Lacking crystallinity.

Amorphous Domain *n* (and crystalline domains) Amorphous or noncrystalline portions of a solid polymer; conversely, crystalline a domains are nonamorphous. A single crystalline polymer chain can possess amorphous regions (domains), e.g., polyethylene.

Amorphous Nylons *n* (1)When unsymmetrical monomers are use to synthesize polyamides (Nylons), the normal ability of the chains to crystallize can be disrupted and amorphous (often transparent) polymers are formed. (2) Transparent aromatic polyamide thermoplastics. Produced by condensation of hexamethylene diamine, isophthalic and terephthalic acid (Sepe MP (1998) Dynamic mechanical analysis. Plastics Design Library, Norwich, New York).

Amorphous Polymer *n* Amorphous polymers are polymers having noncrystalline or amorphous supramolecular structure or morphology. Amorphous polymers have some molecular order but usually are substantially less ordered than crystalline polymers and subsequently have inferior mechanical properties. Materials in this class do not have a detectable melting point. Examples are PVC, acrylic, and polycarbonate.

Amorphous Silica *n* SiO_2. A naturally occurring or synthetically produced pigment, characterized by the absence of pronounced crystalline structure, and which has no sharp peaks in its X-ray diffraction pattern. It may contain water of hydration or be an anhydrous type. It is used as an extender pigment, fatting agent, and as a desiccant in metal flake and metal powder coatings.

Amorphous Solid *n* A substance with the external appearance and characteristics of a solid, but with the irregular structure typical of a liquid; a highly supercooled liquid; a *glass*.

AMP *n* Abbreviation for ▶ 2-Amino-2-Methyl-1-Propanol. Used as a pigment dispersant or as a pH modifier.

Ampere \ˈam-ˌpir, -per\ *n* [André-Marie *Ampère*] (1881) (A) The primary electrical unit of the SI system, upon which all other electrical units are based. The ampere itself is defined as that current which, if maintained in two long, parallel, fine wires located 1 m apart in a vacuum, will produce between these conductors a force of 2×10^{-7} newton per meter of length. Practically, an ampere is the current that flows between two points connected with an electric resistance of 1 ohm when their potential difference is 1 V.

Ampere's Rule *n* A positive charge moving horizontally is deflected by a force to the right if it is moving in a region where the magnetic field is vertically upward. This may be generalized to currents in wires by recalling that a current in a certain direction is equivalent to the motion of positive charges in that direction. The force felt by a negative charge is opposite to that felt by a positive charge.

Amphibole \ˈam(p)-fə-ˌbōl\ *n* [F, fr. LL *amphibolus*, fr. Gr *amphibolos*] (ca. 1823) A group of asbestos minerals.

Amphoprotism *n* The ability of a species either to gain or to lose a proton.

Amphoteric \ˌam(p)-fə-ˈter-ik\ *adj* [ISV, fr. Gk *amphoteros* each of two, fr. *amphō* both] (ca. 1849) Designating an element or a compound that can behave either as an acid or a base, i.e., as an electron donor or an electron acceptor. Polymerization emulsifiers having both anionic and cationic groups are called *amphoteric emulsifiers*.

Amphoterism *n* The ability of a substance to react either as an acid or as a base.

Amplitude \ˈam-plə-ˌtüd, -ˌtyüd\ *n* (1542) The maximum value of the displacement in an oscillatory motion.

AMU *n* The atomic mass unit (amu), a unit of mass equal to 1/12 the mass of the carbon atom of mass number 12. On the atomic mass scale $^{12}C \equiv 12$.

$$1 \text{ amu} = 931.4812(52) \text{ MeV}$$
$$= 1.660531(11)10^{-27} \text{ kg (SI units)}$$
$$= 1.660531(11) \, 10^{-24} \text{ g (cgs units)}$$

The numbers in parentheses are the standard deviation uncertainties in the last digits of the quoted value, computed on the basis of internal consistency.

Amyl \ˈa-məl\ *n* [L *amylum* + E *–yl*] (1850) The radical C_5H_{11}–, also known as *pentyl*. The amyl radical occurs in six isomeric forms, and the term amyl usually refers to any mixture of the isomers.

Amylaceous *adj* Pertaining to, or of the nature of, starch; starchy.

Amyl Acetate *n* (ca. 1881) (banana oil, pear oil, amylacetic ester) $CH_3COOC_5H_{11}$. A commercial solvent for several resins, including the cellulosics, vinyls, acrylics, polystyrene, and uncured alkyds and phenolics. It has a strong, fruity odor (hence its nicknames), and its main constituent is isoamyl acetate, but other isomers such as normal- and secondary-amyl acetates are present in amounts determined by the grade and origin. It is a medium boiling solvent. The commercial product has a boiling range of 120–145°C, a sp gr of approximately 0.876, and a flp of 31°C (87°F). Syn: ▶ Banana Oil.

Amyl Alcohol *n* (1863) $C_5H_{11}OH$. Commercial alcohol having a boiling range of 128–132°C; sp gr of 0.816/15°C; a flp of 43°C (110°F); and a vapor pressure of 10 mmHg at 20°C. It has some application in the manufacture of cellulose lacquers as a high boiling diluent.

Amyl Benzoate *n* $C_6H_5COOC_5H_{11}$. High boiling solvent-plasticizer, with a bp of 261°C, and sp gr of 0.994.

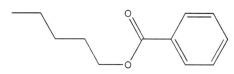

Amyl Borate *n* $(C_5H_{11})_3BO_3$. High boiling solvent-plasticizer. The commercial liquid has a boiling range of 250–260°C.

Amyl Citrate See ▶ Citrates.

Amyl Ether *n* $(C_5H_{11})_2O$. Amyl ether, or diamyl ether as it is correctly called, is a useful high boiling solvent of the ether series, with a boiling range of 170–190°C.

Amyl Formate *n* $HCOOC_5H_{11}$. A solvent for resins and cellulose derivatives. It is a medium boiling solvent with a bp of 13°O, and flp of 27°C.

Amyl Lactate $CH_3CH(OH)COOC_5H_{11}$. Water white to pale yellow colored liquid used as an ester solvent. It plasticizes in lacquers and has a bp of 210°C.

Amyl Nitrate *n* (ca. 1881) $C_5H_{11}NO_2$. A pale yellow, pungent flammable liquid ester, of commercial amyl alcohol and nitrous acid.

Amyl Oleate Solvent and plasticizer for cellulosic and vinyl resins.

Amyl Propionate *n* $CH_3CH_2COOC_5H_{11}$. Very strong solvent, similar in properties and uses to amyl acetate, but possessing a higher boiling range, 140–170°C. Its flp is about 41°C.

Amyl Salicylate See ▶ Isoamyl Salicylate.

Amyl Stearate *n* $C_{17}H_{35}COOC_5H_{11}$. Pale yellow colored substance used as a high-boiling plasticizer. It has a mp of 30°C and a bp of 360°C.

Amyl Tartrate *n* Plasticizer with strong solvent power and a bp of about 400°C.

Amyrin *n* $C_{30}H_{50}O$ for both α and β modifications. Amyrins, isolated in α and β forms, are constituents of elemi resin. They are believed to be alcoholic in character.

Anacardic Acid *n* $C_{22}H_{32}O_3$. Ortho pentadecadienylsalicylic acid. Principal constituent of cashew nutshell liquid.

Anacardol *n* $C_{21}H_{32}O$. Monohydroxy phenol, which closely resembles the anacardic acid from which it is readily obtained by decarboxylation when heated. The substituent side chain is the same for both compounds.

Anadonis Green See ▶ Chromic Oxide Green.

Anaerobic \ˌa-nə-ˈrō-bik; ˌan-ˈa(-ə), -ˈe(-ə)-\ *adj* (ca. 1881) Free of oxygen and/or air. Used in connection with bacteria, developing without air. Important for anaerobic corrosion (tanks, ship bottoms).

Anaerobic Adhesive *n* An adhesive that cures only in the absence of air after being confined between assembled parts. An example is dimethacrylate adhesive, used for bonding assembly parts, locking screws and bolts, retaining gears and other shaft-mounted parts, and sealing threads and flanges.

Analyzer \ˈa-nᵊl-ˌīz\ *vt* [prob. irreg. fr. *analysis*] (1587) A second polarizing element inserted above a preparation. When its vibration direction is at right angles to the vibration direction of the polarizer, the field becomes black if no anisotropic specimen is on the stage.

Anamorphosis *n* Distorted painting which appears normal when viewed from the side. A form of "trick" painting fashionable in the sixteenth and seventeenth centuries.

Anatase \ˈa-nə-ˌtās, -ˌtāz\ *n* [F, fr. Gk *anatasis* extension, fr. *anateinein* to extend, fr. *ana-* + *teinein* to stretch] (ca. 1828) (octahedrite) A crystalline ore of Titanium Dioxide. See ▶ Titanium Dioxide, Anatase.

Anchorage \ˈaŋ-k(ə-)rij\ *n* (15c) Part of an insert that is molded inside of a plastic part and held fast by shrinkage of the plastic onto the insert's knurled surfaces. (1) Property or profile of metal or wood substrate to enhance the adhesion of a coating. (2) Adhesion of rubber to fiber, fabric, metal, or other material to which the rubber compound is applied by calendering, welding, cement spreading, or other means.

Andrade Creep *n* A type of creep behavior in which the compliance of the sample is proportional to the cube root of the time under stress. A variety of polymers exhibit this behavior.

Anelasticity \ˌa-nᵊl-as-ˈti-sə-tē\ *n* (1947) The dependence of elastic strain on both stress and time, resulting in a lag of strain behind stress. In materials subjected to cyclic stress, the anelastic effect causes ▶ Damping.

Angel's Hair *n* Fibrous strands of material pulled away from thermoplastic films, particularly polypropylene, in heat-sealing and cutting operations that employ hot knives or wires. The angel's hair accumulates on the cutting mechanism, eventually affecting performance and requiring removal.

Angle \\ˈaŋ-gəl\\ *n* [ME, fr. MF, fr. L *angulus*] (14c) The ratio between the arc and the radius of the arc. Units of angle, the radian, the angle subtended by an arc equal to the radius; the degree $\frac{1}{360}$ part of the total angle about a point. Dimensions, –a numeric.

Angle Head (offset head, crosshead) An extruder head so designed that the principal direction of the extrudate makes an angle with the (extended) axis of the screw. See also ▶ Crosshead.

Angle of Contact *n* Associated with the phenomenon of wetting. When a drop of liquid contacts a solid surface it can remain exactly spherical, in which case no wetting occurs; or it can spread out to a perfectly flat film, and in that latter case complete wetting occurs. The angle of contact is the angle between the tangent to the periphery of the drop at the point of contact with the solid, and the surface of the solid. When the drop spreads to a perfectly flat film, the angle of contact is zero. When it remains exactly spherical, the area of contact is a point only, and the angle of contact is 180°.

Angle of Incidence *n* Angle between the axis of an impinging light beam and the perpendicular to the object surface.

Angle of Reflection *n* Angle between the axis of a reflected light beam and the perpendicular to the object surface.

Angle of Repose *n* (angle of rest) The maximum angle that a conical pile of particles makes with the horizontal surface on which it rests. No ASTM test is listed for this important property of plastic powders and pellets. The smaller the angle of repose, the more easily does the material flow through hoppers and constrictions.

Angle of Viewing Angle between the axis of a detected light beam and the perpendicular to the object surface.

Angle-Ply Laminate *n* A laminate in which equal numbers of plies are oriented at equal plus and minus angles from the plies in the length direction, making the laminate orthotropic. The most commonly chosen angles are ±60°, giving nearly equal strengths in all directions in the plane of the laminate. See also ▶ Cross Laminate.

Angle Press *n* A hydraulic molding press equipped with horizontal and vertical rams, used in the production of complex moldings containing deep undercuts or side cavities.

Angocopalolic Acid *n* $C_{23}H_{36}O_3$. A constituent acid of Angola copal. It is an unsaturated monobasic acid and has a melting point of 85°C.

Angocopaloresenes *n* Identified as both α and β forms, these are constituents of Angola copal. The former has a mp of about 64°C and the latter about 222°C; α angocopaloresene corresponds with $C_{30}H_{54}O_6$, and β angocopaloresene with $C_{25}H_{38}O_4$.

Angola Copal *n* Fossil copal of African origin.

Angora \\aŋ-ˈgōr-ə\\ *n* (1852) (1) The hair of the Angora goat. The long, fine fibers are so smooth and soft that they must be combined with other fibers in weaving. (2) The hair of the Angora rabbit. The fine, lightweight hair is warm, and it is often blended with wool to decrease price and to obtain novelty effects in weaving. By law, the fiber must be described as Angora rabbit hair.

Angstrom Unit (Å) \\ˈaŋ-strəm\\ *n* [Anders J. *Angstrom*] (1892) {G Angströmeinheit *f*, F unité f Angtröm, S unidad *f* Angström} A unit of length equal to 1×10^{-10} m. A unit of linear measure named after A. J. Ångström, used especially in expressing the length of light waves, equal to one 10,000th of a micron, or 100 millionth of a centimeter (1×10^{-8} cm). It has been replaced by the nanometer (nm). 1 Å = 0.1 nm.

Angular Acceleration *n* The time rate of change of angular velocity either in angular speed or in direction of the axis of rotation (precession). Cgs unit, 1 radian per second per second. Dimensions, $[t^{-2}]$. If the initial angular velocity is ω_t, the angular acceleration,

$$\alpha = \frac{\omega_t - \omega_o}{t}$$

The angular velocity after time t,

$$\omega_t = \omega_o + \alpha_t$$

The angle swept out in time t,

$$\theta = \omega_o t + \frac{1}{2}\alpha t^2$$

The angular velocity after movement through the arc θ,

$$\omega = \sqrt{w_o^2 + 2\alpha\theta}$$

In the above equations, for angular displacement in radians, angular velocity will be in radians per second and angular acceleration in radians per second.

Angular Aperture *n* The angular aperture of an objective is the largest angular extent of wave surface which it can transmit.

Angular Aperture *n* AA The largest angle between the image forming rays collected or transmitted by a lens system, e.g., objective or condenser of a microscope.

Angular Harmonic Motion or Harmonic Motion of Rotation *n* Periodic, oscillatory angular motion in which the restoring torque is proportional to the angular displacement. Torsional vibration.

Angular Momentum or Moment of Momentum Quantity of angular motion measured by the product of the angular velocity and the moment of inertia. Cgs unit, unnamed, its nature is expressed by gram centimeter square per second. Dimensions, $[m\ l^2 t^{-1}]$. The angular momentum of a mass whose moment of inertia is I, rotating with angular velocity ω, is $I\omega$.

Angular Velocity n Time rate of angular motion about an axis. Cgs unit, 1 radian per second. Dimensions, $[t^{-1}]$. If the angle described in time t is θ, the angular velocity,

$$\omega = \frac{\theta}{t}$$

θ in radians and t in seconds gives ω in radians per second.

Angular Welding See ▶ Friction Welding.

Anhedral *adj* Anhedral crystals are those whose growth has been impeded by adjacent crystals growing simultaneously, so that the development of plane faces is inhibited. Also crystals eroded, partly dissolved, or mechanically deformed to the point where nearly all traces of crystal faces have been removed. All anhedral crystals are irregularly shaped and do not have plane faces, cf., euhedral.

Anhydride \(ˌ)an-ˈhī-ˌdrīd\ *n* (1863) (1) A compound from which water has been extracted. (2) An oxide of a metal (basic anhydride) or of a nonmetal (acidic anhydride) that forms a base or an acid, respectively, when united with water. (3) An organic compound made (conceptually) by the union of two acid molecules with the elimination of a molecule of water. In practice, organic anhydrides are usually produced by other reactions.

Anhydrite \-ˌdrīt\ *n* [Gr *Anhydrit*, fr. Gk *anydros*] (ca. 1823) The mineralogical name for native anhydrous calcium sulfate which is often associated in nature with calcium sulfate dehydrate or gypsum. It occurs occasionally as an impurity in gypsum and plaster of Paris. See ▶ Calcium Sulfate, Anhydrous.

Anhydrous \-drəs\ *adj* [Gk *anydrods*, fr. *a-* + *hydōr* water] (1819) Perfectly dry; containing no water. It is generally accepted to mean free from water of any kind, whether as such in the free-state, or as water of crystallization or other tightly held water.

Anidex Fiber *n* A manufactured fiber in which the fiber-forming substance is any long chain synthetic polymer composed of at least 50% by weight of one or more esters of a monohydric alcohol and acrylic acid (CH$_2$=CH–COOH) (FTC definition).

Anilides *n* Acyl derivatives from aniline. Possibly the commonest anilide in the paint trade is acetanilide, C$_6$H$_5$NHCOCH$_3$.

Aniline \ˈa-nᵊl-ən\ *n* [Gr *Anilin*, fr. *Anil* indigo, fr. FP, fr. Arabic *an-nīl* the indigo plant. fr. Sanskrit fr. *nīlī* indigo, fr. feminine of *nīla* dark blue] (1850) (phenylamine, aminobenzene) C$_6$H$_5$NH$_2$. A colorless, oily liquid made by the reduction of nitrobenzene with iron chips and an acid catalyst. It is used in the production of aniline-formaldehyde resins and certain catalysts and antioxidants. See ▶ Dyes.

Aniline Cloud Point See ▶ Aniline Point.

Aniline Dye *n* Large class of synthetic dyes made from intermediates based upon or made from aniline.

Aniline-Formaldehyde Resin *n* An aminoplastic that is made by condensing formaldehyde and aniline in an acid solution. The resins are thermoplastic and are used in making molded and laminated insulating materials with high dielectric strength and good chemical resistance. See also ▶ Amino Resin.

Aniline-Furfural Resin *n* Furfural yields resinous compounds not only with aniline but with aromatic amines generally. Rosin modified resins have also been produced. Amine-furfural resins are dark in color. Bleaching occurs on exposed to light. They possess the unique property of compatibility with cellulose acetate.

Aniline Ink *n* A fast-drying ink used for printing on cellophane, polyethylene, etc. Aniline inks were first made from solutions of coal-tar dyes in organic solvents, hence the name. Modern inks generally employ pigments rather than dyes. See ▶ Flexographic Ink.

Aniline Number See ▶ Aniline Point.

Aniline Pigments *n* Made from distillation products of coal tar, a by-product of coke and coal gas manufacture. *Also known as Coal-Tar Colors.*

Aniline Point *n* The minimum temperature at which a hydrocarbon solvent is completely soluble in an equal volume of freshly distilled aniline. Below this point, the mixture is cloudy and separates into two layers. It is used as a measure of solvent power of hydrocarbon solvents. Refer to ASTM D 1012. Syn: ▶ Aniline Cloud Point and ▶ Aniline Number.

Aniline Printing See ▶ Flexography.

Anilox* Roller Mechanically engraved intaglio from roller used in flexo presses to transfer a controlled film of ink from the contacting rubber-covered fountain roller to the rubber plate (or rubber-covered roller) which prints the web *Registered Trade Name, Inmont Corp.

Animal Black *n* (animal char, animal charcoal, boneblack) A form of charcoal derived from animal bones, used as a pigment. Three types are known: drop black; bone black; and ivory black. Drop black and bone black are produce by the calcinations of bones, but ivory black in its original cake form is virtually extinct. See also ▶ Carbon Black, Bone Black and Drop Black.

Animal Fats *n* Include such products as lard, bone fat, tallow, and butter fat. They are semisolid at ordinary air temperature, and consist chiefly of the glycerides of oleic and stearic acids.

Animal Fibers *n* Fibers of animal origin such as wool, alpaca, camel hair, and silk.

Animal Oils *n* Normally restricted to the animal foot oils and lard oil, and quite distinct from the fish oils. Neatsfoot oil is typical of the foot oils. The animal oils are characterized by the almost complete absence of acids more unsaturated than oleic acid, and by the presence of cholesterol. The presence of this latter compound enables them to be distinguished from the nondrying vegetable oils.

Animal Waxes *n* Obtained from a great variety of sources and have little in common, except absence of glycerides.

Animi *n* Fossil copal, sometimes described as "gooseskin" animi because of the characteristic markings on its surface. It originated in Zanzibar and hence its alternative name, Zanzibar copal. *Also known as Gum Animi and Mombassa Gum.*

Anion \ˈa-ˌnī-ən\ *n* [Gk, neut. of *aniōn*, prp. of *anienai* to go up, fr. *ana-* + *ienai* to go] (1834) An atom, molecule, or radical that has gained an electron to become negatively charged. Anions in a liquid subjected to electric potential collect at the positive pole or anode.

Anion Exchange Resin *n* Certain synthetic resins with the property of absorbing anions from aqueous salt solutions. See ▶ Ion-Exchange Resin.

Anionic \ˌa-(ˌ)nī-ä-nik\ *adj* (ca. 1920) Pertaining to a negatively (−) charged atom, radical, or molecule, or to any compound or mixture having negatively charged groups.

Anionic-Cationic Polymerization *n* A method of preparing multi-block copolymers using a combination of anionic and cationic) (ion coupling) methods.

Anionic Polymerization *n* Polymerization using an anionic initiator (− charge), e.g., n-butyl lithium initiator to form polyisoprene. See ▶ Ionic Polymerization {G anionische Polymerisation f, F polymérisation anionique, polymérisation f, S polimerización aniónica, polimerización f, I polimerizzazione anionica, polimerizzazione f}.

Anionic Surfactant *n* Surfactants which give negatively charged ions in an aqueous solution.

Anisole *n* $C_6H_5OCH_3$. Mixed aromatic-aliphatic ether. It has good solvent properties and a bp of 154°C.

Anisotropic \ˌa-ˌnī-sə-ˈträ-pik\ *adj* (1879) (1) A transparent particle having different refractive indices depending on the vibration direction of light. (2) Said of materials whose properties, e.g., strength, refractive index, thermal conductivity, are unequal in different directions. Oriented thermoplastics and unidirectionally fiber-reinforced resins are typically anisotropic. Not isotropic. See ▶ Isotropic.

Anisotropy *n* The quality of being anisotropic; having directionally dependent properties. Exhibiting different optical properties when tested along axes in different directions. See ▶ Isotropic {G Anisotropie f, F anisotropie f, S anisotropía f, I anisotropia f}.

ANLAB Color Difference Equation *n* Abbreviation used primarily in Britain for Adams-Nickerson L, a, b Color Difference Equation. The normalizing factor is generally 40 and b is calculated as $0.4(V_z − V_Y)$. See ▶ Adams Chromatic Value Color Difference Equation.

ANM *n* Copolymers from acrylic ester and acrylonitrile.

Annealing \ə-ˈnē(ə)l\ *v* [ME *anelen* to set on fire, fr. OE *onîlan*, fr. *on* + *îlan* to set on fire, burn, fr. *āl* fire; akin to OE *îled* fire, ON *eldr*] (1664) The process of relieving stresses in molded plastics, metal, or glass by heating to a predetermined temperature, maintaining this temperature for a set period of time, and slowly cooling the articles. Sometimes the articles are placed in jigs to prevent distortion as internal stresses are relieved during annealing.

Annual Growth Ring Growth layer of a tree put on in a single growth year. This includes springwood and summerwood.

Annular Stop *n* The opaque ring-shaped stop with a central opening placed usually in the objective back focal plane to give annular stop dispersion staining.

Anode \ˈa-ˌnōd\ *n* [Gk *anodos* way up, fr. *ana-* + *hodos* way] (1834) (1) The positive terminal of an electrical source to which electrons and negatively charged ions travel, negative for an electrolytic cell and positive for a galvanic cell. (2) The electrode at which corrosion (oxidation) occurs, electron collection electrode.

Anprolene See ▶ Ethylene Oxide.

ANS *n* Abbreviation for Adams-Nickerson-Stultz Color Difference Equation.

ANSI *n* Abbreviation for ▶ American National Standards Institute, which is the U.S. member of ISO.

Anthophyllite \ ˌan(t)-thə-ˈfi-ˌlīt, (ˌ)an-ˈthä-fə-\ *n* [Gr for *Anthophyllit*, fr. NL *anthophyllum*, fr. Gk *anthos* + *phyllon* leaf] (ca. 1828) A type of ▶ Asbestos, the major source which is in Finland. Anthophyllite is a natural magnesium iron silicate, formerly used as a filler in polypropylene to provide heat stability.

Anthracene \ ˈan(t)-thrə-ˌsēn\ *n* (1862) C_6H_4: $(CH)_2C_6H_4$. Yellow, crystalline solid obtained from the distillation of coal tar.

Anthracene Oil *n* Fraction obtained from the distillation of coal tar, with a boiling range in the region of 270°C, and containing a substantial proportion of anthracene and similar hydrocarbons.

Anthranilic Acid \ ˌan(t)-thrə-ˈni-lik-\ *n* [ISV *anthra*cene + *ani*line] (1881) See ▶ A-Acid.

Anthraquinone Dyes \ ˌan(t)-thrə-kwi-ˈnōn, -ˈkwi-ˌnōn\ *n* [prob. fr. Fr, fr. *anthracene* + *quinone*] (1869) Dyes that have anthraquinone as their base and the carbonyl group (>C = O) as the chromophore. Anthraquinone-based dyes are found in most of the synthetic dye classes. See ▶ Dyes.

Antibacterial Finish *n* A treatment of a textile material to make it resistant to, or to retard growth of, bacteria.

Antibiotic \ ˌan-ti-bī-ˈä-tik, -ˌtī-; ˌan-ti-bē-\ *adj* (1894) A substance derived from a living organism capable of killing or incapacitating another organism.

Antiblocking Agent (Antiblock) An additive that is incorporated into resins and compounds to prevent surfaces of products (mainly films) from sticking to each other or to other surfaces. The term is not generally used for coatings, dusts, or sprays applied to surfaces for the same purpose, or as ▶ Slip Agents, after products have been formed. Antiblocking agents usually are finely divided, solid, infusible materials, such as silica, but an be minerals or waxes. They function by forming minute, protruding asperities that maintain separating air spaces that interfere with adhesion.

Antibonding Orbital *n* The molecular orbital in which electrons have higher energies than in the unbonded atoms; an orbital characterized by a region of low electron probability density between the bonded atoms, producing a destabilizing effect on the molecule.

Antichlor *n* A chemical, such as sodium thiosulfate, used to remove excess chlorine after bleaching.

Anticondensation Paint *n* Coating designed to minimize the effects of condensation of moisture under intermittently dry and humid conditions. Such a material normally has a matt textured finish and frequently contains heat-insulating material as a filler.

Anticorrosion Paint or Composition *n* Coating used for preventing the corrosion of metals and, more particularly, specially formulated to prevent the rusting of iron, steel, and other metals.

Antifelting Agents *n* Products that prevent or minimize matting and compaction of textile materials.

Antifloating Agent *n* Additive used to prevent floating.

Antiflooding Agent *n* Additive used to prevent flooding.

Antifoaming Agent *n* An additive that reduces the surface tension of a solution or emulsion, thus inhibiting or modifying the formation of bubbles and foam. Commonly used are insoluble oils, dimethyl polysiloxanes and other silicones, certain alcohols, stearates, and glycols. In many polymerizations, these agents prevent foaming altogether. They are also used to delay foaming when producing cellular plastics (Harper CA, (2000) Modern plastics encyclopedia. McGraw-Hill/Modern Plastics, New York).

Antifogging Agent *n* An additive that prevents or reduces the condensation of fine droplets of water on a shiny surface. Such additives function as mild wetting agents that exude to the surface and lower the surface tension of water, thereby causing it to spread into a continuous film. Antifogging agents are much used in PVC wrapping films for meats and other moist foods. Examples are alkylphenol ethoxylates, complex polyol monoesters, polyoxyethylene esters of oleic acid, polyoxyethylene sorbitan esters of oleic acid, and sorbitan esters of fatty acids (Modern plastics encyclopedia. McGraw-Hill/Modern Plastics, New York).

Antifouling Composition *n* Paint-like composition used to prevent the growth of barnacles and other organisms on ships' bottoms. It usually contains substances which are poisonous to such organisms in the early stages of growth. Such compositions are applied over protective paints, e.g., over anti-corrosion paints in the case of steel ships.

Antigelling Agent n An additive that prevents a solution from forming a gel.

Antilivering Agent n Additive used to prevent the livering of a coating. See ▶ Livering.

Antimicrobial Agent See ▶ Biocide.

Antimony Orange Sb_2S_3. Like antimony vermillion, which it resembles closely, it is a sulfide pigment, and can be prepared by precipitation from as aqueous antimony salt solution with hydrogen sulfide ((1996) Kirk-Othmer encyclopedia of chemical technology: pigments-powders. Wiley, New York; Solomon DH, Hawthorne DG (1991) Chemistry of pigments and fillers. Krieger Publishing, New York).

Antimony Oxide n Sb_2O_3. Pigment White 11 (77052). Antimony and oxygen combine to form trioxide, tetraoxide, and pentoxide. The most common antimony trioxide pigment is used as a flame retardant in paints, plastics, and textiles. It contains, in addition, traces of the tetroxide (Sb_2O_4), and of the oxides of lead, arsenic, and iron. Density, 5.3–5.7 g/cm^3 (44.0–47.5 lb/gal); O.A., 11–14; particle size, 1 μm ((1996) Kirk-Othmer encyclopedia of chemical technology: pigments-powders. Wiley, New York; Solomon DH, Hawthorne DG (1991) Chemistry of pigments and fillers. Krieger Publishing, New York). Syn: ▶ Antimony Trioxide, ▶ Antimony White, and ▶ Flower of Antimony.

Antimony Trioxide n (antimony white, flowers of antimony, antimony oxide) Sb_2O_3. A very fine white powder made by vaporizing antimony metal in an oxidizing atmosphere, then cooling and collecting the oxide dust. Available in several ranges of particle size, it is used as a flame retardant and pigment in plastics, usually in synergistic combination with an organo-halogen compound. PVC is the biggest consumer ((1996) Kirk-Othmer encyclopedia of chemical technology: pigments-powders. Wiley, New York; Solomon DH, Hawthorne DG (1991) Chemistry of pigments and fillers. Krieger Publishing, New York).

$$OH_2^{-4}$$
$$Sb^{+5} \quad OH_2^{-4}$$
$$OH_2^{-4}$$

Antimony Vermillion n Sb_2S_3. Pigment varying on color from pale orange to deep crimson, with good opacity and tinting strength. It is prepared either by adding a solution of sodium thiosulfate to a solution of tartar emetic and tartaric acid, or to a solution of another suitable antimony salt. The pigment is not resistant to acids, alkalis, or heat, but, if properly made, is lightfast. It is sometimes referred to as *Crimson Antimony*. ((1996) Kirk-Othmer encyclopedia of chemical technology: pigments-powders. Wiley, New York; Solomon DH, Hawthorne DG (1991) Chemistry of pigments and fillers. Krieger Publishing, New York).

$$S^{--}$$
$$Sb^{+5} \quad S^{--}$$
$$S^{--}$$
Antimony trisulfide

Antimony White n See ▶ Antimony Oxide.

Antimony Yellow n $Pb_3(SbO_4)$. Essentially, lead antimonite, which may be considered to be chemically combined lead and antimony oxides. It varies in color from sulfur yellow to orange-yellow, depending upon the proportion of the two materials ((1996) Kirk-Othmer encyclopedia of chemical technology: pigments-powders. Wiley, New York).

Antinode \ˈan-ti-ˌnōd, ˈan-ˌtī-\ n [ISV] (1882) A region or location of maximum disturbance in a standing wave (Giambattista A, Richardson R, Richardson RC, Richardson B (2003) College physics. McGraw-Hill Science/Engineering/Math, New York).

Anti-Noise Paint n Coating used for its sound deadening effect, on vibrating and naturally noisy machinery. It relies for its effectiveness on its very rough surface, produced by the incorporation of cork and similar granular, or fibrous materials. *Also called Acoustical Paint*.

Antioxidant \ˌan-tē-ˈäk-sə-dənt, ˌan-ˌtī-\ n (1926) A substance that slows down the oxidation of oils, fats, etc., and thus helps to check deterioration: antioxidants are added commercially to foods, soaps, etc. (Merriam-Webster's Collegiate Dictionary (2000) 10th edn. Springfield, MA). Although the term technically applies to molecules reacting with oxygen, it is often applied to molecules that protect from any free radical molecule with unpaired electrons (Chemistry encyclopedia, www.ChemistryAbout.com). A substance incorporated in a material to inhibit oxidation at normal or elevated temperatures. Antioxidants are used mainly with natural and synthetic rubbers, petroleum-based resins, and other such polymers that oxidize readily due to structural unsaturation. However, some thermoplastics, namely polypropylene, ABS, rubber-modified polystyrene, acrylic and vinyl resins, also require protection by antioxidants for some uses. There are two main classes: (1) Those that inhibit oxidation by reacting with chain-propagating radicals, such as hindered phenols that intercept free radicals. These are called primary antioxidants or free-radical scavengers. (2) Those that decompose peroxide into nonradical and stable products; examples are phosphates and various sulfur compounds, e.g., esters of thiodipropionic acid. These are referred to as secondary antioxidants, or peroxide decomposers (Glenz W (ed) (2001) A glossary of plastics terminology in 5 languages, 5th edn. Hanser Gardner Publications, Cinicinnati) {G Ozonschutzmittel n, F anti-ozone m, S antiozonante m, I antiozonante m}; Carley JF (ed) (1993) Whittington's dictionary of plastics. Technomic Publishing).

Antioxidants *n* Agents which deter or retard autooxidation degration such as phenol and arylamines, agents react with the intermediate peroxy radicals. These agents retard the action of oxygen in drying oils and other substances subject to oxidation (Glenz W (ed) (2001) A glossary of plastics terminology in 5 languages, 5th edn. Hanser-Gardner, Cinicinnati; Zaiko GE (ed) (1995) Degradation and stabilization of polymers. Nova Science, New York) {G Antioxidans n, F antioxydant m, S antioxidante m, I antiossidante m}.

Antiozonant \-ˈō-(ˌ)zō-nənt\ *n* (1954) Substance that retards or prevents the action of ozone on elastomers when exposed under tension, either statically or dynamically, to air containing ozone.

Antiozonants Same antioxidants.

Anti-Parallel Spins Two spins which are in opposite directions.

Antique Finish *n* Usually applied to nearly painted furniture or objects for the purpose of giving the appearance of mellowness of age.

Antiquing \(ˌ)an-ˈtēk\ *v* (1923) (1) Treating or finishing of wood to make it look old. (2) Imparting a special coloring effect by the application of a base coat followed by a transparent or semitransparent colored glaze, which is either applied by brush, then partially wiped away or applied by wiping with a cloth.

Anti-Sag Agent *n* Additive used to control sagging of a coating. See ▶ Thickening Agent.

Antiseptic \ ˌan-tə-ˈsep-tik\ *adj* [*anti-* + Gk *sēptikos*] (1751) An agent used to destroy or restrain the growth of microorganisms.

Anti-Septic Wash See ▶ Fungicidal Wash.

Anti-Settling Agent *n* Substance incorporated into pigmented paint to retard settling and to maintain uniform consistency during storage or, in dipping paints, during painting operations. These additives normally function by altering the rheological properties of the paint. Syn: ▶ Suspending Agent.

Anti-Shatter Composition *n* Composition designed to prevent the fragmentation of glass and glass-like materials. The main use of these compositions is in the manufacture of spliterproof glass. They are applied as an adhesive, resilient sandwich between relatively thin sheets of glass. A violet blow may cause cracking of the glass, but the anti-shatter composition is designed to retain the various pieces in position. Properties include good permanent adhesion to the glass, a natural resilience such that it will absorb violet shocks without itself distintegrating, a permanent freedom from color, and absolute transparency.

Anti-Silking Agent *n* Additive used to prevent silking, which is a special case of floating, resulting in the formation of parallel hairlike striations of differing colors, running through the length of the painted surface.

Anti-Skinning Agent *n* Any material added to a coating or printing ink to prevent or retard the processes of oxidation or polymerization which results in the formation of an insoluble skin on the surface of the coating or printing ink in a container.

Anti-Slip Paint *n* Used for application to decks and other surfaces where the conditions conductive to slipping are present. It prevents slipping by providing a tough, rough surface. The roughness is usually achieved by the addition of sand or cork dust. *Also called Nonskid Paint.*

Antisoiling Properties The properties of textile materials whereby they resist deposition of dirt and stains.

Antistaining Properties The ability of a textile to resist the deposition of oil- or water-borne stains.

Anti-Stat \-ˈstat *or* an-ti-stat-ic, -ˌsta-tik\ *adj* (1952) See ▶ Antistatic Agent.

Antistatic Agent *n* (antistat) A chemical (or other material) that imparts slight electrical conductivity to plastics compounds (or other polymeric formulation), preventing the accumulation of electrostatic charges on finished articles. The agent may be incorporated in the materials before molding (most preferable) or applied to their surfaces afterward. Antistatsic agents function either by being inherently conductive or by absorbing moisture from the air. Examples are long-chain aliphatic amines and amides, phosphate esters, quaternary ammonium salts, polyethylene glycols, polyethylene-glycol esters, and ethoxylated long-chain amines ((1986, 1990, 1992, 1993) Modern plastics encyclopedia. McGraw-Hill/Modern Plastics, New York).

Antistatic Properties The ability of a textile material to disperse an electrostatic charge and to prevent the build up of static electricity.

Antitracking Varnish *n* When two points of high electrical potential difference are suitably located in contact with the surface of an organic insulating medium, the surface of the insulator may be degraded to carbon under the imposed electrical stresses, and a carbon track formed between the two points. Phenol-formaldehyde laminated boards are sometimes given a special coating of varnish at their exposed surfaces to prevent such carbonization. Varnishes used for such purposes are described as anti-tracking varnishes (American Society for Testing and Materials, www.astm.org, 100 Barr Harbor Drive, West Conshohocken, PA 19428-2959).

Antiwrinkling Agent *n* Material added to surface coating compositions to prevent the formation of wrinkles in films during drying.

Ant Oil See ▶ Furfural.

Antwerp Blue *n* Essentially, zinc ferrocyanide with a pale greenish-blue color. It is characterized by good opacity and almost complete absence of floating tendency.

AOCS Abbreviation for American Oil Chemists Society.

Apertured Nonwoven Fabric *n* A nonwoven fabric having many small through-holes made by laying the fabric on a perforated plate or screen and applying fluid pressure.

APHA Color Scale *n* Abbreviation for American Public Health Asscoation color scale developed for visually evaluating slight yellowness of solutions. It is based on dilutions of a platinum–cobalt standard solution of 500 APHA. This single-number color scale was originally developed by Hazen, and may also be referred to as the Hazen Color Scale.

API *n* Abbreviation for American Petroleum Institute.

API Gravity *n* Measure of specific gravity of petroleum and petroleum products, defined by the following equation:

$$\text{API Gravity, degrees} = \frac{141.5}{\text{specific gravity } 60/60°F} - 131.5$$

Apochromat \ˌa-pə-krō-ˈmat\ *adj* [ISV] (1887) A term applied to photographic and microscope objectives indicating the highest degree of color correction.

Apochromatic Objective *n* An objective corrected for spherical aberration at two wavelengths and for chromatic aberration at three wavelengths.

Apparent Density The mass per unit volume of material including voids inherent in the material as tested, such as pellets, powders, foams, chopped film, or fiber scrap. Usually expressed as lbs/cubic foot or g/cm^3. The term ▶ Bulk Density is synonymous for particulate materials. ASTM tests are D 1895 for pellets and powders, except PTFE powders, for which D 1457 applies; D 1622 for rigid foams. See also ▶ Density and ▶ Bulk Factor.

Apparent Viscosity *n* At any point in a fluid undergoing laminar shear, the nominal shear stress divided by apparent shear rate. In simple fluids, viscosity is a state property, depending only on composition, temperature, and pressure. In polymer melts and solutions, it is, nearly always, also dependent on the shear rate (or stress), hence the term *apparent viscosity*. The term is also applied to the quotient of the shear rate at the wall, which reduces to $\pi \cdot R^4 \cdot \Delta P/(8 \cdot Q \cdot L)$, where R and L are the radius and length of the tube, ΔP is the pressure drop through the tube, and Q is the volumetric flow rate (Elias HG (1977) Macromolecules, vol 1–2. Plenum Press, New York; Staudinger H, Heuer W (1930) A relationship between the viscosity and the molecular weight of polystyrene (German), *Ber*, 63B, 222–234).

Appearance \ə-ˈpir-ən(t)s\ *n* (14c) Manifestations of the nature of objects and materials through visual attributes such as size, shape, color, texture, glossiness, transparency, opacity, etc.

Appliance Finish *n* Generally, thermoset coatings, which are characterized by their hardness, mar resistance, and good chemical resistance.

Application \ˌa-plə-ˈkā-shən\ *n* [ME *applicacioun*, fr. L *application-*, application inclination, fr. *applicare*] (15c) Process by which surface coating compositions are transferred to a variety of surfaces, such as: brushing; spraying (cold and hot); dipping (simple immersion); dipping (assisted by application of vacuum and/or pressure); roller coating; flushing; and spreading.

Applicator \ˈa-plə-ˌkā-tər\ *n* (1659) Device to deposit a film of a specified thickness, such as doctor blade, wire wound rod, drawdown bar, etc.

Applicator Roll *n* Roll in a roller coater which applies the paint to a continuous strip.

Appliqué \ˌa-plə-ˈkā\ *n* [Fr, pp of *appliquer* to put on, fr. L *applicare*] (1801) Pattern or picture, formed by laying various materials usually on a fabric ground. Especially suitable for dressmaking and needlework. A design or ornament applied to another surface. In wallpaper, cutouts applied to a plain, textured, or figured background.

Appurtenance \ə-ˈpərt-nən(t)s, -ˈpər-tᵊn-ən(t)s\ *n* (14c) Any built-in, nonstructural portion of a building, such as doors, windows, ventilators, electrical equipment, partitions, etc.

Apricot Kernel Mixed Fatty Acids *n* These have a mp of 12–14°C; an acid value of 194; and an iodine value of 103.

Apricot Kernel Oil *n* Nondrying oil with iodine value of 96.

Apron \\ˈā-prən, -pərn\ *n* {*often attributive*} [ME, alteration (resulting fr. false division of *a* napron) of *napron*, fr. MF *naperon*, dim. of *nape* cloth, mod. of L *mappa* napkin] (15c) A paved area, such as the juncture of a driveway with the street or with a garage entrance (Merriam-Webster's Collegiate Dictionary (2004) 11th ed. Merriam-Webster, Springfield, MA).

Apron Mark See ▶ Decating Mark.

Aprotic Solvent \(ˌ)ā-ˈprō-tik\ *adj* [^2a- + *proton* + ^{1}-*ic*] (1931) An organic solvent that neither donates protons to or accepts them from a substance dissolved in it. Benzene, C_6H_6, is such a solvent (Wypych G (ed) (2001) Handbook of solvents. Chemtec Publishing, New York).

Aprotic Substance *n* A substance that can act neither as an acid nor as a base (Goldberg DE (2003) Fundamentals of chemistry. McGraw-Hill Science/Engineering/Math, New York).

Aptitude, Color See ▶ Color Aptitude.

A. Pullulans See ▶ Color Aptitude.

Aqua Regia *n* A powerful oxidizing and complexing solvent mixture composed of about three parts concentrated HC1 to one part concentrated HNO_3.

Aquarelle \ˌa-kwə-ˈrel, ˌä-\ *n* [F, fr. obs. I *acquarella* (now *acquerello*), fr. *acqua* water, fr. L *aqua*] (1869) (1) Drawing or print which has been colored with transparent water color washes. (2) Color that is made workable when mixed with water.

Aquatint \ˈa-kwə-ˌtint, ˈä-\ *n* [I *acqua tinta* dyed water] (1782) Method of engraving which, like etching, involves the use of a mordant acid for biting a metal plate, but differing in that aquatint is used to render tonal instead of linear effects.

Aqueous \ˈa-kwē-əs, ˈa-\ *adj* [ML *acqueus*, fr. L *aqua*] (1646) Water-containing or water-based.

Aqueous Acrylic See ▶ Latex.

Aqueous Phase of Emulsion The aqueous phase of an aqueous emulsion is water mixed with surfactant sometimes initiator and catalyst.

Aqueous Thermoplastic Emulsions *n* Water-based acrylic emulsions with noncrosslinked latex particles.

Aquo Complex *n* A complex in which water molecules serve as ligands.

AR *n* Abbreviation for ▶ Acrylic Rubber.

Arabic *n* (14c) See ▶ Acacia Gum.

Arachis Oil \ˈar-ə-kəs-\ *n* [NL *arachis*, genus that includes the peanut, fr. Gk *arakis*, dim. of *arakos*, a legume] (ca. 1889). See ▶ Peanut Oil.

Arachne Machine *n* (1) A machine for producing loop-bonded nonwovens. The fabric is formed by knitting a series of warp yarns through a fiber web processed on a card. Also see ▶ Bonding. (2) Stitch Bonding.

Aragonite \ə-ˈra-gə-ˌnīt, ˈar-ə-g-\ *n* [Gr *Aragonit*, fr. *Aragon*, Spain] (1803) See ▶ Calcium Carbonate.

$$Ca^{++}$$

$$\begin{array}{c} O^- \\ | \\ ^-O-C \\ \parallel \\ O \end{array}$$

Aralac Albumin fiber. Manufactured by National Dairy Products, U.S.

Araldite *n* Epoxide resins. Manufactured by CBA, Switzerland.

Aramid \ˈar-ə-məd, -ˌmid\ *n* [*ar*omatic poly*amide*] (1972) Acronym for *ar*omatic poly*amide*, currently available in fiber foam only, having at least 85% of the amide groups bonded to two aromatic rings. DuPont's Kevlar® is poly(*p*-phenylene terephthalamide). Aramid fibers exhibit low flammability, high strength, and high modulus. Fabrics made from aramid fibers maintain their integrity at high temperatures. Such fabrics are used extensively in hot-air filters. Aramids are also found in protective clothing, ropes and cables, and tire cord.

Aramid Fiber *n* Any of a family of high-strength, high-modulus fibers made from aramid resin. DuPont's Kevlar® –49 and –29 are the best known. K–49's ultimate strength is 3.4 GPa (500 kpsi), modulus is 131 GPa (19 Mpsi), ultimate elongation is 2.4%. K–29 has about equal strength, but half the modulus and twice the elongation. Density of either is 1.44 g/cm^3. Strength per density is higher for either of these fibers than for any others except some whiskers.

Archimedes Principle \ˌär-kə-ˈmē-dēz-\ *n* A body wholly or partly immersed in a fluid is buoyed up by a force equal to the weight of the fluid displaced. A body of volume V cm^3 immersed in a fluid of density ρ grams per cubic centimeter is buoyed up by a force in dynes,

$$F = \rho g V$$

where g is the acceleration due to gravity. A floating body displaces its own weight of liquid. (Serway RA, Faugh JS, Bennett CV (2005) College physics. Thomas, New York)

Architectural Coatings *n* Coatings intended for on-site application to interior or exterior surfaces of residential, commercial, institutional, or industrial – as opposed to industrial coatings. They are protective and decorative finishes applied at ambient temperatures (Wicks ZN, Jones FN, Pappas SP (1999) Organic coatings science and technology, 2nd edn. Wiley-Interscience, New York).

Architecture \ˈär-kə-ˌtək-chər\ *n* (1555) The art and science of designing and building structures, or large groups of structures, in keeping with aesthetic and functional criteria, and structures built in accordance with such principles.

Arcing \ˈär-kiŋ\ *v* (1893) Swinging spray gun away from the perpendicular.

Arco Microknife Instrument designed for testing scratch hardness and adhesion of coatings. A diamond point is weighted until it can penetrate a film to the metal substrate in two retracing steps. The weight necessary to achieve this cutting force for films of standard thickness is a measure of hardness.

Arc Resistance *n* The ability of a plastic material to maintain low conductivity along the path of exposure to a high-voltage electrical arc, usually stated in terms of the time required to render the material electrically conductive. Failure of the specimen may be caused by heating to incandescence, burning, tracking, or carbonization of the surface (Dissado LA, Fothergill CJ (eds) (1992) Electrical degradation and breakdown of polymers. Institution of Electrical Engineering (IEE), London; Emerson JA, Torkelson JM (eds) (1991) Optical and electrical properties of polymers: materials research society symposium proceedings, vol 24. Materials Research Society, Warrendale, PA; Ku CC, Liepins R (1987) Electrical properties of polymers. Hanser Publishers, New York).

Arc Tracking See ▶ Tracking.

Ardil *n* Fiber from peanut protein. Manufactured by ICI, Great Britain.

Area, Unit of The square centimeter. The area of a square whose sides are 1 cm in length. Other units of area are similarly derived. Dimensions, $[l^2]$.

Arene See ▶ Aromatic Hydrocarbon.

Argillaceous \ˌär-jə-ˈlā-shəs\ *adj* (ca. 1731) Composed primarily of clay or shale; clayey.

Argyle also ar-gyll \ˈär-ˌgī(ə)l, är-ˈ\ *n* [*Argyle, Argyll*, branch of the Scottish clan of Campbell, fr. whose tartan the design was adapted] (1899) A pattern consisting of diamond shapes of different colors knit in a fabric.

Arithmetic Mean *n* (Arithmetic average, mean, $\bar{x}$) (1) In statistics, the average of a set of measurements found by summing the measurements and dividing the sum by the number of measurements (▶ Number-Average Molecular Weight is an arithmetic average). (2) The conceptual mean, μ, of the population from which a set of measurements was drawn, rarely known exactly. The sample mean, $\bar{x}$, is the most efficient estimator of the population mean, μ.

Arnaudon's Green See ▶ Chromium Oxide Green.

Arnite Poly(ethylene terephthalate) (as plastic). Manufactured by AKU, The Netherlands.

Aromatic Compounds *n* A class of organic compounds containing a resonant, unsaturated ring of carbon atoms. Included are benzene, naphthalene, anthracene, and their derivatives. The term *aromatic* stems from the fact that many of these compounds have an agreeable odor (Morrison RT, Boyd RN (1992) Organic chemistry, 6th edn. Prentice-Hall, Englewood Cliffs, NJ).

Aromatic Content Percent aromatic hydrocarbons present in a solvent mixture or in a compound.

Aromatic Electrophilic-Substitution *n* Typical reactions of the benzene ring since the ring is electron rich (a base) and reactants are electron poor; typical reactions include nitration, halaogenation, and sulonation (Morrison RT, Boyd RN (1992) Organic chemistry, 6th edn. Prentice-Hall, Englewood Cliffs, NJ).

Aromatic Hydrocarbon *n* (1) A compound of carbon and hydrogen whose molecular structure contains one or more rings of six carbon atoms, with at least one of the rings containing alternating, resonant single and double bonds. Benzene, which is the simplest of the aromatic hydrocarbons, has the molecular formula C_6H_6. The family includes many solvents for plastics. (2) An aromatic hydrocarbon (AH) or arene (sometimes aryl hydrocarbon) is a hydrocarbon with a conjugated cyclic molecular structure that is much more stable than the hypothetical localized structure. Aromatic was assigned before the physical mechanism determining aromaticity was discovered, and because many of the compounds have a sweet scent. The configuration of six carbon atoms in aromatic compounds is known as a benzene ring, after the simplest possible such hydrocarbon–benzene. Aromatic hydrocarbons can be monocyclic or polycyclic. Some nonbenzene-based compounds called heteroarenes, which follow Hückle's rule, are also aromatic compounds. In these compounds, at least one carbon atom is replaced by one of the heteroatoms oxygen, nitrogen, or sulfur. Nonbenzene compounds with aromatic properties include furan, a heterocyclic compound with a five-membered ring that includes an oxygen atom, and pyridine, a heterocyclic compound with a six-membered ring containing one nitrogen atom. Reference: Morrison

RT, Boyd RN (1992) Organic chemistry, 6th edn. Prentice-Hall, Englewood Cliffs, NJ.

Aromatic Petroleum Residues *n* See ▶ Diolefin Resins.

Aromatic Polyamide *n* See ▶ Aramid.

Aromatic Polyester *n* (1) A Polyester that has aromatic rings in its chain, e.g., ▶ Polyethylene Terephthalate; (2) engineering thermoplastic prepared by polymerization of aromatic polyol with aromatic polyol with aromatic dicarboxylic anhydride. They are tough with somewhat low chemical resistance. Processed by injection and blow molding, extrusion, and thermoforming. Drying is required. Used in automotive housing, and trim, electrical wire jacketing, printed circuit boards, and appliance enclosures (Sepe MP (1998) Dynamic mechanical analysis. Plastics Design Library, Norwich, New York).

Aromatic Polyester Estercarbonate *n* A thermoplastic block copolymer of aromatic polyester with polycarbonate. Has higher heat distortion temperature tan regular polycarbonate.

Aromatic Polymer *n* Aromatic polymers, the backbone of which consist of repeating aromatic ring units. Aromatic rings in a unit may be single, fused, or joined by a chemical bond, bridging atom, or a group of atoms. Aromatic rings are six carbon rings containing three double bonds and are typified by benzene. Some hydrogen atoms in these rings may be substituted by other atoms or atom groups.

Aromatics *n* (15c) Compounds containing at least one benzene ring. Benzene, toluol, Xylol, or aromatic solvents.

Aromatic Solvents *n* Organic liquids having a cyclic or ring hydrocarbon structure and KB values over 40. They are good solvents for printing ink resins. Examples are toluol and xylol (Cf., Aliphatic Solvents). Use of aromatic solvents is severely restricted by Rule 66. Aromatic solvents containing less than 80% aromatic compounds are frequently designated as partial aromatic solvents.

Aromatization \ə-ˌrō-mə-tə-ˈzā-shən\ *vt* (15c) Process of converting saturated or aliphatic hydrocarbons to aromatic hydrocarbons.

Arrhenius Equation \ə-ˈrē-nē-əs, -ˈrā-\ *n* (1) A classical equation describing how rates of chemical reactions increase with rising absolute temperature:

$$r = Ae^{-E/(R \cdot T)}$$

in which

r = reaction rate (in appropriate units),
A = the collision factor (in the same units as r),
e = 2.71828...,
E = the activation energy of the reaction, J/mol,
R = the universal molar-energy constant, 8.3144 J/(mol·K), and
T = absolute temperature, K.

(Atkins PW, De Paula J (2001) Physical chemistry. W.H. Freeman, New York).

(2) An almost identical form, differing only in that the sign of the exponent is positive (viscosities *decrease* with rising temperature), has been used with good results to represent the temperature dependence of liquid viscosities, including those of polymer solutions and melts (Patton TC (1979) Paint flow and pigment dispersion: a rheological approach to coating and ink technology. Wiley, New York; Miller ML (1966) Structure of polymers. Reinhold Publishing, New York). It, too, is often referred to as an Arrhenius equation. This form has also been successful in modeling the temperature dependence of creep failure and property retention during heat aging. For viscosity work, the logarithmic form of this equation is more convenient.

$$\ln(\mu/\mu_o) = E(T - T_o)/(R \cdot T \cdot T_o).$$

Here E, R, and T have the same meanings as above; μ_o represents the viscosity at a reference temperature, T_o, which is usually chosen to be within the temperature range over which the viscosities have been measured (Elias HG (1977) Macromolecules, vol 1–2. Plenum Press, New York).

Arrhenius Plot *n* A plot of the logarithm of the specific rate constant for a reaction against the reciprocal of the absolute temperature; useful for determining the activation energy of the reaction (Atkins PW, Atkins P, De Paula J (2001) Physical chemistry. W.H. Freeman, New York).

Arrhenius Theory of Electrolytic Dissociation *n* This theory states that the molecule of an electrolyte can give rise to two or more electrically charged atoms or ions (Atkins PW, Atkins P, De Paula J (2001) Physical chemistry. W.H. Freeman, New York).

Arsenic Orange \ˈärs-nik ˈär-inj\ *n* See ▶ Realgar.

Artificial Aging \ˌär-tə-ˈfi-shəl\ *n* The accelerated testing of plastics to determine their changes in properties such as dimensional stability, water resistance, resistance to chemicals and solvents, light stability, and fatigue resistance.

Artificial Daylight *n* Term loosely applied to light sources, frequently equipped with filters, which are claimed to reproduce the color and spectral distribution of daylight. A more specific definition of the light source is to be preferred. See ▶ Correlated Color Temperature,

▶ Color Rendering Index, and ▶ Spectral Power Distribution Curve.

Artificial Stone *n* Special concretes and tiles, artificially colored to simulate natural stone, obtained by mixing stone dust aggregate and chips with Portland cement.

Artificial Turf *n* A manufactured carpet having the appearance of grass. It is used to replace grass in sports arenas, yards, etc. Also see ▶ Recreational Surfaces.

Artificial Ultramarine *n* Synthetic ultramarine blue. See ▶ Ultramarine Blue.

Artificial Weathering *n* A process of simulating weathering conditions such sunlight and rain using accelerated methods (e.g., Salt Fog Spray ASTM B 117). Artificial weathering is a method of utilizing exposure to cyclic laboratory conditions involving changes in temperature, relative humidity, and radiant energy, with or without direct water spray, in an attempt to produce changes in the material similar to those observed after long-term continuous outdoor exposure (American Society for Testing and Materials, www.astm.org, 100 Barr Harbor Drive, West Conshohocken, PA 19428-2959; American Society for Testing and Materials, www.astm.org, 100 Barr Harbor Drive, West Conshohocken, PA 19428-2959).

Artists' Colors *n* Various paint media used by artists, such as oil paints, watercolors, gouache, tempera, encaustic, fresco, silicate esters, and latex.

Art Linen *n* A plain-weave, softly finished fabric used either bleached or unbleached as a base fabric for needlework.

Aryl \\ar-əl\ *adj* [ISV *ar*omatic + *-yl*] (1906) Pertaining to monovalent aromatic groups, e.g., C_6H_5, phenyl.

ASA *n* See ▶ Acrylate Styrene Acrylonitrile Polymer.

ASA See ▶ American National Standards Institute.

Asbestine See ▶ Magnesium Silicate, Nonfibrous.

Asbestos \as-bes-təs, az-\ *n* [ME *albestron* mineral supposed to be inextinguishable when set on fire, prob. fr. MF, fr. ML *asbeston*, alter. of L asbestos, fr. Gk, unslaked lime, fr. *asbestos* inextinguishable, fr. *a-* + *sbennynai* to quench] (1607) The commercial term for a family of fibrous mineral silicates comprising some 30 known varieties, of which six were, for many years, commercially important. They are of two general types, *serpentine* and *amphibole* (Hibbard MJ (2001) Mineralogy. McGraw-Hill, New York). The serpentine type contains chrysotile that was most widely used as a reinforcement in thermosetting resins and laminates and, in finer form, as a filler in polyethylene, polypropylene nylons, and vinyls. Abestos used in flooring sheet and tiles. The amphiboles provided good chemical resistance and lower water absorption, but the outstanding properties provided all asbestos types were resistance to heat, fire retardance, and resistance to chemicals and for reinforcing plastics, and some brake linings for vehicles may still contain it. Extreme precautions must be taken in handling asbestos and asbestos-filled materials. The OSHA limit for such workers is 2 fibers per cubic meter of air, averaged over an 8-h shift. The use of asbestos products has declined in the United States due to health hazards associated with it (Carley JF (ed) (1993) Whittington's dictionary of plastics. Technomic Publishing).

Asbestos-Cement *n* A dense, rigid, board containing a high proportion of asbestos fibers bonded with Portland cement; resistant to fire, flame, and weathering; has low resistance to heat flow. Used as a building material in sheet form and corrugated sheeting (Harris CM (2005) Dictionary of architecture and construction. McGraw-Hill, New York).

Asbestos-Cement Board *n* A dense, rigid, board containing a high proportion of asbestos fibers bonded with Portland cement; resistant to fire, flame, and weathering; has low resistance to heat flow. Used as a building material in sheet form and corrugated sheeting (Harris CM (2005) Dictionary of architecture and construction. McGraw-Hill, New York).

Asbestos Plaster *n* A fireproof insulating material generally composed of asbestos with Bentonite as the binder (Harris CM (2005) Dictionary of architecture and construction. McGraw-Hill, New York).

Asbestos Sheeting *n* A dense, rigid, board containing a high proportion of asbestos fibers bonded with Portland cement; resistant to fire, flame, and weathering; has low resistance to heat flow. Used as a building material in sheet form and corrugated sheeting (Harris CM (2005) Dictionary of architecture and construction. McGraw-Hill, New York).

Asbestos Wallboard *n* A dense, rigid, board containing a high proportion of asbestos fibers bonded with Portland cement; resistant to fire, flame, and weathering; has low resistance to heat flow. Used as a building material in sheet form and corrugated sheeting.

Ascaridole (1,4-peroxide-*p*-menthene-2) $C_{10}H_{16}O_2$. A naturally occurring peroxide with uses as a polymerization initiator.

Aseptic Packaging *n* A package that has been sterilized with gas, heat or radiation after it was sealed.

Ash \ash\ *n* [ME *asshe*, fr. OE *æsc*, akin to OHGr *ask* ash, L *ornus* mountain ash] (before 12c) The mineral residue left after burning or decomposition of a sample of a substance.

Ash Content The solid residue remaining after a substance has been incinerated or heated to a temperature sufficient to drive off all volatile or combustible substances.

Ashing *n* A finishing process used to produce a satin-like finish on plastic articles, or to remove cold spots or teardrops from irregular surfaces which cannot be reached by wet sanding. The part is applied to a loose muslin disk loaded with wet ground pumice, rotating at a lineal speed of about 20 m/s.

ASM International *n* (American Society of Materials) A material science and engineering society headquartered in Materials Park, OH 44073, www.asminternational.org.

Aspartic Acid \ə-ˈspär-tik-\ *n* [ISV, irreg. fr. L *asparagus*] (1863) COOHCH$_2$CH(NH$_2$)COOH. Amino acid constituent of proteins of both animal and vegetable origin. In particular it is present in casein. It is a dibasic acid.

Aspect Ratio *n* (1907) (1) The ratio of length to diameter of a fiber or yarn bundle. (2) In tire production, the ratio of the height of the tire to its width. (3) In a rectangular structure, the ratio of the longer dimension to the shorter ((2000) Complete textile glossary, Celanese Acetate LLC, Three Park Avenue, New York).

Asphalt \ˈas-ˌfólt *also* ˈash-, *esp British* -ˌfalt\ *n* [ME *aspalt*, fr. LL *asphaltus*, fr. Gk *asphaltos*] (14c) (bitumen) (1) A dark brown or black, bituminous, viscous material which gradually liquefies when heated. The predominating constituents are bitumens, all of which occur in the solid or semisolid form in nature or are obtained by refining petroleum, or which are combinations of the bitumens mentioned with each other or with petroleum or derivatives thereof. (2) A similar material obtained artificially in refining petroleum; used in built-up roofing systems as a water proofing agent. (3) A mixture of such substances with an aggregate for use in paving (Usmani AM (1997) Asphalt science and technology. Marcel Dekker, Wallingford, UK).

Asphalt Cut Back *n* Asphalt plus thinner; asphalt solution; asphalt coating formed by dissolving asphalt (Usmani AM (1997) Asphalt science and technology. Marcel Dekker).

Asphalt Emulsion *n* A suspension or emulsion of ordinary asphalt in water. Such emulsions have attained wide use because, unlike straight asphalt, they do not have to be heated to be applied. The suspended asphalt is spread on in the usual way; and after the water has evaporated, the asphalt hardens into a continuous mass. Uses: highways; cement water-proofing; roofing compounds; and the like. *Also called Emulsified Asphalt*.

Asphaltenes *n* Highly condensed hydrocarbon compounds present in bitumens and asphaltums. Little is known of their chemical structure, but they are usually characterized by their insolubility in low boiling aliphatic hydrocarbons. Their physical condition is solid or semisolid (Usmani AM (1997) Asphalt science and technology. Marcel Dekker).

Asphaltic Bitumens *n* May be either naturally-occurring materials or otherwise obtained from the distillation of asphaltic base petroleum; they are readily soluble in aliphatic and aromatic hydrocarbons and have good film-forming properties (Usmani AM (1997) Asphalt science and technology. Marcel Dekker).

Asphaltic Pyrobitumens *n* Include elaterite, impsonite, and wurtzlite. They are characterized by their hydrocarbon nature, infusibility, and insolubility in carbon disulfide. Oxygen-containing compounds are present only in small amounts (Usmani AM (1997) Asphalt science and technology. Marcel Dekker).

Asphalt Paint *n* A liquid asphaltic product sometimes containing small amounts of other materials such as lampblack, aluminum flakes, and mineral pigments (Ash M, Ash I (1996) Handbook of paint and coating raw materials: trade name products – chemical products dictionary with trade name cross-references. Ashgate Publishing, New York).

Asphalt Paper *n* A paper sheet material that has been coated, saturated, or laminated with asphalt to increase its toughness and its resistance to water (Ash M, Ash I (1996) Handbook of paint and coating raw materials: trade name products – chemical products dictionary with trade name cross-references. Ashgate Publishing, New York).

Asphalt Seal Coat *n* A bituminous coating, with or without aggregate, applied to the surface of a pavement to waterproof and preserve the surface and to improve the texture of a previously applied bituminous surface (Ash M, Ash I (1996) Handbook of paint and coating raw materials: trade name products – chemical products dictionary with trade name cross-references. Ashgate Publishing, New York).

Asphalt Tile Sealer *n* Resin or plastic-type finish usually containing a water or alcohol carrier (Ash M, Ash I (1996) Handbook of paint and coating raw materials: trade name products – chemical products dictionary with trade name cross-references. Ashgate Publishing, New York).

Asphaltum (or Asphalt) \as-ˈfól-təm, *esp British* - ˈfal-\ *n* [ME *asphalt*, fr. LL *aspaltus*, fr. Gk *asphaltos*] (14c) U.S. paint industry's term for asphalt. In Britain, the term is reserved for natural asphalt (Usmani AM (1997) Asphalt science and technology. Marcel Dekker).

Asplit *n* Phenoplast. Manufactured by Hoechst, Germany.

ASR See ▶ Alkali-Soluble Resins.

Assembly \ə-ˈsem-blē\ *n* [MD *assemblee*, fr. MF, fr. OF, fr. *assembler*] (14c) A group of materials or parts, including adhesive, which has been placed together for bonding or which has been bonded together.

Assembly Adhesive See ▶ Adhesive, Assembly.

Assembly Glue See ▶ Adhesive, Assembly.

Assembly of Plastics *n* Plastic parts may be joined to others by many methods. Self-tapping screws are made with special thread designs to suit specific resins. Threaded inserts to receive mounting screws may be molded in or installed by press-fitting or by means of self-tapping external threads. Press-fitting may be employed to join plastics to similar or dissimilar materials. Snap-fit joints are made by molding or machining an undercut in one part, and providing a lip to engage this undercut in the mating component.

Assembly Time *n* Elapsed time after the adhesive is spread and until the pressure becomes effective. See ▶ Time, Assembly.

Associative Thickener A thickener which obtains its efficiency presumably by association between thickener molecules or thickener and latex particles rather than through high molecular weight or chain stiffness of the thickener molecules themselves. See also ▶ Thickener.

A-Stage *n* An early stage in the preparation of certain thermosetting resins in which the material is still fusible and soluble in certain liquids. Sometimes referred to as a *resol*. See also ▶ C-Stage.

A-Stage Thermosetting Resins *n* The first stage of novolac (phenolic) formation before crosslinking or curing.

Astigmatism \ə-ˈstig-mə-ˌti-zəm\ *n* (1849) An error of spherical lenses peculiar to the formation of images by oblique pencils. The image of a point when astigmatism is present will consist of two focal lines at right angles to each other and separated by a measurable distance along the axis of the pencil. The error is not eliminated by reduction of aperture as is spherical aberration.

ASTM American Society For Testing Materials. ASTM headquarters is located at 1916 Race St., Philadelphia, PA 19103, may be the largest nongovernmental, standards-writing body in the world, with 33,000 members. All information, standards, etc., for adhesives, coatings, inks, plastics, and sealants are available on the official ASTM website, www.ASTM.org.

ASTM D256 *n* An American Society for Testing of Materials (ASTM) standard method for determination of the resistance to breakage by flexural shock of plastics and electrical insulating materials, as indicated by the energy extracted from standard pendulum-type hammers in breaking standard specimens with one pendulum swin. The hammers are mounted on standard machines of either Izod or Charpy type. **Note:** Impact properties determined include Izod or Charpy impact energy normalized per width of the specimen. Also called ASTM method D256–84. See also ▶ Impact Energy.

ASTM D256-84 See also ▶ ASTM D256.

ASTM D412 *n* An American Society for Testing of Materials (ASTM) standard methods for determining tensile strength, tensile stress, ultimate elongation, tensile set, and set after break of rubber at low, ambient, and elevated temperatures using straight, dumbbell, and cut-ring specimens.

ASTM D638 *n* An American Society for Testing of Materials (ASTM) standard method for determining tensile strength, elongation, and modulus of elasticity of reinforced or unreinforced plastics in the form of sheet, plate, moldings, rigid tubes, and rods. Five (I–V) types, depending on dimensions, of dumbbell-shaped specimens with thickness not exceeding 14 mm are specified. Specified speed of testing varies depending on the specimen type and plastic rigidity. **Note:** Tensile properties determined include tensile stress (strength) at yield and at break, percentage elongation at yield or at break and modulus of elasticity. Also called ASTM method D638-84. See also ▶ tensile strength.

ASTM D638-84 See also ▶ ASTM D638.

ASTM method D638-84 See also ▶ ISO 75.

ASTM D671 *n* An American Society for Testing of Materials (ASTM) standard test method for determination of the flexural fatigue strength of rigid plastics subjected to repeated flexural stress of the same magnitude in a fixed-cantilever type testing machine, designed to produce a constant-amplitude-of-force on the test specimen each cycle. The test results are presented as a plot (S–N curve) of applied stress vs. number of stress cycles required to produce specimen failure by fracture, softening, or reduction in stiffness by heating caused by internal friction (damping). The stress corresponding to the point when the plot becomes clearly asymptotic to a horizontal (constant-stress) line is reported as fatigue strength in pascals, along with corresponding number of cycles. Also called ASTM D671-71B.

ASTM D671–71B See ▶ ASTM D671.

ASTM D696 *n* An American Society for Testing of Materials (ASTM) standard test method for the measurement of the coefficient of linear thermal expansion of plastics by using a vitreous silica dilatometer. The test is carried out under conditions excluding any significant creep or elastic strain rate and effects of moisture, curing, loss of plasticizer, etc. The specimen is placed at the bottom of the outer dilatometer tube and the tube is immersed in a liquid bath at a desired temperature.

ASTM D746 *n* An American Society for Testing of Materials (ASTM) standard method for determining brittleness temperature of plastics and elastomers by impact. The brittleness temperature is the temperature at which 50% of cantilever beam specimens fail on impact of a striking edge moving at a linear speed of 1.8–2.1 m/s and striking the specimen at a specified distance from the clamp. The temperature of the specimen is controlled by placing it in a heat-transfer medium, the temperature of which (usually sub-freezing) is controlled by a thermocouple.

ASTM D785 *n* An American Society for Testing of Materials (ASTM) standard test method for determination of indentation hardness of plastics by a Rockwell tester. The hardness number is derived from the net increase in the depth of impression as the load on a ball indenter is increased from a fixed minor load (10 kg) to a major load and then returned to the minor load. This number consists of the number of scale divisions (each corresponding to 0.002 mm vertical movement of the indentor) and scale symbol. Rockwell scales, designated by a single capital letter of English alphabet, vary depending on the diameter of the indentor and the major load.

ASTM D1708 *n* An American society for Testing of Materials (ASTM) standard method for determining tensile properties of plastics using microtensile specimens with maximum thickness 3.2 mm and minimum length 38.1 mm, including thin films. Tensile properties include yield strength, tensile strength, tensile strength at break, elongation at break, etc., determined per ASTM D638.

ASTM D2240 *n* An American Society for Testing of Materials (ASTM) standard method for determining the hardness of materials ranging from soft rubbers to some rigid plastics by measuring the penetration of a blunt (type A) or sharp (type D) indenter of a durometer at a specified force. The blunt indenter is used for softer materials and the sharp indenter – for or rigid materials.

ASTM D3763 *n* An American Society for Testing of Materials (ASTM) standard method for determination of the resistance of plastics, including films, to high-speed puncture over a broad range of test velocities using load and displacement sensors. **Note:** Puncture properties determined include maximum load, deflection to maximum load point, entegy to maximum load point, and total energy. Also called ASTM method D3763-86. See also ▶ Impact Energy.

ASTM method D3763-86 See ▶ ASTM D3763.

Astrakhan Cloth \ˈas-trə-kən, -ˌkan\ *adj* A thick knit or woven fabric with loops or curls on the face. The base yarns are usually cotton or wool and the loops are made with fibers such as mohair, wool, and certain manufactured fibers. The face simulated the pelt of the astrakhan lamb (Vincenti R (ed) (1994) Elsevier's textile dictionary. Elsevier Science and Technology Books, New York).

Asymmetric \ˌa-sə-ˈme-tri-kəl\ *adj* [Gk *asymmetria* lack of proportion, fr. *asymmetros* ill-proportioned, fr. *a-* + *symmetrical*] (1690) Of such a form that no point, line or plane exists about which opposite portions are congruent. The opposite of *symmetrical* (Merriam-Webster's Collegiate Dictionary (2004) 11th edn. Merriam-Webster, Springfield, MA).

Asymmetry *n* In chemistry, a molecular arrangement in which a particular carbon atom is joined to four different groups.

Atactic \(ˌ)ā-ˈtak-tik\ *adj* [ISV ^{2}*a-* + *-tactic*] (1957) Pertaining to a polymer in which the pendant side groups, as –CH$_3$ in polypropylene, are randomly located around the main chain. It can also be defined as the term for polymer tacticity to indicate unsymmetrical and alternating substituent groups along a polymer chain {G ataktisch, F atactique, S atáctico, I atattico}.

Atactic Block *n* A block of chain units in a polymer of copolymer that has a random distribution of equal numbers of the possible configurational base units (Odian GC (2004) Principles of polymerization. Wiley, New York; Mark JE (ed) (1996) Physical properties of polymers handbook. Springer, New York).

Atactic Configuration See ▶ Atactic.

Atactic Polymer *n* A polymer with molecules in which substituent groups or atoms are arranged at random around the backbone chain of atoms. The opposite of a *stereospecific polymer* and *isotactic polymer*. Also see

▶ Isotactic Polymer, ▶ Syndiotactic Polymer, and ▶ Tactic Polymer.

ATE *n* Abbreviation for Aluminum Triethyl (▶ Triethyl-Aluminum), a polymerization catalyst for olefins.

Atecticity *n* The degree of random location that the side chains exhibit off the backbone chain of a polymer (Rosato DV (ed) (1992) Rosato's plastics encyclopedia and dictionary. Hanser-Gardner, New York).

Athermal Transformation *n* A reaction that proceeds without thermal (not dependent on heat, enthalpy $\Delta H = 0$) activation as contrasted to isothermal transformation which occurs at constant temperature. An *athermal* mixture of liquids or polymer in a solvent involves no enthalpy ($\Delta H = 0$) change in an ideal solution; and the Gibbs Free energy ($-\Delta G$) is always negative for the solution (polymer dissolved in solvent) to occur while the entropy (ΔS) increases (Barton AFM (1983) Handbook of solubility parameters and other cohesion parameters. CRC Press, Boca Raton, FL).

Atlac *n* Polyester resin, manufactured by Atlas, U.S.

Atmosphere (Standard Atmosphere) A unit of pressure; 1 atm = 1.013×10^5 pascals (Pa) = 760 mmHg.

Atmospheric Conditions *n* In general, the relative humidity, barometric pressure, and temperature existing at a given time.

Atmospheric Fading See ▶ Gas Fading.

Atom \ˈa-təm\ *n* [ME, fr. L *atomus*, fr. Gk *atomos*, fr. *atomos* indivisible, fr. a- + *temnein* to cut] (15c) The smallest particle of an element which can enter into a chemical *combination*. All chemical compounds are formed of atoms, the difference between compounds being attributable to the nature, number, and arrangement of their constituent atoms (Goldberg DE (2003) Fundamentals of chemistry. McGraw-Hill Science/Engineering/Math, New York).

Atomic Absorption *n* An analytic method of measuring qualitatively/quantitatively elements that is based on absorption of specific wave-lengths of radiation.

Atomic Emission Spectroscopy *n* Same as ▶ Atomic Absorption, except radiation is emitted and measured as being indicative of specific atomic spectral characteristics (Willard HH, Merritt LL, Dean JA (1974) Instrumental methods of analysis. D. Van Nostrand, New York).

Atomic Energy *n* (1) The constitutive internal energy of the atom which was absorbed when it was formed. (2) Energy derived from the mass converted into energy in nuclear transformations.

Atomic Mass Unit *n* (1942) (1) The mass of a neutral atom of a nuclide. It is usually expressed in terms of the physical scale of atomic masses, that is, in atomic mass units (amu) See ▶ AMU. (2) A unit of mass; 1 atomic mass unit (amu) is defined as $\frac{1}{12}$ the mass of one $^{12}_{6}C$ atom (Weast RC (ed) Handbook of chemistry and physics, 52nd edn. CRC Press, Boca Roton, FL).

Atomic Number *n* (1821) The number (Z) of protons within the atomic nucleus. The electrical charge of these protons determines the number and arrangement of the outer electrons of the atom and thereby the chemical and physical properties of the element.

Atomic Structure *n* According to the currently accepted view, the atom consists of a central part, called nucleus, and a number of *electrons* (called orbital or planetary electrons) circling about the latter, like planets about the sun (Russell JB (1980) General chemistry. McGraw-Hill, New York). The nucleus is of a high specific weight; it contains most of the mass of the entire atom (its mass is considered equal to the atomic mass) and is composed of positively charged particles, called *protons* (the number of which always equals the atomic number, and particles of 0 charge, called neutrons. The diameter of the nucleus is between 10^{-13} and 10^{-12} cm, and the relatively vast distance in which the orbital electrons circle about it is illustrated by the fact that this nuclear diameter is only 10^{-4}–10^{-5} of the entire atomic diameter. While the nucleus carries an integral number of positive charges (an integral number of protons) each of 1.6×10^{-19} coulomb, each electron carries one negative charge of 1.6×10^{-19} coulomb, and the number of orbital electrons is equal to the number of protons in the nucleus (i.e., to the atomic number, Z), so that the atom as a whole has a net charge of 0. The electrons are arranged in successive shells around the nucleus; the maximum number of electrons in each shell is determined by natural laws, and the extra nuclear electronic structure of the atom is characteristic of the element. The electrons in the inner shells are tightly bound to the nucleus; this inner structure can be altered by high-energy particles, γ-rays or radium, or X-rays. The electrons in the outer shells are responsible for the chemical properties of the elemen (Whitten KW, Davis RE, Davis E, Peck LM, Stanley GG (2003) General chemistry. Brookes/Cole, New York).

Atomic Theory *n* (ca. 1847) All elementary forms of matter are composed of very small unit quantities called atoms. The atoms of a given element all have the same size and weight. The atoms of different elements have different sizes and weights. Atoms of the same or different elements unite with each other to form very small unit quantities of compound substances.

Atomic Unit *n* A mass equal to 1/12 the mass of an atom of carbon-12. The unit is deprecated by the SI system.

Atomic Weight *n* (1820) (atomic mass) The mass of an elemental isotope relative to that of the C-12 isotope of carbon, whose mass has been set at exactly 12.0000 atomic units (Daltons). For most elements, the tabulated atomic weight is the average, weighted by natural mass abundance, over all the element's isotopes, so is never an integer. The actual mass-fraction of a gram-of one au is the reciprocal of Avogadro's number $1/(6.02283 \times 10^{23})$.

ATRP *n* Atom-transfer radical polymerization. Odian G (2004) Principles of polymerization. Wiley-Intersciene, New York.

Attapulgite *n* $(Mg,Al)_5Si_8O_{20}(OH)_2 \cdot 8H_2O$. A clay mineral, with the ideal formula, in which there is considerable replacement of magnesium by aluminum. Electron micrographs show single laths or needles and bundles of laths or needles oriented in random fashion, which structure is generally believed to account for the high degree of porosity and absorptivity characteristic of this mineral. The name is derived from the place name, Attapulgus, Georgia. See ▶ Hydrated Magnesium Aluminum Silicate.

Attenuation \ə-ˈten-yə-ˌwāt\ *n* [L *attenuatus*, pp of *atfenuare* to make thin, fr. *ad-* + *tenuis*] (15c) (1) The process for making slim and slender, for example, the formation of fibers from molten glass (Merriam-Webster's Collegiate Dictionary (2000) 10th edn. Springfield, MA). (2) The gradual diminution of intensity or amplitude of a damped vibration with distance or time (Giambattista A, Richardson R, Richardson RC, Richardson B (2003) College physics. McGraw-Hill Science/Engineering/Math, New York).

atto- \ˈa-(ˌ)tō\ [ISV, fr. D or N *atten* eighteen, fr. ON *āttjān*; akin to OE *eahtatīene* eighteen] The SI prefix meaning $\times 10^{-18}$.

Attrition Mills *n* Machines for reducing materials into smaller particles by grinding down by friction. Equipment used in shredding pulp prior to acetylation in the manufacture of acetate and triacetate fibers, equipment used in shredding pulp prior to acetylation.

AU *n* Abbreviation for ▶ Polyurethane Elastomers with polyester segments. See ▶ Angstrom Unit.

Aufbau Procedure *n* The procedure of starting at the lowest of a set of energy levels and gradually adding electrons to progressively higher levels; "building-up" procedure.

Auger Electron Spectroscopy (AES) This technique is most powerful for providing analysis of the first few

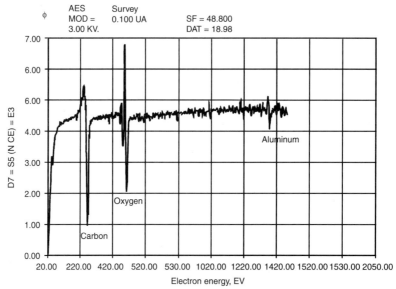

Auger electron spectroscopy spectrum of alumina, Al_2O_3.

atom layers (10 Angstroms or less) on the surface of the sample (AES explores the the electronic energy levels in atoms. Ther term Auger-process has come to denote any electron de-excitation in which de-excitation energy is transferred to a second electron, the "Auger electron," (Thompson M, Baker MD, Tyson JF, Christie A (1985) Auger electron spectroscopy. Wiley, New York). An AES spectrum of alumina is shown.

Aureobasidium Pullulans *n* A common ubiquitous, omnivorous, and highly polymorphic species of fungus which is often the primary organism in defacement of paint films (black yeast). *A Pullulans* is heterocaryotic since cells of most strains contain one to eight nuclei. It is now thought that *Pullularia Pullulans* and *A. Pullulans* are one and the same organism (Kirk PM, Cannon PF (2001) Fungi, 9th edn. CABI, New York).

Aureolin *n* Potassium cobaltinitrite yellow pigment. Syn: ▶ Cobalt Yellow.

Austrian Cinnabar See ▶ Chrome Orange.

Autoacceleration *n* (1) In some vinyl polmerizations, as the reaction approaches completion and the viscosity of the reaction medium rises, there is a rising rate of increase of molecular weight of the chains that have not yet been terminated. This rising increase is called autoacceleration, or the Trommsdorff effect, or gel effect. In similar terms, the acceleration of a reaction, such polymerization which continues to increase without external stimulus, i.e., gel effect and Trommsdorff effect. (2) Autoacceleration (Gel Effect, Trommsdorff–Norrish Effect) is a dangerous reaction behavior that can occur in free radical polymerization systems. It is due to the localized increases in viscosity of the polymerizing system that slow termination reactions. The removal reaction obstacles therefore causes a rapid increase in the overall rate of reaction, leading to possible reaction runaway and altering the characteristics of the polymers produced (Alger, 1989). Autoacceleration (Gel Effect, Trommsdorff–Norrish Effect) is a dangerous reaction behavior that can occur in free radical polymerization systems. It is due to the localized increases in viscosity of the polymerizing system that slow termination reactions. The removal reaction obstacles therefore causes a rapid increase in the overall rate of reaction, leading to possible reaction runaway and altering the characteristics of the polymers produced (Chekal, 2002). Autoacceleration of the overall rate of a free radical polymerization system has been noted in many bulk polymerization systems. The polymerization of methyl methacrylate, for example, deviates strongly from classical mechanism behavior around 20% conversion; in this region the conversion and molecular weight of the polymer produced increases rapidly. This increase of polymerization is usually accompanied by a large rise in temperature if heat dissipation is not adequate. Without proper precautions, autoacceleration of polymerization systems could cause metallurgic failure of the reaction vessel or, worse, explosion (Chekal, 2002). Norrish and Smith, Trommsdorff, and later, Schultz and Harborth, concluded that autoacceleration must be caused by a totally different polymerization mechanism. They rationalized through experiment that a decrease in the termination rate was the basis of the phenomenon. This decrease in termination rate, k_t, is caused by the raised viscosity of the polymerization region when the concentration of previously formed polymer molecules increases. Before autoacceleration according to Alger (2002), chain termination by combination of two free radical chains is a very rapid reaction that occurs at very high frequency (about one in 10^4 collisions). However, when the growing polymer molecules with active free radical ends are surrounded in the highly viscous mixture consisting of a growing concentration of "dead" polymer, the rate of termination becomes limited by diffusion Flor, 1953; Chekal, 2002). The Brownian motion of the larger molecules in the polymer "soup" are restricted, therefore limiting the frequency of their effective (termination) collisions. With termination collisions restricted, the concentration of active polymerizing chains and simultaneously the consumption of monomer rises rapidly. Assuming abundant unreacted monomer, viscosity changes affect the macromolecules but do not prove high enough to prevent smaller molecules – such as the monomer – from moving relatively freely. Therefore, the propagation reaction of the free radical polymerization process is relatively insensitive to changes in viscosity (Chekal, 2002). This also implies that at the onset of autoacceleration the overall rate of reaction increases relative to the rate of unautoaccelerated reaction given by the overall rate of reaction equation for free radical polymerization. Approximately, as the termination decreases by a factor of 4, the overall rate of reaction will double. The decrease of termination reactions also allows radical chains to add monomer for longer time periods, raising the weight average molecular weight dramatically. However, the number average molecular weight described by Flory (1953) only increases slightly, leading to broadening of the molecular weight distribution (high polydispersity index (PDI), very polydispersed product). Alger, M (1989) Polymer science dictionary. Elsevier Applied Science, New York; 28 Chekal BP (2002) Understanding the roles of chemically-controlled and diffusion-limited processes in determining the severity of autoacceleration behavior in free radical polymerization. Diss. Northwestern; Flory PJ (1953) Principles of polymer chemistry. Cornell UP, Ithaca, 124–129.

Autoadhesion *n* (tackiness) The ability of two contiguous surfaces of the same material, when pressed together, to form a strong bond that prevents their separation at the place of contact.

Autocatalytic Degradation *n* A type of breakdown in which the initially generated products accelerate the rate at which later degradation proceeds.

Autoclave \ˈȯ-tō-ˌklāv\ *n* [F, fr. *aut-* + L *clavis* key] (1876) A strong pressure vessel with a quick-opening door and means for heating and applying pressure to its contents. Autoclaves are widely used for bonding and curing reinforced-plastic laminates such as polyesters, epoxies, and phenolics. They are closed vessels for conducting a chemical reaction, sterilization or other operation under pressure and heat.

Autoclaveable *adj* Capable of being sterilized in steam at two to three times standard atmospheric pressure with no change in properties.

Autoclave Molding See ▶ Bag Molding.

Autodissociation *n* (Self-dissociation) The production of cations and anions by dissociation of solvent molecules without interaction with other species (Whitten KW, Davis RE, Davis E, Peck LM, Stanley GG (2003) General chemistry. Brookes/Cole, New York).

Auto-Flex Die *n* Trade name for a type of sheet-extrusion die with a flexible lip in which each lip-adjusting bolt, which can either push against the lip or pull it, is paired with a nearby cartridge heater. When a signal from the ▶ Beta-Ray Gauge that is constantly traversing the width of the sheet indicates that the sheet is too thick or too thin at a given point, the heater voltage at the relevant bolt is raised or lowered, expanding or contracting the bolt and decreasing or increasing the lip opening at that point (Shenoy AV (1996) Thermoplastics melt rheology and processing. Marcel Dekker; Carley JF (ed) (1993) Whittington's dictionary of plastics. Technomic Publishing, Lancaster, Pennsylvania).

Autogenous \ȯ-ˈtä-jə-nəs\ *adj* [Gk *autogenēs*, fr. *aut-* + *genēs* born, produced] (1846) A system processing independent of external influences, equivalent to autothermal extrusion (Shenoy AV (1996) Thermoplastics melt rheology and processing. Marcel Dekker).

Autohesion Term which refers to the ability of two contiguous surfaces of the same material to form a strong bond which prevents their separation at the place of contact. Also known as *Self-Adhesion* (Skeist I (ed) (1990, 1977, 1962) Handbook of adhesives. Van Nostrand Reinhold, New York).

Autoignition Temperature *n* The temperature at which a combustible material will ignite and burn spontaneously under specified conditions (Troitzsch J (2004) Plastics flammability handbook: principle, regulations, testing and approval. Hanser-Gardner, New York; Babrauskas V (2003) Ignition handbook. Fie Science Publishers, New York; Wypych G (ed) (2001) Handbook of solvents. Chemtec Publishing, New York; (1997) Tests for comparative flammability of liquids, UI 340. Laboratories Incorporated Underwriters, New York; Nelson G (1990) Fire and polymers: hazards identification and prevention. Oxford University Press, UK).

Automatic Control *n* In processing, control achieved by instruments that **Automatic Mold** A mold for compression, transfer, or injection molding that is equipped to perform all operations of the molding cycle, including ejection of the molded parts, in a completely automatic manner without human assistance.

Automatic Press *n* Hydraulic press for compression molding or an injection machine which operates continuously, being controlled mechanically, electrically, hydraulically, or by a combination of any of these methods.

Automatic Profile Control *n* In film and sheet extrusion, a system for controlling the uniformity of thickness across the sheet. The main components are a traversing thickness sensor such as a ▶ Beta-Ray Gauge, a computer and program that uses the sensor's signals to direct a mechanism that rotates the die-lip-adjusting bolts.

Automatic Unscrewing Mold *n* A mold for making threaded products – bottle caps are typical – that incorporates a mechanism for unscrewing the product from the mold core (or *vice versa*) as the mold opens, thereby releasing the product (Strong AB (2000) Plastics materials and processing. Prentice-Hall, Columbus, OH).

Autoxidation or Autooxidation \ȯ-ˌtäk-sə-ˈdā-shən\ *n* (1883) Oxidation by direct combination with molecular oxygen (as in air) at ordinary temperatures as described in Smith MB, March J (2001) Advanced organic chemistry, 5th edn. Wiley, New York. The reaction of diatomic oxygen (O_2) with π-bonds to form intermediate peroxides, etc. and permanent chemical bonds that are responsible for drying or curing of oils in alkyd resins (e.g., the drying of linseed oil). A reaction in which one substance acts simultaneously as an oxidizing agent and a reducing agent; also called *disproportionation* (Morrison RT, Boyd RN (1992) Organic chemistry, 6th edn. Prentice-Hall, Englewood Cliffs, NJ). The area of interest to materials scientists is the polymerization of oils, oil based resins such as vegetable oil based alkyd resins and inks (Muizebelt WJ, Donkerbroek MWF, Nielsen JB, Hussem and Biedmond MEF (1998) Oxidative crosslinking of alkyd resins studied with mass spectroscopy and NMR using model compounds. J Coatings Technol 70(876):83–92).

Autoxidation Inhibitors *n* Chemical agents which inhibit the autoxidation (or oxidative polymerization) reactions, i.e., phenol and arylamines (Bart J (2005) Additives in polymers: industrial analysis and applications. Wiley, New York; Zaiko GE (ed) (1995) Degradation and stabilization of polymers. Nova Science Publishers, New York).

Autoxidative Polymerization *n* The reaction of oxygen with fatty acids, oils or other reactive materials to form higher molecular weight polymers (e.g., autoxidative polymerization of vegetable oils and emulsified vegetable oils) (Gooch JW (2002) Emulsification and polymerization of alkyd resins. Kluwer Academic/Plenum Publishers, New York).

Autoxidation of Polymers *n* The reaction of oxygen with polymers to form peroxides, etc., which can result in crosslinking and/or degradation (Zaiko GE (ed) (1995) Degradation and stabilization of polymers. Nova Science Publishers, New York).

Autothermal Extrusion *n* (adiabatic or autogenous extrusion) In screw extruders, a steady state of operation in which the increase in enthalpy of the plastic from feed throat to die entry is equal to the net energy furnished by the drive to the screw (Strong AB (2000) Plastics Materials and Processing. Prentice-Hall, Columbus, OH).

Auxochrome See ▶ Chromophore.

Average Degree of Polymerization *n* $\overline{DP}$ Average degree of polymerization is equal to or twice for a chain growth polymerization, disproportionation or combination (Coleman MM, Strauss S (1998) Fundamentals of polymer science: an introductory text, CRC Press, Boca Raton, FL).

Average Molecular Weight (Viscosity Method) *n* The molecular weight of polymeric materials determined by the viscosity of the polymer in solution at a specific temperature. This gives an average molecular weight of the molecular chains in the polymer independent of specific chain length (Slade PE (2001) Polymer molecular weights, vol 4. Marcel Dekker, New York; Billmeyer FW, Jr. (1984) Textbook of polymer science, 3rd edn. Wiley-Interscience, New York; Staudinger H, Heuer W (1930) A relationship between the viscosity and the molecular weight of polystyrene (German), *Ber*, 63B, 222–234). See ▶ Number-Average, ▶ Viscosity-Average, and ▶ Weight-Aver-Age Molecular Weight.

Average Stiffness The ratio of change in stress to change in strain between two points on a stress–strain diagram, particularly the points of zero stress and breaking stress (Brown R (1999) Handbook of physical polymer testing, vol 50. Marcel Dekker, New York). Also see ▶ Modulus.

Average Toughness See ▶ Toughness.

Avogadro's Law \ˌa-və-ˈgä-(ˌ)drō, ˌä-\ *np* (1811) Equal volumes of different gases at the same pressure and temperature contain the same number of molecules.

Avogadro's Number \ˌa-və-ˈgä-(ˌ)drōz-\ *np* [Count Amedeo *Avogadro*] (1924) The number of atoms in exactly 12 g of $^{12}_{6}C$; 6.02×10^{23}. The number of units in 1 mole of units (Whitten KW, Davis RE, Davis E, Peck LM, Stanley GG (2003) General chemistry. Brookes/Cole, New York).

Avogadro's Principle *np* (or theory) The numbers of molecules present in equal volumes of gases at the same temperature and pressure are equal.

Avrami Equation \ˈäv-rä-mē\ [M. Avrami, European, published 1939–1941] *n* The time dependence of the overall crystallization of a polymer is described by the Avrami equation, where specific volume is related to crystalline and amorphous regions within the same polymer, then

$$(v_t - v_f)/(v_o - v_f) = \exp(-Kt^n)$$

where, v_o and v_f are initial and limiting values of the specific volume, and v_t is the specific volume of at a time t, and K is a kinetic constant. This equation describes the rate of polymer crystallization at a temperature. The crystallinity is expressed as the volume fraction (ϕ) of the crystalline material (Elias, H-G (2003) An introduction to plastics. Wiley, New York; Avrami M Kinetics of phase change I: general theory. J Chem Phys 7:1103; Kinetics of phase change II: transformation-time relations for random distribution of nuclei, 8:212; Kinetics of phase change III: granulation, phase change and microstructures, 9:177; Elias H-G (2003) An introduction to plastics. Wiley, New York).

Axial Yarn *n* A system of Axis, Crystallographic. One of several imaginary lines assumed in describing the positions of the planes by which a crystal is bounded, the positions of the atoms in the structure of the crystal and the directions associated with vectorial and tensorial physical properties longitudinal yarns in a triaxial braid that are inserted between bias yarns (Vincenti R (ed) (1994) Elsevier's textile dictionary. Elsevier Science and Technology Books, New York).

Axminster Carpet *n* A machine-woven carpet in which successive weft-wise rows of pile are inserted during weaving according to a predetermined arrangement of colors. There are four main types of Axminster looms: Spool, Gripper, Gripper-Spool, and Chenille (Complete textile glossary, Celanese Corporation, Three Park Avenue, New York, NY).

Azelaic Acid *n* (nonanedioic acid, 1,7-heptanedicarboxylic acid) $HOOC(CH_2)_7COOH$. A yellowish to white crystalline powder, derived from a fatty acid such as oleic acid by oxidation with ozone. It is an intermediate used in the production of plasticizers, polyamides, and alkyd resins. Acid value, 595.5; mol wt, 188.2; mp, 106°C; bp, 237°C/15 mmHg ((2001) Merck index, 13th edn. Merck and Company, Whitehouse Station, NJ). For plasticizers derived from azelaic acid. See ▶ Dicyclohexyl−, ▶ Di(2-Ethylbutyl)−, ▶ Di (2-Ethylhexyl)−, ▶ Di(2-Ethylhexyl)-4-Thio−, ▶ Di-*N*-Hexyl−, ▶ Diisobutyl−, and ▶ Diisooctyl Azelates.

Azeotrope \ˈā-zē-ə-ˌtrōp\ *n* [*a*- + *zeo*- (fr. Gk *zein* to boil) + -*trope*] (1938) A liquid mixture that is characterized by a constant minimum or maximum boiling point which is lower or higher than any of the components, and that distills without change in composition (Goldberg DE (2003) Fundamentals of chemistry. McGraw-Hill Science/Engineering/Math, New York; Weast RC (ed) Handbook of chemistry and physics, 52nd edn. CRC Press, Boca Raton, FL).

Azeotropic Copolymer *n* A copolymer in which the relative numbers of the different mer units are the same as in the mixture of monomers from which the copolymer was obtained. During a copolymerization reaction, the copolymer has the same composition as the monomer feed mixture, and a polymeric product of constant composition is formed throughout the copolymerization reaction (Kricheldorf HR, Swift G, Nuyken O, Huang SJ (2004) Handbook of polymer synthesis. CRC Press, Boca Raton, FL; Odian GC (2004) Principles of polymerization. Wiley, New York).

Azimuthal Quantum Number, *l* \ˌa-zə-ˈmə-thəl-\ *n* A quantum number which specifies a subshell for an electron in an atom.

2(1-Aziridinyl)ethyl Methacrylate *n* A vinyl monomer that combines a reactive vinyl group with an aziridinyl functional group. It can be polymerized alone or with other vinyl monomers to yield polymers with pendant aziridinyl groups that promote adhesion of coatings to the polymer (Odian GC (2004) Principles of polymerization. Wiley, New York).

Azlon Fiber *n* A manufactured fiber in which the fiber-forming substance consists of any regenerated naturally occurring proteins (FTC definition). Azlon is not currently produced in the United States.

Azobisformamide *n* (ABFA, azodicarbonamide) $H_2NCON=NCONH_2$. An aliphatic azo compound widely used as a chemical blowing agent in PVC, polystyrene, polyolefins, many other plastics, and in natural and synthetic rubbers. It is nontoxic, odorless, nonstaining, and, unlike other organic blowing agents, it is self-extinguishing and does not support combustion. Since ABFA in the pure state decomposes at temperatures above 216°C, when used with heat-sensitive plastics such as PVC, an activator that lowers its decomposition temperature is added to the compound. Such activators are compounds of cadmium, zinc, and lead, which also act as heat stabilizers, either directly or synergistically with other stabilizers (Wickson EJ (ed) (1993) Handbook of polyvinyl chloride formulating. Wiley, New York; Carley JF (ed) (1993) Whittington's dictionary of plastics. Technomic Publishing).

Azobis(isobutyronitrile) *n* A blowing agent developed in Germany for use in rubber and PVC. It is nonstaining and yields white PVC foam of fine uniform cell structure. However, its decomposition product, tetramethyl succinonitrile, is toxic and must be eliminated from the expanded product. For this reason the material is not used commercially in the U.S. as a blowing agent. It *is* used as a polymerization initiator (Odian GC (2004) Principles of polymerization. Wiley, New York; Elias HG (1977) Macromolecules, vol 1–2. Plenum Press, New York).

Azo-Compound Initiators *n* Azo compounds (R–N=N–R) which decompose to form free radicals (R + N$_2$ + R·) which are capable of initiating polymerization.

Azodicarbonamide See ▶ Azobisformamide.

Azo Dye *n* Any of an important family of dyes containing the –N=N– group, produced from amino compounds by the processes of diazotization and coupling. By varying the composition it is possible to produce acidic, basic, triazo, and tetrazo types, depending on the number of –N=N– groups in the molecule (Herbst W, Hunger K (2004) Industrial organic pigments. Wiley).

Azo Group *n* The structural grouping, –N=N–.

Azo Pigment *n* See ▶ Benzidine Yellows.

Azure Blue *n* See ▶ Cobalt Blue.

Azurite *n* \ˈa-zhə-ˌrīt\ *n* [F, fr. azur] (ca. 1868) Natural blue pigment derived from the mineral, azurite, a basic copper carbonate. The mineral occurs in various parts of the world in secondary copper ore deposits where it is frequently associated with malachite, a green basic carbonate of copper (Perkins D (2001) Mineralogy. Prentice-Hall, New York; (1996) Kirk-Othmer encyclopedia of chemical technology: pigments-powders. Wiley, New York; Lewis PA (ed) (1985–1990) Pigment handbook, 2nd edn. vols 1–4. Wiley-Interscience, New York). Also known as *Mountain Blue*.

B

b n \ˈbē\ SI abbreviation for ▶ Barn.

"b" or "β" n Yellowness-blueness coordinate in color spaces, Δb, the difference in "b", between a specimen and a standard reference color, normally used with "a" or "α" as part of the chromaticity difference (McDonald R (1997) Colour physics for industry, 2nd edn. Society of Dyers and Colourists, West Yorkshire). If "b" is plus, there is more yellowness than blueness; if "b" is minus, there is more blueness than yellowness. The exception is Adams Chromatic Value. See also ▶ Uniform Chromaticity Scale Diagram and ▶ Color Difference Equations.

"b", Kubelka-Munk n Mathematical constant characteristic of a color at complete opacity; dependent on the optical constants K and S. $b = [2(K/S) + (K/S)^2]^{1/2}$. See ▶ "a", ▶ Kubelka-Munk Equation. (McDonald R (1997) Colour physics for industry, 2nd edn. Society of Dyers and Colourists, West Yorkshire)

B (1) Chemical symbol for the element boron. (2) Symbol for magnetic induction.

Ba n Chemical symbol for the element barium.

Babo's Law The addition of a non-volatile solid to a liquid in which it is soluble lowers the vapor pressure of the solvent in proportion to the amount of substance dissolved. (Whitten KW, Davis RE, Davis E, Peck LM, Stanley GG (1994) General chemistry, Brookes/Cole, New York)

Back Coating n The application of latex or adhesive to the back of a carpet to anchor the tufts, usually followed immediately by addition of a secondary backing material such as woven jute or nonwoven polypropylene. (Complete textile glossary (2000) Celanese Acetate LLC, Three Park Avenue, New York).

Back Draft n (back taper, counterdraft) A slight undercut or tapered area in a mold tending to prevent removal of the molded part. See also ▶ Undercut. (Complete textile glossary (2000) Celanese Acetate LLC, Three Park Avenue, New York).

Backed Cloth n A material with an extra warp or filling added for weight and warmth. Satin-weave and twill-weave constructions are frequently used in the design of backed cloth because they are relatively resistant to the passage of air. (Complete textile glossary (2000) Celanese Acetate LLC, Three Park Avenue, New York).

Back Filling n A solution composed of varying amounts of cornstarch, China clay, talc, and tallow that is applied to the back side of low-grade, low-cost cloth to change its hand, improve its appearance, and increase its weight. (Complete textile glossary (2000) Celanese Acetate LLC, Three Park Avenue, New York)

Back Focal Plane n A "plane" normal to the axis of a lens in which all back focal points lie. (Moller KD (2003) Optics. Springer, New York; Freir GD (1965) University physics. Appleton-Century-Crofts, New York).

Backing n (1) A general term for any system of yarn which interlaces on the back of a textile material. (2) A knit or woven fabric or plastic foam bonded to a face fabric. (3) A knot or woven fabric bonded to a vinyl or other plastic sheet material. (4) See ▶ Carpet Backing. (Kadolph SJJ, Langford AL (2001) Textiles. Pearson Education, New York; Complete textile glossary (2000) Celanese Acetate LLC, Three Park Avenue, New York).

Backing Away From Fountain n A condition in which an ink lacking in flow will not keep in contact with the fountain roller so that the latter can transfer ink to the ductor roller. Eventually the prints become uneven, streaky, and weak. (Leach RH, Pierce RJ, Hickman EP, Mackenzie MJ, Smith HG (eds) (1993) Printing ink manual, 5th edn. Blueprint, New York).

Backing Coat n In coil coating, the finish on the back side of continuous strip. (Tracton AA (ed) (2005) Coatings technology handbook. Taylor and Francis, Inc. New York).

Backing Plate n In injection molding, a plate used as a support for the cavity blocks, guide pins, bushing, etc. Sometimes called *Support Plate*.

Back Paint See ▶ Back Priming.

Back Pressure n In extrusion, the head pressure. In screw-injection molding, the head pressure just before the valve opens to make the shot.

Back-Pressure Relief Port n A side channel in the head of an extruder, usually leading to a ▶ Rupture Disk through which the melt can escape if the pressure exceeds a safe limit.

Back Priming n The process of applying a coating to the back surface of construction materials prior to installation, usually for protection against the weather.

Back Putty See ▶ Bed Putty.

Back Sizing See ▶ Filler.

Backstitch n (1611) See ▶ Purl.

Back Taper See ▶ Back Draft.

Back Warp n The warp which, along with the back filling, actually forms the second face (back) of double, triple, or quadruple fabrics.

Back Winding n (1) Rewinding yarn or fiber from one type of package to another. (2) Winding yarn as it is deknit.

Bacteria \bak-ˈtir-ē-ə\ *n* [plural of *bacterium*] (1884) Any of the numerous microscopic, spherical, rod-shaped, or spiral organisms of the class, *Schizomycetes*. (Black JG (2002) Microbiology, 5th edn. Wiley, New York).

Bacterial Corrosion *n* A corrosion which results from substances (e.g., ammonia or sulfuric acid) produced by the activity of certain bacteria. (Baboian R (2002) Corrosion engineer's handbook, 3rd edn. NACE International – The Corrosion Society, Houston).

Bactericidal \bak-ˌtir-ə-ˈsī-dᵊl\ *adj* (1878) Capable of causing the death of bacteria. (Black JG (2002) Microbiology, 5th edn. Wiley, New York).

Bactericidal Fiber *n* Fiber used for medical applications, socks, shoe liners, etc., in which bactericides are introduced directly into the fiber matrix as opposed to fiber simply having a bactericidal finish applied.

Bactericidal Paint *n* Coating which discourages the multiplication of bacteria.

Bactericide *n* (1878) An agent capable of destroying bacteria. See also ▶ Biocide. (Black JG (2002) Microbiology, 5th edn. Wiley, New York).

Bacteriostat \-ˈtir-ē-ō-ˌstat\ *n* (1920) An agent that, when incorporated in a plastics compound, will prevent the growth of bacteria on surfaces of articles made from the compound. (Black JG (2002) Microbiology, 5th edn. Wiley, New York).

Bacteriostatic *n* (1920) Inhibiting the growth of bacteria, but not bactericidal.

Bacteriostats *n* (1920) Chemicals which inhibit the growth of bacteria cells, i.e., bacteriocidal chemicals.

Baddeleyite See ▶ Zirconium Oxide.

Baekeland, Dr *n* Inventor of the oldest family of phenolformaldehyde polymers (1907). (Odian GC (2004) Principles of polymerization. Wiley, New York).

Baffle \ˈba-fəl\ *vt* [prob. alter. of ME (Sc) *bawchillen* to denounce, discredit publicly] (ca. 1590) A plug or other device inserted in a flow channel to restrict the flow or change its direction. A metal piece so placed in a rollercoater pan to direct the stream or flow of paint to the pick-up roller. Also used to direct air currents in an oven. (Perry RH, Green DW (1997) Perry's chemical engineer's handbook, 7th edn. McGraw-Hill, New York).

Bagasse \be-ˈgas\ *n* [Fr] (ca. 1826) (megass) A tough fiber derived from sugar cane, remaining after the sugar juice has been extracted. It is used as reinforcement in some laminates and molding powders. (Vincenti R (ed) (1994) Elsevier's textile dictionary. Elsevier Science and Technology Books, New York).

Bagging \ˈba-giŋ\ *n* (1732) (1) A fabric woven in cylindrical or tubular form on an ordinary cam loom and used for grain bags, etc. (2) Fabric bulging caused by extension at the knees, elbows, etc., of a garment lacking dimensional stability.

Baggy Cloth *n* A fabric that does not lie flat, caused by sections of tight or loose yarns in either the warp or the filling.

Baggy Selvage See ▶ Slack Selvage.

Bagillumbang Oil *n* Drying oil obtained from the species *Aleurites trisperma*, which grows in the Philippines.

Bag Molding *n* A method of forming and curing reinforced plastic laminates employing a flexible bag or mattress to apply pressure uniformly over one surface of the laminate. A perform comprising a fibrous sheet impregnated with an A- or B-state resin is placed over or in a rigid mold forming one surface of the article. The bag is applied to the upper surface, then pressure is applied by vacuum, in an autoclave, in a press, or by inflating the bag. Heat may be applied by steam in the autoclave, or through the rigid half of the mod. When an autoclave is used, the process is sometimes called *autoclave molding*.

Bail \ˈbā(ə)l\ *n* [ME *beil, baile*, prob. of Scand origin; akin to Swedish *bygel* bow, hoop; akin to OE *būgan* to bend] (15c) Semicircular handle of a kettle, pail, or can.

Bailing Machine *n* Mechanical device designed to automatically attach bails to containers.

Bake Hardness Increased hardness on heating due to elimination of retained solvent in thermoplastic coatings and thermal crosslinking in thermosetting coatings.

Bakelite \ˈbā-kə-ˌlīt, -ˌklīt\ *n* (1) A trade name derived from the name of Leo H. Baekeland, a pioneering Belgian chemist who developed phenolic resins in the early 1900s. They are the oldest family of phenolformaldehyde polymers. The trade name was long used by the Bakelite Corporation, later absorbed by Union Carbide, who still uses the name for some of its resins. (2) Phenolformaldehyde resins, manufactured by Bakelite Inc., USA.

Bake System *n* Set of prime, intermediate, and/or top coats which require baking to effect a cured or dried film, and which together have been determined to yield the desired properties for a particular purpose.

Baking *n* Process of drying or curing a coating by the application of heat in excess of 65°C (150°F). When below this temperature, the process is referred to as Forced Drying. Syn: ▶ Stoving.

Baking Finish *n* Paint or varnish that requires baking at temperatures above 65°C (150°F), for the development of desired properties.

Baking Schedule *n* Set of related baking temperatures and baking times which will yield a cured or dried film from a coating materials such as a powder coating for maximum performance. The baking schedule is specific for each material because powder coatings are made from different materials and require different periods of flow time (fusion) and compete cure time (reaction) or program of time comprising temperature held at specific periods of time and minimum and maximum temperatures. (Wicks ZN, Jones FN, Pappas SP (1999) Organic coatings science and technology, 2nd edn. Wiley-Interscience, New York).

Baking Temperature *n* Temperature above 65°C (150°F) at a specified time in a baking schedule which has been determined to produce a cured or dried film having optimum desired properties for the coating.

Baking Time *n* Time at a specified temperature, above 65°C (150°F), in a baking schedule which has been determined to produce a cured or dried film having optimum desired properties for the coating.

Baking Vehicle (Varnish) *n* Vehicle specifically formulated for coatings which are intended to be baked in order to effect a cured or dried film.

Balanced Cloth *n* A term describing a woven fabric with the same size yarn and the same number of threads per inch in both the warp and the filling direction.

Balanced Design *n* In reinforced plastics, a winding pattern so designed that the stresses in all filaments are equal. (Pittance JC (ed) (1990) Engineering plastics and composites. SAM International, Materials Park).

Balanced Gating *n* In multicavity injection molds, the objective is to fill all of the cavities simultaneously, to fill to the same final pressure, and then to have their gates freeze at the same time. (Strong AB (2000) Plastics materials and processing. Prentice Hall, Columbus).

Balanced Laminate *n* A composite structure in which fiber layers laid at angles to the main axis occur in pairs, at equal ± angles, that may or may not be adjacent. (Pittance JC (ed) (1990) Engineering plastics and composites. SAM International, Materials Park).

Balanced Reaction *n* A reaction in which a state of equilibrium has been reached, and it can be made to proceed in one direction or another by adjusting the conditions. Conditions which affect the direction of a reaction are the concentrations of the reactants involved and the temperature and the pressure. (Whitten KW, Davis RE, Davis E, Peck LM, Stanley GG (1994) General chemistry, Brookes/Cole, New York).

Balanced Runners *n* In a multicavity injection mold, the runners are balanced when the injected melt reaches all the cavity gates at the same instant after the start of injection. In practice, with identical cavities whose shape, size, number, and layout permit, all runner branches are given equal cross sections and corresponding branch lengths are made equal. Uniform metal temperature throughout is assumed.

Balanced Solvents *n* Combination of solvents designed to give a specified performance. (Wypych G (ed) (2001) Handbook of solvents. Chemtec Publishing, New York).

Balanced Twists *n* In a plied yarn or cord, an arrangement of twist which will not cause the yarn or cord to twist on itself of kink when held in an open loop.

Balata \bə-ˈlä-tə\ *n* [Sp, fr. Carib] (1860) (1) Dried juice of a West Indian tree (*Mimusops globosa*). It has rubbery characteristics and is sometimes used as a substitute of gutta-percha. It resembles both this and ordinary rubber in that it consists of isoprene molecules (C_5H_8), but polymerized in its own particular way. Impure form of trans-polyisoprene. (2) *Trans*-1,4-Poly (isoprene). (Langenheim JH (2003) Plant resins: chemistry, evolution ecology and ethnobotany. Timber, Portland).

Balata, Natural *n* A material identical in properties and composition to ▶ Gutta-Percha, obtained from trees in South America.

Balata, Synthetic *n* A stereospecific rubber, the trans isomer of polyisoprene, made by catalyzed addition polymerization of isoprene.

Bald Spot Area or patch, usually in a wrinkle finish film, which has failed to wrinkle or give the desired optical effect.

Bale *n* [ME, fr. MF, of Gr origin; akin to OHGr *balla* ball] (14c) Quantity of compressible articles or materials assembled into a shaped unit and bound with cord or metal ties.

Bale-and-Ring Test *n* (ring-and-ball test) A method of determining the softening temperature of resins (see www.astm.org). A specimen is cast or molded in a metal ring of 16-mm inside diameter and 6.4-mm depth. This ring is placed upon a metal plate in a liquid bath heating at a controlled range, and a steel ball 9.5 mm in diameter weighting 3.5 g is placed in the center of the specimen. The softening point is considered to be the temperature of the liquid when the ball penetrates the specimen and touches the lower plate.

Baling Machine *n* Device used to form and fasten compressible materials into a shaped unit (bale).

Ball \ˈbȯl\ [ME *bal*; akin to OE *bealluc* testis, OHGr *balla* ball, OE *bl* āwan to blow] (13c) When rubbing down paint, varnish films, or the like, with abrasive paper, the material removed by the abrasive action may be either in the form of a dry powder, or, if it has a tendency to softness or stickiness, will collect as relatively large lumps or balls. A film, which exhibits this latter phenomenon is said to ball. (Paint/coatings dictionary

(1978). Federation of Societies for Coatings Technology, Philadelphia).

Ball and Pebble Mills *n* A rotating mill for reducing size of materials (e.g., pigments and fillers) or pulverize them using hard metal or ceramic balls. (Perry RH, Green DW (1997) Perry's chemical engineer's handbook, 7th edn. McGraw-Hill, New York).

Ball and Ring Method *n* System of testing the melting and softening temperatures of asphalt, waxes, resins, and paraffins. A ring 15.875 mm in diameter is filled with the substance to be tested, and a steel ball 9.5 mm in diameter is placed upon the substance in the ring. The end point is the temperature at which the substance softens sufficiently to allow the ball to fall through the ring to the bottom of the beaker. See also ▶ Softening Point. (Paint: pigment, drying oils, polymers, resins, naval stores, cellulosics esters, and ink vehicles, vol. 3. (2001) American Society for Testing and Material; Usmani AM (ed) (1997) Asphalt science and technology. Marcel Dekker).

Ball Charge *n* Volume of porcelain or steel balls loaded in a ball mill. It is generally one third the total volume of the ball mill.

Ball Mill *n* (pebble mill) A cylindrical or conical shell rotating horizontally about its axis, partly filled with a grinding medium such as natural flint pebbles, ceramic pellets, or hard metal balls. The material to be ground is added to just fill, or slightly more than fill, the voids between the balls. Water may or may not be added. The shell is rotated at a speed that causes the balls to cascade, thus reducing the particle sizes by repeated impacts. The operation may be batchwise or continuous. In the plastics industry, the term ball mill is reserved by some persons for mills containing metallic grinding media, and the term pebble mill for nonmetallic media. For paint, steel balls are frequently used. (Perry RH, Green DW (1997) Perry's chemical engineer's handbook, 7th edn. McGraw-Hill, New York; Weismantal GF (1981) Paint handbook. McGraw-Hill Corporation, Inc., New York).

Ball Milling *n* A method of grinding and mixing material, with or without liquid, in a rotating cylinder or conical mill partially filled with grinding media such as balls or pebbles. (Perry RH, Green DW (1997) Perry's chemical engineer's handbook, 7th edn. McGraw-Hill, New York).

Balloon \bə-ˈlün\ The curved paths of running yarns about the take-up package during spinning, downtwisting, plying, or winding, or while they are being withdrawn over-end from packages under appropriate yarnwinding conditions. (Humphries M (2000) Fabric glossary, Prentice-Hall, Upper-Saddle River; Vincenti R (ed) (1994) Elsevier's textile dictionary. Elsevier Science and Technology Books, New York).

Balloon Fabric *n* A plain-weave cloth having the same breaking strength in each direction. This fabric is made from fine (60s to 100s) combed yarn woven to constructions of 92 × 108 to 116 × 128. Vulcanized balloon fabric is used for air cells in planes and barrage balloons. (Vincenti R (ed) (1994) Elsevier's textile dictionary. Elsevier Science and Technology Books, New York).

Ballotini *n* Glass beads used in reflective paints. See ▶ Beaded Paint and ▶ Traffic Paint. (Paint /coatings dictionary (1978). Compiled by Definitions Committee of the Federation of Societies for Coatings Technology)

Ball Punch Impact Test *n* Measure of the resistance of a coating to impact or shock. A ball punch is used for this test. (Paint and coating testing manual (Gardner-Sward Handbook) (1995) MNL 17, 14th edn. ASTM, Conshohocken).

Ball Rebound Test *n* A method for measuring the resilience of polymeric materials by dropping a steel ball on a specimen from a fixed height and observing the height of rebound. The difference between the two heights is proportional to the energy absorbed. By conducting tests over a range of temperature, results can indicate temperature of first- and second-order transitions, and effects of additives and plasticizers. (Shah V (1998) Handbook of plastics testing technology. Wiley, New York; Pittance JC (ed) (1990) Engineering plastics and composites. SAM International, Materials Park; Harper CA (ed) (2002) Handbook of plastics, elastomers and composites, 4th edn. McGraw-Hill, New York; see relevant test methods from American Society for Testing and Materials, 100 Barr Harbor Drive, West Conshohocken, 19428–2959, www.astm.org).

Ball-Up *n* A term used in adhesive circles to describe the tendency of an adhesive to stick to itself.

Ball Viscometer See ▶ Falling-Ball Viscometer.

Ball Wrap *n* Parallel threads in the form of a twistless rope wound into a large ball. When wound mechanically with quick traverse a ball warp may be made in the form of a large cylindrical package.

Balmer Series *n* (of spectral lines) The wavelengths of a series of lines in the spectrum of hydrogen are given in angstroms by the equation

$$\lambda = 3646 \frac{N^2}{N^2 - 4}$$

where N is an integer having values greater than 2. (Weast RC (ed) Handbook of chemistry and physics, 52nd edn. The Chemical Rubber Co., Boca Raton)

Balsa \ˈbȯl-sə\ *n* [Sp] (ca 1600) Wood from the tree *Ochroma lagopus*, Ecuador. Its density is only 0.12–0.2 g/cm^3, yet it has good strength, especially end-grain compressive strength, so it has found application as an interlayer in reinforced-plastics (Hoadley RB (2000) Understanding wood. The Taunton, Newtown). See ▶ Sandwich Structures.

Balsam \ˈbȯl-səm\ *n* [L *balsamum*, fr. Gk *balsamon*, prob. of Semitic origin; akin to Hebrew *bāshām* balsam] (before 12c) Oleoresinous exudations from plants which are characterized by softness or a semi liquid consistency. They consist of mixtures of resin, essential oils, and other compounds. Typical balsams are Canada, gurjun, Peru, tolu, and storax balsams. (Merriam-Webster's collegiate dictionary, 11th edn (2004). Merriam-Webster, Inc., Springfield, Massachusetts). See ▶ Oleoresin.

Baluster \ˈba-lə-stər\ *n* [Fr *balustre*, fr. It *balaustro*, fr. *balaustra* wild pomegranatye flower, fr. L *balaustium*, fr. Gk *balaustion;* fr. its shape] (1602) (1) One of a number of short vertical members, often circular in section, used to support a stair handrail or a coping. (2) A post in a balustrade. Also called *Banister*.

Balustrade \ˈba-lə-strār\ *n* [Fr, fr. It *balaustrata,* fr. *balaustro*] (1644) An entire railing system (as along the edge of a balcony) including a top rail and its balusters, and sometimes a bottom rail.

Banana Liquid A solution of nitrocellulose in amyl acetate or similar solvent.

Banana Oil See ▶ Amyl Acetate.

Banbury *n* Compounding apparatus composed of a pair of contra rotating blades which masticate the materials to form a homogeneous blend. An internal type heavy duty mixer. See ▶ Internal Mixer.

Banbury Mixer *n* An intensive mixer originally used for rubber, and for many years used for mixing plastics such as cellulosics, vinyls, polyethylene and others. It consists of two counter rotating, spiral-shaped blades encased in segments of cylindrical housing, the housing halves joined along internal ridges between the blades. Blades and housing may be cored for circulating heating or cooling liquids. A recent adaptation of the design, with connections to feed and discharge screws, permits continuous operation. (Strong AB (2000) Plastics materials and processing. Prentice Hall, Columbus; Juran R (ed) (1986) Modern plastics encyclopedia. McGraw-Hill/Modern Plastics, New York; 1990, 1992, 1993 editions).

Band Heater See ▶ Heater Band.

Banding, Heavy Tow *n* Nonuniform distribution of filaments across towband width.

Bandle *n* A coarse homespun linen made on narrow hand looms in Ireland.

Band Spectrum *n* Lines produced by molecular vibrations, which are so close together that they appear to be continuous.

Band Wire *n* A wire attached between dispensing and collecting vessels in order to equalize the electrical potential between them to dissipate electrostatic charge.

Banister \ˈba-nəs-tər\ *n* [alter. of *bahuster*] (1667) See ▶ Baluster.

Bank (1) In calendering and roll-milling, a cylindrical accumulation of working material in the nip of the rolls at the feed point. (2) Another name for a yarn creel.

Bank, Rubber Mill *n* A relatively small amount of unvulcanized rubber compound rolling in the space between two mill rolls, while the major portion of the compound is bonded around the first roll. A pencil bank is a small, smooth rolling amount of compound between a calendar roll and fabric surface being surface or friction coated. (Vincenti R (ed) (1994) Elsevier's textile dictionary. Elsevier Science and Technology Books, New York).

Bar *n* [Gr fr. Gk *baros*] (1910) A deprecated unit of pressure, long used in meteorological work as approximately one atmosphere, and actually equal to 0.987 standard atmosphere, i.e., 100 kPa.

Barathea \ˈbar-ə-ˈthē-ə\ *n* [fr. *Barathea*] (1862) (1) A silk, rayon, or manufactured fiber necktie fabric with a broken rib weave and a characteristic pebbly appearance. (2) A fine, dress fabric with a silk warp and worsted filling, woven in a broken filling rib which completely covers the warp. (3) A smooth-faced worsted uniform cloth with an indistinct twilled basket weave of fine two-ply yarns.

Barbender Plastograph *n* (PlastiCorder®) An instrument that continuously measures the torque exerted in shearing a polymer or compound specimen over a wider range of shear rates and temperatures, including those conditions anticipated in actual processing. The instrument records torque, time and temperature with a computer generated graphical representation referred to as a *plastigram*, from which one can infer processability. It shows the effect of additives and fillers, measures and records lubricity, plasticity, scorch, cure, shear and heat stability, and polymer consistency. The modern instrument is equipped with interchangeable heads for bench-scale studies of mixing and extrusion, and with computer control and data acquisition. (Strong AB (2000) Plastics materials and processing. Prentice Hall, Columbus; Shah V (1998) Handbook of plastics testing technology. Wiley, New York).

Barcol Hardness *n* The resistance of a material to penetration by a sharp steel point under a known load with an instrument called the Barcol Impressor. Direct readings are obtained on a scale from 0 to 100. The instrument has often been used as a way of judging the degree of cure of thermosetting resins and harder materials. The ASTM test, www.astm.org, is referred to as "Indentation Hardness of Rigid Plastics". See ▶ Indentation Hardness.

Barefoot Resin See ▶ Neat Resin.

Bare Glass n Glass (yarns, rovings, fabrics) from which the sizing or finish has been removed: also, such glass before the application of sizing or finish.

Barex Copolymer from acrylonitrile and methyl methacrylate (3:1). Manufactured by Vistron, USA.

Bargeboard \❘bärj- ❘bōrd, - ❘bórd\ n (1833) An often ornatmented and sloping board along a gable, covering the ends of roof timbers (Merriam-Webster's collegiate dictionary, 11th edn (2004) Merriam-Webster, Inc., Springfield, Massachusetts.
Syn: ▶ Vergeboard and ▶ Gableboard.

Barite \❘bar- ❘īt, ❘ber-\ n [Gr *barytēs* weight, fr. *barys*] (1868) See ▶ Barium Sulfate.

Barium \❘bar-ē-əm, ❘ber-\ n [NL, fr. *bar-*] (1808) A silver-white malleable toxic bivalent metallic element of the alkaline-earth group that occurs only in combination. (Whitten KW, Davis RE, Davis E, Peck LM, Stanley GG (1994) General chemistry, Brookes/Cole, New York).

Barium Carbonate n $BaCO_3$. A white compound, insoluble in water, occurring as a mineral or made by direct precipitation. It has a sp gr of 4.275 and a mp of 1360°C. It is used in ceramics, fillers, and extenders. (Ash M, Ash I (1998) Handbook of fillers, extenders and dilutents. Synapse Informtion Resources, Inc., New York). Syn: ▶ Witherite.

Barium Chromate See ▶ Barium Yellow. (Ash M, Ash I (1998) Handbook of fillers, extenders and dilutents. Synapse Informtion Resources, Inc., New York)

Barium Chrome See ▶ Barium Yellow.

Barium-Extended Titanium Dioxide n Analogous to calcium-extended titanium dioxide, prepared by coprecipitating the $BaSO_4$ with TiO_2 or alternately made by a physical blending. In either case, the product contains 30% TiO_2 and 75% $BaSO_4$. It is no longer commercially available. (Ash M, Ash I (1998) Handbook of fillers, extenders and dilutents. Synapse Informtion Resources, Inc., New York)

Barium Ferrite See ▶ Ferrite.

Barium Fillers n Barium sulfate or other compound of barium which serves as a filler for plastics compounding. (Ash M, Ash I (1998) Handbook of fillers, extenders and dilutents. Synapse Informtion Resources, Inc., New York)

Barium Hydroxide Monohydrate n (barium monohydrate) $Ba(OH)_2 \cdot H_2O$. A white powder used in the production of phenol-formaldehyde resins and barium soaps.

Barium Lithol Red See ▶ Lithol Red.

Barium Lithol Toner n Barium salt of 2-amino-1-naphthalene sulfonic acid coupled to 2-naphthol.

Barium Metaborate Modified n $BaB_2O_4 \cdot H_2O$. White crystalline pigment prepared by precipitation from aqueous solution; sp gr of 3.3, density of 27.5 lb/gal refractive index of 1.55–1.60, O.A. of 30 lb/100 lb, and pH of 9.8–10.3 (Ash M, Ash I (1998) Handbook of fillers, extenders and dilutents. Synapse Informtion Resources, Inc., New York).

Barium Naphthenate n Metallic naphthenate used as a wetting agent for certain pigments, a hardener for some alkyds, and a thickener for oils.

Barium Oleate n Barium salt of oleic acid. Used for the prevention of chalking, the maintenance of color of titanium paints, and the dispersion of pigments in media during grinding.

Barium Peroxide n BaO_2 or $BaO_2 \cdot 8H_2O$. An oxidizing catalyst used in some polymerization reactions.

Barium Plaster *n* A special mill-mixed gypsum plaster containing barium salts; used to plaster walls of X-ray rooms.

Barium Ricinoleate *n* Ba(OOCC$_7$H$_4$CH=CHCH$_2$CHOHC$_5$H$_{10}$CH$_3$)$_2$. A heat stabilizer imparting good clarity, used most often in vinyl plastisols and organosols.

Barium Stearate Ba(C$_{18}$H$_{35}$O$_2$)$_2$*n* White crystalline solid; insoluble in water or alcohol; mp of 160°C, and sp gr of 1.145. Used as a light and heat stabilizer in plastics. As a heat stabilizer, it is used particularly when sulfur staining is to be avoided. Also used as a lubricant where high temperatures are to be encountered.

Barium Sulfate *n* (1903) (1) BaSO$_4$. Natural Pigment White 22 (77120). Mineral consisting essentially of barium sulfate. It may contain, in addition, small amounts of sulfates of calcium and strontium, common chalk, calcium fluoride, silica, and iron oxide. Used in paints as a filler or extender. Density, 4.5 g/cm^3 (37.5 lb/gal); O.A., 9; particle size 2–30 μm. (Ash M, Ash I (1998) Handbook of fillers, extenders and dilutents. Synapse Informtion Resources, Inc., New York).

Barium Sulfates *n* (barites, blanc fixe, heavy spar, permanent white, terra ponderosa) BaSO$_4$. A white powder obtained from the mineral barite or synthesized chemically. One of the synthetic varieties, *blanc fixe*, is made by mixing aqueous solutions containing sulfate and barium ions. As a filler in plastics and rubbers, barium sulfate imparts opacity to X rays but only a low order of optical opacity. Thus it is useful as a filler when it is desired to increase specific gravity without adversely affecting the tinctorial power of pigments. (Ash M, Ash I (1998) Handbook of fillers, extenders and dilutents. Synapse Informtion Resources, Inc., New York).

Barium White *n* See ▶ Barium Sulfate and ▶ Synthetic.

Barium Yellow *n* Pigment known also as barium chromate, barium chrome, or lemon yellow. It is essentially barium chromate made by adding a solution of sodium dichromate to barium chloride solution. (Ash M, Ash I (1998) Handbook of fillers, extenders and dilutents. Synapse Informtion Resources, Inc., New York).

Barking *n* The removal of bark from wood prior to pulping.

Bar Mold A mold in which the cavities are arranged in rows on separate bars that may be individually removed to facilitate stripping.

Barn (b) *n* A miniscule area unit commensurate with the cross sections of atomic nuclei. 1 barn = 10^{-28} m^2. Nuclear cross sections range from about 0.01 to 1.5 b.

Barnacle \ ˈbär-ni-kəl\ *n* [ME *barnakille*, alter. of *bernake*, *bernekke*] (15c) Any of certain crustaceans of the group *cirripedia*, as the goose barnacles, the stalked species which cling to ship bottoms and floating timber, and the rock barnacles, the species which attach themselves to marine rocks.

Barre \ ˈbär\ *n* [Fr, fr. ML *barra*] (1936) A defect characterized by bars or streaks, fillingwise in woven fabrics or coursewise in weft-knit fabrics, caused by uneven tension in knitting, defective yarn, improper needle action, or other similar factors.

Barrel *n* [ME *barel*, fr. MF *baril*] (14c) (1) The tubular main cylinder of an extruder, within which the screw rotates. (2) A container, agitated by rotation or vibration, used for tumbling moldings to remove flash and sharp edges. Also used for mixing of dry solids, e.g., pigments with resin pellets. (3) Standard unit of liquid volume in the petroleum industry. It is equal to 42 U.S. gallons, approximately 35 Imperial gallons; approximately 160 liters.

Barrel Finishing See ▶ Tumbling.
Barrelling See ▶ Tumbling.
Barrel Mixing See ▶ Tumbling.
Barrel Polish See ▶ Tumbling.
Barrier Coat n Coating used to isolate a paint system from the surface to which it is applied in order to prevent chemical or physical interaction between them, e.g., to prevent the paint solvent attacking the underlying paint or to prevent bleeding from underlying paint or material. (Wicks ZN, Jones FN, Pappas SP (1999) Organic coatings science and technology, 2nd edn. Wiley-Interscience, New York) See ▶ Tie Coat and ▶ Primer.
Barrier Layer n In multiplayer films, coextruded sheet, and blow-molded containers, a layer of polymer having very low permeability to the gases and/or vapors of interest for the application of the film, sheet, or container.
Barrier Plastics n Thermoplastics with low permeability to gases and/or vapors. Most important commercially are ▶ Nitrile Barrier Resins. Several others, however, are based on various copolymers, some of which are more permeable than nitrile resins but are easier to process. A major application is bottles for carbonated beverages.
Barrier Screw See ▶ Solids-Draining Screw.
Barrier Sheet n An inner layer of a laminate, placed between the core and an outer layer.
Bar Stock n Standard lengths of plastics extrusion having simple cross-sectional shapes such as circular (rod stock), square, hexagonal, and rectangular of low aspect ratio, used in fabricating plastics parts by machining, welding, fastening, and adhesive bonding.
Barye n Cgs pressure unit = one dyne/cm^2.
Barytes \ˈbar-ˌīt, ˈber-\ See also ▶ Blanc Fixe. Natural barium sulphate used as an ink pigment and a white extender. It is considerably more abrasive and gritty than precipitated barium sulphate.
Base n (1) The opposite of ▶ Acid. Any molecule or ionic substance that can combine with a proton to produce a differnt substance (salts, soaps) (Whitten KW, Davis RE, Davis E, Peck LM, Stanley GG (1994) General chemistry, Brookes/Cole, New York. (2) Reinforcing material (glass fiber, paper, cotton, asbestos, etc) which is impregnated with resin in the forming of laminates and is used as an insulating support for an electrical printed pattern. (3) Metallic salt upon which coloring matter is precipitated to form the insoluble pigments called lakes. (4) A paste or liquid which is to be thinned or tined. (5) In ink manufacture, a dispersion containing usually only one coloring matter, pigment or dye, properly dispersed in a vehicle. This is subsequently used for mixing to produce the desired end product, (Leach RH, Pierce RJ, Hickman EP, Mackenzie MJ, Smith HG, (eds) (1993) Printing ink manual, 5th edn. Blueprint, New York). (6) See ▶ Baseboard. (7) See ▶ Substrate and ▶ Ground.
Baseboard n (1853) (1) Board along the base of a wall usually finished with moldings. (2) A molding that conceals the joint between an interior wall and floor. Syn: for ▶ Base, ▶ Skirting Board, ▶ Base Plate, ▶ Mop Board, ▶ Scrub Board, ▶ Kick Board, and ▶ Washboard.
Basebox n (1) In the metal-coating trade, a unit of area equal to 0.4861 m^2. (2) In can coating, the standard unit equal to 31,360 square inches (20.23 m^2).
Base Coat n (1) All plaster applied before the finish coat; may be a single coat or a scratch coat and a brown coat. (2) The first coat applied to a surface, as paint; a prime coat. (3) An initial coat applied to a wood surface before staining or otherwise finishing it.
Base Fabric n In coated fabrics, the underlying substrate.
Base Plate See ▶ Baseboard.
Base Unit n The smallest possible repeating unit of a polymer. (Odian GC, (2004) Principles of polymerization, Wiley, New York).
Basic n Of an alkaline nature. Capable of uniting with an acid to form a salt.
Basic Carbonate of Lead See ▶ Carbonate White Lead.
Basic Dye n Having a slightly basic property due to the presence of aniline or a similar group. (Industrial dye: chemistry, properties and applications (2003). Wiley, New York)
Basic Fuchsin Magenta See ▶ Fuchsin.
Basic Lead Acetate n Compound derived from lead oxide (litharge) by treatment with insufficient acetic acid for complete neutralization. It is used in the manufacture of acetate chromes. (Gooch JW, (1993) Lead based paint handbook. Plenum, New York)
Basic Lead Carbonate n $2PbCO_3 \cdot Pb(OH)_2$. A very effective heat stabilizer, used where toxicity is of no concern as in electrical-insulating compounds. Its use is limited because of its tendency to form blisters during processing and to cause spew when exposed to weather, also by rising concern about lead in the environment (Gooch JW (1993) Lead based paint handbook. Plenum, New York). See ▶ Carbonate White Lead.
Basic Lead Chromate n $PbCrO_4 \cdot PbO$. Normal lead chromates are represented by the formula $PbCrO_4$, but if these are treated with alkalis, the color can be changed from the original bright yellow to an orange color, and

basic chromates are produced. (Gooch JW (1993) Lead based paint handbook. Plenum, New York)

Basic Lead Nitrate *n* $Pb(NO_3)_2$. Compound prepared in a similar manner to basic lead acetate, but nitric acid is used in place of acetic acid to produce nitric chromes (Gooch JW (1993) Lead based paint handbook. Plenum, New York).

Basic Lead Silico Chromate *n* Calcined basic lead chromate – basic lead silicate complex on a silica core, used as a corrosion-inhibiting pigment. Sp gr of 4.1; particle size of 7 μm; O.A. of 10–18 g/100 g; CrO_3 – 5.4 wt. %; SiO_2 – 47.6 wt. %; PbO – 47.0 wt. %. (Gooch JW (1993) Lead based paint handbook. Plenum, New York).

Basic Lead Silicate *n* $3\ PbO \cdot 2\ SiO_2 \cdot H_2O$. Pigment White 16 (77625). Pigment consisting of an adherent surface layer of basic lead silicate "cemented" to silica. Compositions and constants vary depending on commercial grade. (Gooch JW (1993) Lead based paint handbook. Plenum, New York).

Basic Lead Sulfate *n* $PbSO_4$, Pigment White 2 (77633). Pigment containing normal lead sulfate ($PbSO_4$) in "combination" with a proportion of basic lead which is taken to be an oxide (PbO). It is a white pigment which is used in considerable amounts in weather-resisting undercoats and finishes, usually in conjunction with carbonate white lead and/or zinc oxide. Its O.A. is 12–18 g/100 g and its sp gr is 6.4 (Gooch JW (1993) Lead based paint handbook. Plenum, New York). Syn: ▶ Sublimed White Lead.

Basic Pigment *n* Any pigment which is capable of reacting with fatty acids to form soaps.

Basic Solution *n* An aqueous solution in which the concentration of hydroxide ions exceeds that of hydrogen (hydronium) ions.

Basis Weight *n* The weight in pounds of a ream (500 sheets) of paper cut to a given standard size for that grade: 25 × 38 for book papers, 20 × 26 for cover papers, 22½ × 28½ or 22½ × 35 for bristols, 25½ × 30½ for index. E.g., 500 sheets 25 × 38 of 80-lb. coated will weigh eighty pounds. (2) The weight of a unit area of fabric. Examples are ounces per square yard and grams per square centimeter.

Basic Zinc Chromate *n* $ZnCrO_4 \cdot 4\ Zn(OH)_2$. Yellow pigment used primarily for its corrosion-inhibiting properties.

Basket Stitch (Weave) *n* In this knit construction, purl and plain loops are combined with a preponderance of purl loops in the pattern courses to give a basket-weave effect. ()

Bas-relief (Basso-rilieva) *n* Sculpture in low relief.

Basswood Oil *n* Semidrying oil. Sp gr of 0.938/15°C; iodine value of 111.0.

Bast Fiber *n* Any of a group of fibers taken from the inner barks of plants that run the length of the stem, is surrounded by enveloping tissue, and is cemented together by pectic gums. Included in the group are jute, flax, hemp, and ramie, some of which are used to reinforce plastics.

Batavia Dammar or Damar \bə-ˈtä-vē-ə ˈda-mər\ *n* (1698) Gum or resin exported from Batavia (now called Jakarta), Indonesia. Batavia dammar has two subgrades: Padang or Sumatra.

Batch \ˈbach\ *n* [ME *bache*; akin to OE *bacan* to bake] (15c) In industry, a unit or quantity of production used in one complete operation. (Merriam-Webster's collegiate dictionary, 11th edn (2004). Merriam-Webster, Inc., Springfield, Massachusetts).

Batch Adjustment Record *n* A record of materials added to a paint batch to adjust viscosity, color, and such.

Batch Polymerization *n* The polymerization in a single and static, noncontinuous volume of monomers. (Odian GC (2004) Principles of polymerization. Wiley, New York).

Batch Process *n* Any process in which the charge is added intermittently in definite portions, or batches, the operations of he process and the removal of products being completed on each portion before the addition of the next; as opposed to continuous process.

Batch Reactors *n* A reaction vessel for batch reactions as opposed to continuous reactors. (Smith JM (1981) Chemical engineering kinetics. Mc-Graw-Hill Co, New York).

Batch Record *n* A record of all materials and proportions used to produce a batch of paint.

Bathrobe Blanketing *n* A double-faced fabric woven with a tightly twisted spun warp and two sets of soft spun filling yarns. The fabric is thick and warm and its filling yarns are frequently napped to produce a soft surface. Today's blankets are made of spun polyester, acrylic, or polyester/cotton blends. (Kadolph SJJ, Langford AL (2001) Textiles. Pearson Education, New York).

Batik \bə-ˈtēk, ˈba-tik\ *n* [Japanese *batik*] (1880) See ▶ Dyes.

Batiste \bə-ˈtēst, ba-\ *n* [F] (1697) (1) A sheer, woven, mercerized fabric of combed cotton or polyester/cotton resembling nainsook, only finer, with a lengthwise streak. (2) A rayon fabric decorated with dobby woven striped and jacquard florals. (3) A smooth, fine, woven fabric, lighter that challis and very similar to nun's veiling. (Kadolph SJJ, Langford AL (2001) Textiles. Pearson Education, New York).

Batt \ˈbat\ *n* (1871) In textile jargon, insulation in the form of a blanket, rather than loose filling. ()

Batten \ˈba-tən\ *n* [alter. of ME *batent*, *bataunt* finished board, fr. MF *batant*, fr. pre part of *battre*] (1658) Small thin strips covering joints between wider boards on exterior building surfaces. (Harris CM (2005) Dictionary of architecture and construction, McGraw-Hill Co.)

Battery \ˈba-t(ə-)-rē\ *n* [MF *batterie*, fr. OF, fr. *battre* to beat, fr. L *battuere*] (1531) (1) A set of galvanic cells connected in series; (2) In common usage, a single commercial galvanic cell. (Whitten KW, Davis RE, Davis E, Peck LM, Stanley GG (1994) General chemistry, Brookes/Cole, New York; Merriam-Webster's collegiate dictionary, 11th edn (2004). Merriam-Webster, Inc., Springfield, Massachusetts).

Batting \ˈba-tiŋ\ *n* (1773) A soft, bulky assembly of fibers, usually carded. Battings are sold in sheets or rolls and used for warm interlinings, comforter stuffings, and other thermal or resiliency applications.

Batu *n* East Indian semifossil resin, resembling the dammars. Soluble in aromatic or mixed aromatic-aliphatic hydrocarbons. Dissolves readily in warm vegetable oil. It is almost neutral. Acid value averages bout 30.

Baudouin Test \bō-ˈdwaⁿ-\ Specific test for sesame oil, which gives a strong carmine coloration when a dilute alcoholic solution of furfural is added to a mixture of the oil and hydrochloric acid, and the whole violently shaken.

Baumé (Bé) \bō-ˈmā\ *adj* (1877) A floating hydrometer method of measuring density of a liquid. The scale for floating hydrometer methods used to measure the specific gravity of a liquid (suspension, etc.). The depth of immersion is a linear function of the inverse of the density, www.photonics.com/dictionary. The Baume' scale is linear in inverse density. A dual transformation of specific gravity (S) for liquids devised by the French chemist Antoine Baumé for the graduation of hydrometers for the purpose of measuring percent salt (brine) in water; and, like Twaddell and API, becoming obsolete because they are empirical and have no scientific basis. Letting S equal the ratio of the density of the subject liquid at 15.6°C to that of water at the same temperature, the Baumé transformations are:
For liquids more dense than water, °Bé = 145 [1 − (1/S)] and for liquids less dense than water, °Bé = (140/S) − 130. The hydrometer is used for determining the specific gravities of liquids, engraved, not directly in units of specific gravity, but in a scale consisting of Baumé degrees. These Baumé degrees, often written (Bé), can be converted to true specific gravities by reference to conversion tables. (See www.monashscientific.com)

Bauxite \ˈbȯk-ˌsīt, ˈbäk-\ *n* [F *bauxite*, from Les *Baux*] (1861) $Al_2O_3 \cdot 2H_2O$. Aluminum mineral, consisting chiefly of a hydrated aluminum oxide. The color of native bauxite may be white, brownish red, or grey. Used as the source of many aluminum compounds. Crude powdered bauxite has been used to some extent as filler for paints.

Bay *n* Space between columns or supports of a building. A bay window is placed between such supports and usually projects outward. If curved or semicircular, it is called a bow window.

Bayardere *n* A very broad term for stripes that run crosswise in a knit or woven fabric.

BBP See ▶ Butyl Benzyl Phthalate.

BCF Yarns *n* Bulked continuous filament yarns for carpet trade, usually nylon, polypropylene, or polyester.

Be *n* Chemical symbol for the element beryllium.

Bead \ˈbēd\ *n* [ME *bede* prayer, prayer bead, fr. OE *bed*, *gebed* prayer, akin to OE *biddan* to entreat, pray] (before 12c) (1) Heavy accumulation of a coating which occurs at the lower edge of a panel or other vertical surface as the result of excessive flowing. (2) Solidified droplet of an oil-resin mix, withdrawn from a varnish or resin mix, for testing purposes. Instructions frequently include a statement that the oil-resin mix shall be heated until a hard or tough bead is obtained. (3) Glass spheres. See ▶ Ballotini.

Bead and Butt *n* Framed work in which the panel is flush with the framing and has a bead run on two edges in the direction of the grain; the ends are left plain. *Also known as Bead, Butt, and Bead Butt Work.*

Bead, Butt and Square *n* Similar to bead and butt, but having the panels flush on the beaded face only, and showing square reveals on the other.

Beaded Paint *n* Traffic or marking paint with reflecting beads (ballotini) applied after striping or manufactured with beads and applied.

Beaded Velvet *n* Velvet with a cut-out pattern or a velvet pile effect, made on a Jacquard loom. This fabric is used primarily for evening wear.

Beader *n* A device for rolling beads on the edges of thermoplastic sheets or cylinders.

Bead Polymer *n* A polymer in the form of nearly spherical particles about 1 mm in diameter.

Bead Polymerization *n* A type of polymerization identical to ▶ Suspension Polymerization, except that the monomer is dispersed as relatively large droplets in water or other suitable inert diluents by vigorous agitation.

Bead Polymerization, Suspension *n* Through agitating or stirring with the aid of a dispersion agent, water insoluble monomers can be dispersed in water as fine droplets, oil soluble free radical initiators start polymerization in the droplets, the droplets grow and convert into beads or pearls.

Beads, Reflective See ▶ Beaded Paint and ▶ Ballotini.

Beam \ˈbēm\ n [ME *beem*, fr. OE *bēam* tree, beam; akin to OHGr *boum* tree] (before 12c) (1) Timber or girder supporting a floor, roof, or ceiling, usually supported at each end by a pier, wall, post, or girder. A structural member whose prime function is to carry transverse loads, as a joist, girder, rafter, or purlin. (2) A cylinder of wood or metal, usually with a circular flange on each end, on which warp yarns are wound for slashing, weaving, and warp knitting.

Beam Ceiling n (1) A ceiling, usually of wood, made in imitation of exposed floor beams with the flooring showing between. (2) The underside of a floor, showing the actual beams, and finished to form a ceiling.

Beam Dyeing Machine n A machine for dyeing warp yarns or fabrics that have been wound onto a special beam, the barrel of which is evenly perforated with holes. The dye liquor is forced through the yarn or fabric from inside to outside and vice versa. (Kadolph SJJ, Langford AL (2001) Textiles. Pearson Education, New York; Industrial dye: chemistry, properties and applications (2003). Wiley, New York).

Beaming The operation of winding warp yarns onto a beam usually in preparation for slashing, weaving, or warp knitting. *Also called Warping.*

Beamroll See ▶ Beam.

Bearding n Fuzz on loop pile carpets usually resulting from poor anchorage or fiber snagging

Bearing Strength n The ability of plastics sheets to sustain edgewise loads that are applied by pins, rods, or rivets used to assemble the sheets to other articles.

Beat Frequencies n The beat of two different frequencies of signals on a non-linear circuit when they combine or beat together. It has a frequency equal to the difference of the two applied frequencies.

Beater n (1) The machine which does most of the opening and cleaning work on a fiber picker and opener. Revolving at high speed, it beats against the fringe of fiber as the latter is fed into the machine. (2) A machine used in the paper industry for opening pulp and combining additives.

Beating-Up n The last operation of the loom in weaving, in which the last pick inserted in the fabric is "beat" into position against the preceding picks.

Beat(s) n Two vibrations of slightly different frequencies f_1 and f_2 when added together, produce in a detector sensitive to both these frequencies, a regularly varying response which rises and falls at the "beat" frequency $f_b = [f_1 - f_2]$. It is important to note that a resonator which is sharply tuned to f_b alone will not resound at all in the presence of these two beating frequencies. See ▶ Combination Frequencies.

Beaumontage n A resin, beeswax, and shellac mixture used for filling small holes or cracks in wood or metal. (Paint: pigment, drying oils, polymers, resins, naval stores, cellulosics esters, and ink vehicles, vol. 3. (2001) American Society for Testing and Material).

Beaver Cloth n Made of high-quality wool, this heavy but soft fabric has a deep nap. Beaver cloth is frequently used in overcoats.

Becchi-Millian Test n Sensitive test for cottonseed oil. If cottonseed fatty acids are dissolved in 90% ethyl alcohol and a little dilute silver nitrate solution added, they separate, colored brownish or black, as the alcohol is boiled away on heating. It is claimed that as little as 1% of cottonseed oil is detected in a mixture by this test.

Beck n A vessel for dyeing fabric in rope form, consisting primarily of a tank and a reel to advance the fabric.

Becke Line n The bright halo near the boundary of a transparent particle that moves with respect to that boundary as the microscope is focused through best focus.

Becke Test n The method for determining refractive index of a transparent particle by noting the direction in which the Becke line moves. The halo (Becke lines) will always move to the higher refractive index medium as the focus is raised. The halo crosses the boundary into the lower refractive index medium when the microscope is focused down. The particle must be illuminated with a narrow cone of axial light obtained by closing the aperture diaphragm of the condenser to a small aperture. (Nesse WD (2003) Introduction to optical mineralogy. Oxford University Press, New York).

Beckacite n Phenoplast, manufactured by Reichhold, USA.

Becquerel (Bq) \be-ˈkrel, ˌbe-kə-ˈrel\ The SI unit for rate of disintegration of a radioactive element, equal to one disintegration per second.

Bedford Cord n A rib-weave fabric with raised lengthwise cords produced by using stuffing threads in the warp. Since the fabric is strong and wears well, it is used for upholstery, suits, riding habits, and work clothes. (Tortora PG, Merkel RS (2000) Fairchild's dictionary of textiles, 7th edn. Fairchild Publications, New York).

Bed Putty n The glazier's putty under glass, on which the glass is bedded. See also ▶ Face Putty.

Beechnut Oil *n* A nondrying oil. Its iodine value averages 110, and its sp gr, 0.992/15°C.

Beer-Bouguer Law (Beer-Lambert Law) *n* Combined laws expression absorption of radiant energy of a single wavelength as a function of concentration and optical path length of absorbing material:

$$- \log T_i / cl = K$$

where T_i is the internal transmittance, c is the concentration, l is the path length, and K is the unit absorption coefficient. The value $-\log T_i$ is sometimes called the optical density, OD, or simply the density, d; it is sometimes called the absorption, the absorbency, or the extinction. The unit absorption coefficient, K, is sometimes designated A, a, α, E, or ϵ. The units of concentration and path length must be specified. See ▶ Absorption Coefficient. Note – In Europe, it is called the Law of Lambert-Beer. (DeLevie R (1996) Principles of quantitative analysis. Mc-Graw-Hill Higher Education, New York; Willard HH, Merritt LL, Dean JA (1974) Instrumental methods of analysis, D Van Nostrand, Company, New York).

Beer's Law *np* If two solutions of the same colored compound are made in the same solvent, one of which is, say, twice the concentration of the other, the absorption due to a given thickness of the first solution should be equal to that of twice the thickness of the second. Mathematically this may be expressed as $l_1 c_1 = l_2 c_2$ when the intensity of light passing through the two solutions is a constant and if the intensity and wavelength of light intensity and wavelength of light incident upon each solution are the same. See ▶ Absorption Coefficient. (DeLevie R (1996) Principles of quantitative analysis, Mc-Graw-Hill Higher Education, New York; Willard HH, Merritt LL, Dean JA (1974) Instrumental methods of analysis, D Van Nostrand, Company, New York).

Beeswax \ˈbēz-ˌwaks\ *n* (1676) Mixture of crude cerotic acid and myricin separated from honey. It is a soft natural wax, with a mp of 63°C, an acid value usually below 20, a saponification value of 90–95, and a very low iodine value of 6–13.

Beetling \ˈbē-tǝl-iŋ\ *adj* (ca. 1919) A process in which round-thread linen or cotton fabric is pounded to give a flat effect. Beetled linen damask has an increased luster and a leather-like texture. Beetling is also used to give a thready or linen-like appearance to cotton.

Behenic Acid $CH_3(CH_2)_{20}COOH$. Long chain, aliphatic acid. It has an mp of 81°C, a bp of 306°C/60 mmHg and an acid value of 164.8.

Belly Benzoin \-ˈben-zǝ-wǝn\ *n* Benzoin obtained as an exudation from the tree in the later years of its life. Belly benzoin is usually regarded as an inferior grade, obtained during the 9 years following the 10th year of the tree's life. See also ▶ Benzoin.

Benard (Vortex) Cell *n* Hexagon-shaped cell that is produced by the vortex action of solvent evaporation in thin films. During drying, all solvent paints and many varnishes exhibit Benard cell formation, i.e., roughly hexagonal cells, generally with a well-marked center. The whole cell is in movement with currents streaming up the center and flowing down the walls. These current, because they affect pigments differently, lead to segregation and deposition of different components in different parts of the film. This phenomenon manifests itself in color and surface irregularities of the film. (Weldon DG (2001) Failure analysis of paints and coatings. Wiley, New York).

Bending Length *n* A measure of fabric stiffness based on how the fabric bends in one plane under the force of gravity. (Paint/coatings dictionary (1978). Federation of Societies for Coatings Technology, Philadelphia)

Bending Modulus *n* Maximum stress per unit area that a specimen can withstand without breaking when bent. For fibers, the stress per unit of linear fiber weight required to produce a specified deflection of a fiber.

Bending Moment *n* The resultant moment about the neutral axis of a beam or column, at any point along its span, of the system of forces that produce bending.

Bending Properties See ▶ Flexural Properties.

Bending Rigidity See ▶ Flexural Rigidity.

Bending Strength See ▶ Flexural Strength.

Bending Stress See Flexural Stress.

Bend Test See ▶ Flexibility Test.

Benedict's Test *n* Special test designed to detect the pressure of reducing sugars, and it may also be applied to the products of hydrolysis of certain gums, starch, etc. Benedict's reagent consists of a solution of sodium citrate and sodium carbonate in water, to which a small quantity of copper sulfate is subsequently added. An orange precipitate is obtained when a few milliliters of the reagent is added to the solution of the suspected compound, if a reducing sugar is present. (Whistler JN, BeMiller JN (eds) (1992) Industrial gums: polysaccharides and their derivatives. Elsevier Science and Technology Books, New York).

Bengaline \ˈbeŋ-gə-ˌlēn\ n [F, from *Bengal*] A fabric similar to faille, only heavier, with a fine weave and widthwise cords. Originally, bengalines were made of a silk, wool, or rayon warp with a worsted or cotton filling and used for dresses, coats, trimmings, and draperies. Modern bengalines are made with filament acetate or polyester warps. Also, some bengalines have fine spun warps with 2- and 3-ply heavier spun yarns for filling cord effects. (Vincenti R (ed) (1994) Elsevier's textile dictionary. Elsevier Science and Technology Books, New York).

Bengucopalic Acid n $C_{19}H_{30}O_2$. Monobasic acid which has been isolated from Benguela copal. It constitutes about 44% of the copal. The acid is said to resemble the congocopalic acid of Congo copal.

Bengucopalolic n $C_{21}H_{32}O_3$. Major constituent acid of Benguela copal, being present to the extent of about 25% of the whole. It has a mp of 115°C.

Bengucopalresene n Constituent of Benguela copal. Present in two forms, the α and β varieties. The former is present to the extent of about 5% and the latter to the extent of 15%.

Benguela Copal n Fairly hard natural copal obtained from West Africa. Sometimes known as Lisbon copal and, like other similar copals from this part of West Africa, gives satisfactory oleoresinous varnishes after running.

Beni \ˈbā-nē\ [Japanese] A delicate pink or red pigment of vegetable origin.

Beni-ye [Japanese] A print in which beni is the chief color used. Generally employed to describe all those two-color prints which immediately preceded the invention of polychrome printing.

Benne Oil \ˈbe-nē\ n [of African origin; akin to Malinke *bène* sesame] (1769) See ▶ Sesame Oil.

Ben Oil n Nondrying vegetable oil. The physical properties recorded vary widely from different samples; for example, iodine values have been reported from 72 to 112.

Bentonite \ˈben-tᵊn-ˌīt\ n (1898) Very fine-grained clay (a mixture and not a definite mineral type) derived from volcanic ash and consisting largely of montmorillonite mineral. Two classes of Bentonite are recognized: (1) sodium bentonite, which is a swelling type in water, and (2) calcium bentonite or subbentonite, which exhibits little swelling in water. See ▶ Aluminum Silicate. *Also known as Wilkinite*

Bentonite Clay See ▶ Bentonite.

Benzaldehyde \ben-ˈzal-də-ˌhīd\ n [ISV] (1866) (benzoic aldehyde, oil of bitter almonds, benzoÿl hydride, benzene carbonal) C_6H_5CHO. A solvent, particularly for polyester and cellulosic plastics.

Benzene \ˈben-ˌzēn, ben-ˈ\ n [ISV *benz-* + *-ene*] (ca. 1872) (benzol, phene) C_6H_6. The compound and building block of all aromatic organic chemistry. It took almost a century of investigation after its discovery by Faraday in 1823 to establish the structure of this extraordinarily stable ring: its system of resonant, alternating single and double bonds. Benzene is a solvent and intermediate in the production of phenolics, epoxies, STYRENE, and nylon. Hydrogenation of benzene yields cyclohexane, a solvent and raw material for preparing adipic acid, from which nylon is derived. As a solvent, benzene will dissolve ethyl cellulose, polyvinyl acetate, polymethyl methacrylate, polystyrene, coumarone-indene resins, and certain alkyds. Benzene is toxic and has been declared to be a carcinogen, so it requires very careful handling. Pure benzene has a bp of 80°C and a mp of 5.5°C. Its flp is below normal air temperature and its vapor pressure is about 118 mmHg at 30°C. (Wypych G (ed) (2001) Handbook of solvents. Chemtec Publishing, New York; Merck Index, 13th edn. (2001) Merck and Company, Inc., Whitehouse Station, New Jersey).

Benzene Ring n (phenyl ring) The six carbon atoms, diagrammed as a hexagon, joined by alternating single and double bonds, each carbon with an attached hydrogen in the case of benzene itself, or with one or more hydrogens replaced by other atoms or radicals. The alternating double bonds, in either of two possible arrangements, may or may not be shown, depending on the expected audience's knowledge of organic chemistry. The Greek letter, φ, is also used as a symbol for *phenyl–*. (Morrison RT, Boyd RN (1992) Organic chemistry, 6th edn. Prentice Hall, Englewood Cliffs)

Benzenesulfonylbutylamide n $C_6H_5SO_2NHC_4H_9$. A plasticizer for cellulosics and polyvinyl acetate.

Benzenesulfonylhydrazide *n* [4,4'-oxybis(benzenesulfonylhydrazide), OBSH] A blowing agent, a white crystalline solid that melts and begins to decompose near 105°C. It produces a white unicellular foam when incorporated in PVC plastisol but has a strong residual odor that does not evolve when it is used in rubbers. It is also used in epoxy and phenolic foams and serves as a cross-linking agent in rubber compositions and a rubber/resin blends.

Benzenoid \ˈben-ˌzēn, ben-ˈ\ *n* [ISV *benz-* + *-ene*] (ca. 1872) Benzenoid compound is one which has a structural resemblance to benzene.

Benzidine Orange *n* Metal-free diazo pigments based on dichlorobenzidine. They are highly transparent, bright in color, and low in cost due to their high tinctorial strength, but tend to bleed and fade upon exposure to light.

Benzidine Yellow *n* Benzidine or diarylide yellows are disazo pigment dyestuffs. They are approximately twice as strong, much more bleed and heat resistant and markedly inferior in lightfastness versus Hansa yellows.

Benzillic Acid *n* $(C_6H_5)_2C(OH)COOH$. It has a mol wt of 228.17 and a mp of 150°C. *Known also as Diphenyl Glycolic Acid.*

Benzine \ˈben-ˌzēn, ben-ˈ\ *n* [Ger *Benzin*, fr. *benz-*] (1835) (Deprecated) (1) This term is outdated and misleading, should not be used. (2) European term for gasoline. See ▶ Ligroin.

Benzoate Fiber \ˈben-zə-ˌwāt\ Fiber with a silk-like hand made from a condensation polymer of *p*-(Bhydroxyethoxy) benzoic acid.

Benzofuran \ˌben-zō-ˈfyur-ˌan, -fyu-ˈran\ *n* (1946) A compound C_8H_6O found in coal tar and polymerized with indene to form thermoplastic resins used especially in adhesives and printing inks. See ▶ Coumarone.

Benzofuran Resin See ▶ Coumarone-Indene Resin.

Benzoguanamine *n* (2,4'-diamino-6-phenyl-1,3,5-triazine) $C_6H_5N_3(NH_2)_2$. A crystalline compound that reacts with formaldehyde to give thermosetting resins with resistance to heat and alkalies, and gloss generally superior to hose of melamine-formaldehyde resins. Benzoguanamine resins are used for protective coatings, paper additives and finishes, laminating agents, textile finishes, and adhesives.

Benzoguanamine Resins *n* Resins based on benzoguanamine.

Benzoic Acid \ben-ˈzō-ik-\ *n* [ISV] (1791) (carboxybenzene, benzene carboxylic acid, phenylformic acid) C_6H_5COOH. A white, crystalline compound occurring naturally in benzoin gum and some berries, also synthesized from phthalic acid or toluene. It is used in making plasticizers such as 2-ethylhexyl-*p*-oxybenzoate, diethyleneglycol dibenzoate, dipropyleneglycol dibenzoate, ethyleneglycol dibenzoate, triethyleneglycol dibenzoate, polyethyleneglycol-(200)- and -(600)-dibenzoate, and benzophenone. It has a mp of 121°C, a by of 249°C, and an acid value of 460. It is a useful peptizing agent for metallic soaps, either in their solid form or in hydrocarbon solutions. It is sometimes effective in reducing seeding, especially in spirit enamel containing small amounts of basic pigments.

Benzoic Ether \ben-ˈzō-ik-\ *n* [ISV, fr. *benzoin*] (1791) See ▶ Ethyl Benzoate.

Benzol \ˈben-ˌzól, -ˌzōl\ *n* [Gr, fr. *benz-* + *-ol*] (1838) Commercial solvent derived from the distillation of coal. It is a crude mixture of hydrocarbons, consisting chiefly of benzene, although it may contain in addition some toluene, xylene, carbon disulfide, thiophene, pyridine, acetonitrile, and paraffinic hydrocarbons. A 90%, or 90°, benzol is one from which 90% by volume can be distilled below 100°C.

Also can be spelled *Benzole*.

Benzoline *n* (1) Gum benzoin, as it is more commonly known, is a spirit-soluble resin which appears on the market as Siam, Palembang, Padang, Sumatra and Penang benzoins. It has a very pleasant "vanilla" smell, and it is used partly on account of this, and partly as a plasticizing resin in spirit varnishes. Benzoin is obtained as an exudation from a tree as a result of deliberate incision. See also ▶ Belly Benzoin. (2) (Deprecated) Another name for petroleum benzine.

Benzoperoxide See ▶ Benzoyl Peroxide.

Benzophenone \ben-zō-fi-ˈnōn, -ˈfē-ˌnōn\ *n* [ISV *benz-* + *phen-* + *-one*] (1885) (1) (diphenylketone) An involatile solvent and chemical intermediate. (2) Any of a family of UV stabilizers based on substituted 2-hydroxybenzophenone ("B"). Typical members are 4-methoxy-B, 4-octyloxy-B, 4-dodecyloxy-B, 2,2′-dihydroxy-4-methoxybenzophenone, and 2,2′-dihydroxy-4,4′-dimethoxybenzophenone. They function both as direct UV absorbers and, in the case of polyolefins, also as energy-transfer agents and radical scavengers. It is an aromatic ketone having excellent general compatibility. It is a solid at ordinary temperature, has a mp of 48°C, and a bp of 305°C. It is insoluble in water.

Known also as *Diphenyl Ketone*.

p-Benzoquinone \-kwi-ˈnōn\ (1,4-benzoquinone, chinone) A yellow crystalline compound used, along with many of its derivatives, as an inhibitor in unsaturated polyester resins to prevent premature gelation during storage.

Benzotriazole *n* (1) $C_6H_5N_3$. A double ring compound, parent to many derivatives. (2) Any of a family of UV stabilizers, derivatives of 2-(2′-hydroxyphenyl) benzotriazole, that function primarily as UV absorbers. Typical examples are 2-(2′-hydroxy-5′-methylphenyl) benzotriazole and the corresponding 5′-*t*-octylphenyl analog. The benzotriazoles offer intense and broad UV absorption with a fairly sharp wavelength cutoff close to the visible region. The higher alkyl derivatives are less volatile and therefore more suitable for processing at higher temperatures.

Benzoyl \ˈben-zə-ˌwil, -ˌzóil\ *n* [Gr, from *Benzoësäure* benzoic acid + Gk *hylē* matter] (ca. 1855) C_6H_5CO-. Monovalent aryl radical.

Benzoyl Peroxide *n* (1924) (dibenzoÿl peroxide, DBP) A catalyst employed in the polymerization of polystyrene, styrene, vinyl, and acrylic resins. It is also a curing agent for polyester and silicone resins, usually used together with an accelerator such as dimethylaniline. It can be dispersed in diluents or plasticizers to diminish the explosion hazard associated with the dry product. Its natural state is colorless crystals of mp of 103°C. *Known also as Benzoperoxide*. (Odian GC, (2004) Principles of polymerization, Wiley, New York)

Benzyl \ˈben-ˌzēl, -zəl\ *n* [ISV *benz-* + *-yl*] (1869) (α-tolyl) The radical $C_6H_5CH_2-$, which exists only in combination.

Benzyl Abietate *n* Very high boiling plasticizer derived from rosin. It is compatible with many other varnish and lacquer constituents. As a rosin derivative, it is susceptible to atmospheric oxidation and yellowing. Sometimes known as Benzyl Resinate.

Benzyl Acetate *n* (phenylmethyl acetate) A colorless liquid with a pleasant aroma, a solvent for cellulosic resins.

Benzyl Alcohol *n* (α -hydroxytoluene, phenylcarbinol) Has excellent solvent properties for a wide range of varnish and lacquer constituents. It has some application as a high boiling solvent, more particularly in baking finishes. It has a bp of 205°C, a flp of 300°C, and a vp of less than 1 mmHg at ordinary temperature.

Benzyl Benzoate *n* A water-white liquid used as a plasticizer that freezes at room temperature. It has a mp of 19°C, and a bp of 323°C.

Benzyl Butyl Phthalate *n* High melting solid, with a bp of 208–288°C/20 mmHg.

Benzyl Butyrate *n* A liquid with a heavy, fruity odor, a plasticizer.

Benzyl Cellosolve *n* High boiling solvent with a bp of 256°C and sp gr of 1.070/20°C.

Benzyl Cellulose *n* A benzyl ether of cellulose, it is a cellulosic plastic used in lacquers. It also may be formulated for making films and compounds for molding and extrusion.

Benzyl Formate *n* A solvent for cellulosic resins.

Benzyloctyl Adipate *n* (BOA) A plasticizer for polystyrene, vinyl, and cellulosic resins.

Benzyl Phthalate *n* A solid plasticizer: mp of 42°C, and bp of 277°C/15 mmHg. *Also known as Dibenzyl Phthalate.*

Benzyl Resinate See ▶ Benzyl Abietate.

Benzyltrimethylammonium Chloride *n* A quaternary ammonium salt, a solvent for cellulosics, and a catalyst for phenolic resins.

Ber *n* One of the host trees to which the lac insect affixes itself.

Berlin Blacks *n* Low-cost form of air-drying blacks based on solutions of bitumens to which black pigments or fillers are added. They possess excellent opacity, but reduced gloss. Some Berlin blacks are semi matt.

Berlin Blue *n* A term used for any of the variety of iron-based blue pigments; Prussian blue. See ▶ Iron Blue.

Berlin Red A pigment consisting essentially of red iron oxide.

Bernoulli's Theorem \bər-ˈnü-lēz-\ *n* [Daniel *Bernoulli* † 1782 Swiss physicist] At any point in a tube through which a liquid is flowing the sum of the pressure energy, potential energy, and kinetic energy is constant. If p is pressure; h, height above the reference plane; d, density of the liquid, and v, velocity of flow,

$$p + hdg + \frac{1}{2}dv^2 = \text{a constant}.$$

(Serway RA, Faugh JS, Bennett CV (2005) College physics. Thomas, New York)

Bertholet Principle of Maximum Work *n* Of all possible chemical processes which can proceed without the aid of external energy, that process which always takes place is accompanied by the greatest evolution of heat. This law holds good for low temperatures only and does not account for endothermic reactions. (Russell JB (1980) General chemistry. McGraw-Hill, New York)

Bertrand Lens *n* An auxiliary, low-power lens which may be inserted in the bodytube of the microscope between the eyepiece and the objective for observing the back focal plane of the objective; useful for conoscopic observations in crystal optics and for checking microscope illumination quality.

Beryllium \bə-ˈri-lē-əm\ *n* [NL, from Gk *bēryllion*] (ca. 1847) A steel-gray light strong brittle toxic bivalent metallic element used chiefly ads a hardening agent in alloys.

Beryllium Copper *n* Copper containing about 2.7% beryllium and 0.5% cobalt, used for blow molds and insertable injection-mold cavities. The small percentages of Be and Co greatly increase the strength and hardness of the copper whole preserving its high thermal conductivity and corrosion resistance. Beryllium copper is easily pressure cast and hobbed into mold cavities.

BESA Abbreviation for British Engineering Standards Association.

BET Abbreviation for Brunauer, Emmett and Teller, applied to an equation and method for determining the surface area of an absorbent, such as carbon.

Beta- *adj* A prefix, usually abbreviated as the Greek letter β and usually ignored in alphabetizing compound names, signifying that the so-labeled substitution is on the second carbon atom away from the main functional group of the molecule. See ▶ Alpha-.

Beta Cellulose *n* One of the three forms of cellulose. It has a lower degree of polymerization than the alpha form. With gamma cellulose it is known as hemicellulose. Also see ▶ Alpha Cellulose and ▶ Gamma Cellulose.

Beta Gage (Beta-ray Gage) *n* Gage consisting of two facing elements, a β-ray-emitting source, and a β-ray detector. When a sheet material (e.g., plastic) is passed between the elements, some of the β-rays are absorbed, according to the area density or the thickness of the sheet. Signals from the detecting element can be used to control equipment that automatically regulates the thickness. The most usual sources are krypton 85 and strontium 90. Also used for particular applications are cesium 137, promethium 147, and ruthenium 106. See also ▶ Thickness Gauging.

Beta-minus (β-) Particle *n* A radioactive emission consisting of a high-energy electron, $_{-1}^{0}e$. *Also known, simply, as a Beta Particle.*

Beta Particle *n* (1904) A subatomic particle created at the instant of emission from a decaying radioactive atomic nucleus, having a mass (at rest) of 9.1095×10^{-28} g. A negatively charged beta particle is identical to an ordinary electron, and a positively charged one is identical to a positron. A stream of beta particles is called a *beta ray*. Such rays are use in equipment for measuring and controlling the thickness of plastics films, sheets, and other extrudates. (Serway RA, Faugh JS, Bennett CV (2005) College physics. Thomas, New York).

Beta (b)-Particle, (Beta Ray) *n* (1902) One of the particles which can be emitted by a radioactive atomic nucleus. It has a mass about $\frac{1}{1837}$ that of the proton. The negatively charged beta particle is identical with the ordinary electron, while the positively charged type (positron) differs from the electron in having equal but opposite electrical properties. The emission of an electron entails the change of a neutron into a proton inside the nucleus. The emission of a positron is similarly associated with the change of a proton into a neutron. Beta particles have no independent existence inside the nucleus, but are created at the instant of emissions. See ▶ Neutrino.

Beta-pinene See ▶ Pinene.

Beta-plus (β+) Particle *n* A radioactive emission (*particle*) consisting of a positron $_{-1}^{0}e$.

Beta-Ray Gauge *n* (beta gauge) A device for measuring the thickness of plastics films, sheets, or extruded shapes, consisting of a source of beta rays and a detecting element. When material is passed between the source and the detector, some of the rays are absorbed, the percent absorbed being a measure of the thickness of the material. Signals from the detecting element can be used to control equipment that automatically regulates the thickness. The most usual sources are krypton 85 and strontium 90. Also used for particular applications are cesium 137, promethium 147, and ruthenium 106. See also ▶ Thickness Gauging.

Betatron \ˈbā-tər-ˌträn\ *n* [ISV] (1941) An accelerator that uses an electrostatic field to impart high velocities to electrons. Energies of 5 to 6 MeV will produce X rays equivalent in energy to gamma radiation of 12–20 g of radium.

Beton *n* A kind of concrete; a mixture of lime, sand, and gravel.

Bevatron A six or more billion electron volt accelerator of protons and other atomic particles. Energies of 5–6 MeV will produce X rays equivalent in energy to gamma radiation of 12–20 g of radium. Makes use of a Cockcroft-Walton transformer cascade accelerator and a tinear (q.v.) as well as an electromagnetic field in the build-up.

Bevel Siding See ▶ Clapboard.

BHT *n* Abbreviation for ▶ Butylated Hydroxytoluene. See ▶ Di-*tert*-Butyl-*p*-Cresol.

Bias Fabric *n* A two-dimensional fabric that when oriented in the XY plane contains fibers that are aligned in a different direction, i.e., 45° to the *x*-axis fibers.

Bias Filling *n* A fabric defect in which the filling yarn does not run at a right angle to the warp. The principal cause is improper processing on the tenter frame. Also see ▶ Bow.

Bias Ply *n* A layer of reinforcing fiber, cloth, or sheet oriented at an angle, less than 90° and typically 45°, to the fiber direction in the main reinforcing layers.

Biaxial Braid *n* Braided structure with two yarn systems one running in one direction and the other in the opposite direction.

Biaxial Crystals *n* Anisotropic crystals in the orthorhombic, monoclinic and triclinic systems. They have three principal refractive indices, α, β, and γ.

Biaxial Laminate See ▶ Bidirectional Laminate.

Biaxially Oriented Film *n* Polymeric film (i.e., polyethylene) which as been strained or stretched in one direction which produces optical and physical birefringence or most importantly, high strength one the axial or long direction.

Biaxial Orientation *n* (1) The process of stretching hot plastic film or other article in two perpendicular directions, resulting in molecular alignment. See also ▶ Orientation and ▶ Tentering. (2) The state of the material that has been subjected to such stretching.

Biaxial Winding *n* A method of ▶ Filament Winding in which the helical bands are laid in sequence, side-by-side, with no crossover of fibers.

Bibasic Lead Phthalate *n* A heat and light stabilizer for vinyl insulation, opaque film and sheeting, and foam.

Bicarbureted Hydrogen See ▶ Ethylene.

Bicomponent Fibers See ▶ Composite Fibers.

Bicomponent Yarns *n* Spun or filament yarns of two generic fibers or two variants of the same generic fiber.

Biconstituent Fiber *n* A fiber extruded from a homogeneous mixture of two different polymers. Such fibers combine the characteristics of the two polymers into a single fiber.

Bicyclic *n* Consisting of two rings.

Biddiblack *n* A particular type of mineral black, mined near Bideford, Devonshire, England. See ▶ Mineral Black.

Bidirectional *n* In two directions. (1) The term may be applied to illumination angles in three dimensional space. Thus, the incident angle may be 45° relative to the material surface, but this designation does not describe the portion of the circumferential 45° angle possible. The term "bidirectional illumination" generally implies two incident beams separated circumferentially by 180°. (2) The term may also be applied to goniophotometric measurements, to the illumination and to the viewing angle. In either case, the angular distance is expressed as the degrees (or radians) from the perpendicular (normal) to the surface as 0°. Thus, the grazing angle approaches 90°.

Bidirectional Fabric *n* A fabric having reinforcing fibers in two directions, i.e., in the warp (machine) direction and filling (cross-machine) direction.

Bidirectional Laminate *n* A fiber-reinforced material in which the fibers are laid in two different directions, typically in the length and width directions. In particular such a laminate in which equal volumes of reinforcing fibers are laid in the two directions.

Bidirectional Reflectance Distribution Function (BRDF) *n* The ratio of radiance per unit irradiance, used for describing the geometrical reflectance properties of the surface. See ▶ Bidirectional.

Bierbaum Scratch Hardness See ▶ Scratch Hardness.

Biff See ▶ Monkey.

Bikerman Boundary-Layer Theory *n* The theory that adhesive bonding occurs through the formation of an adsorbed, strong boundary layer on the surface of the adhered.

Bilateral Fibers *n* Two generic fibers or variants of the same generic fiber extruded in a side-by-side relationship.

Billow Forming *n* A variant of ▶ Thermoforming, in which the hot plastic sheet is clamped in a frame and expanded upwards with mild air pressure against a male plug or female die as the plug or die descends into the frame. The process is suitable for thin-walled containers with high draw ratios.

Bimetallic Cylinder *n* In most modern extruders and injection machines, the barrel is lined, by centrifugal casting from the melt, with any of several white irons containing chromium and boron carbides and having hardnesses near Rockwell C65. Aster finish-grinding and polishing, the liner, about 1 mm thick, provides excellent resistance to wear or corrosion or both, depending on the formation. The best known trade name is XALOY®.

Bimetal Plate *n* In lithography, a plate in which the image area is copper or brass, and the non-image area is aluminum, stainless steel, or chromium.

Bimolecular Process *n* An elementary process in which the activated complex is formed as a result of the collision of two particles.

Bin Activator *n* A device that promotes the steady flow of granular or powdered plastics from storage bins or hoppers. Among the many types of equipment are vibrators or mallets acting upon the outside of the container, prodding devices or air jets acting directly on the material, inverted-cone baffles with vibrating means located at the bottom of the hopper, and other "live bottom" devices such as scrapers, rolls, and chains.

Bin Cure *n* Rubber term for partial or complete vulcanization of a mixed compound while stored in a bin or pile waiting for molding or further processing. *Also called Pile Burning or Premature Vulcanization.*

Bin Dischargers *n* Restores flowability to materials which are prone to packing during storage in bins, silos, or surge hoppers, and promotes free flow of materials which tend to bridge or hang up.

Binder *n* (1) Nonvolatile portion of the liquid vehicle of a coating. It binds or cements the pigment particles together and the paint film as a whole to the material to which is it applied. See also ▶ Vehicle. (2) Component of an adhesive composition which is primarily responsible for the adhesive forces which hold two bodies together. (3) Resin or cementing constituent of a plastic compound which holds the other components together. The agent applied to glass mat or performs to bond the fibers prior to laminating or molding. (4) The components in an ink film which holds the pigment to the printed surface. (Wicks ZN, Jones FN, Pappas SP (1999) Organic coatings science and technology, 2nd edn. Wiley-Interscience, New York).

Binder Content *n* The weight of adhesive used to bond the fibers of a web together. Binder content is usually expressed as percent of fabric weight.

Binder Demand *n* That amount of binder needed to completely wet a pigment by displacing the air voids. This is determined primarily by: the particle size, shape, chemical composition, and density of the pigment; and the particle size, degree of polymerization and wetting properties of the binder. (Whereas the binder demand refers to a particular pigment-vehicle system, oil absorption is a numerical value assigned to a pigment based on a specific test method.)

Binder Fibers *n* Fibers that can act as an adhesive in a web because their softening point is relatively low compared with that of the other fibers in the material.

Binder Ratio See ▶ Pigment/Binder Ratio.

Binding Varnish *n* A term used by printers or lithographers to describe a viscous varnish in the ink that is used to toughen the dried film.

Bingham Body *n* Material displaying plastic flow.

Bingham Liquid *n* Liquid exhibiting plastic flow. See ▶ Plastic Flow.

Bingham Plastic *n* A model for flow behavior in which no flow occurs until the shear stress exceeds a critical level called the ▶ Yield Value or yield stress above which shear rate is proportional to the stress (after the yield stress has been surpassed). When a Bingham plastic flows through a circular tube, there is a critical radius r_c at which the shear stress, $\Delta P \cdot r_c/(2 \cdot L)$, equals the yield stress. All actual flow occurs between that radius and the wall radius, R, while from the center to r_c there is a solid plug carried along by the stream. (Parfitt CD, Sing KSW (1976) Characterization of powder surfaces. Academic, London; Patton TC (1979) Paint flow and pigment dispersion: a rheological approach to coating and ink technology. Wiley, New York).

Binodal *n* Having or relating to two modes; especially, having or occurring with two statistical modes

Biochemical Oxygen Demand (B.O.D.) *n* A standard test for estimating the degree of contamination of water supplies. It is expressed as the quantity of dissolved oxygen (in mg/l) required during stabilization of the decomposable organic matter by aerobic biochemical action.

Biocide \ˈbī-ə-ˌsīd\ *n* (1947) An agent incorporated in or applied to the surfaces of plastics to destroy bacteria, fungi, marine organisms, etc. Some plastics, e.g., acetals, acrylics, epoxies, phenoxies, ABS, nylons, polycarbonate, polyesters, fluorocarbons and polystyrene, are normally resistant to attack by bacteria or fungi. Others, e.g., alkyds, phenolics, low-density polyethylene, urethanes, and flexible vinyls can under some circumstances be affected by growth of these organisms on their surfaces. Even though the resins themselves might be resistant, additives such as plasticizers, stabilizers, fillers, and lubricants can serve as food for fungi and bacteria. Examples of biocides are organotins, brominated salicylanilides, mercaptans, quaternary ammonium compounds, and compounds of mercury, copper, and arsenic.

Biocompatibility \-kəm-ˌpa-tə-ˈbi-lə-tē\ *n* (1971) Materials which are compatible with living blood and tissue, which is important for artificial blood vessels, hearts, etc.

Biodegradable \-di-ˈgrā-də-bəl\ *adj* (1961) The process of rapid decomposition as a result of the action of microorganisms.

Biodegradable Surfactant *n* Surfactant which may be decomposed by biological action.

Biodegradation *n* The gradual breakdown of plastics and matter by living organisms such as bacteria, fungi, and yeasts. Most of the commonly used plastics are essentially not biodegradable, exhibiting limited susceptibility to assimilation by microorganisms. An exception is ▶ Polycaprolactam. However, the growing emphasis on environmental aspects of discarded plastics has stimulated research in ways of attaining biodegradability after a predetermined time period. One method is to add a UV-light sensitizer that causes photodegradation after a period of exposure to light, followed by breakup after prolonged exposure to the elements, after which bacteria will finish the job. A third method is the deliberate incorporation of weal links in the polymer chain, temporarily protected by a degradable stabilizer. (Zaiko GE (ed) (1995) Degradation and stabilization of polymers. Nova Science Publishers, Inc., New York) See also ▶ Photodegradation.

Biologically Active Polymers *n* Polymers capable of specifically and reversibly binding to analytes, including molecules and cells. The biologically-active polymers are also capable of releasing substances upon electrical stimulation.

Biomedical Polymers See ▶ Biocompatibility.

Biopolymer \ˌbī-ō-ˈpä-lə-mər\ *n* (1961) (1) A copolymer in which there is irregularity with regard to the placement (relative locations within the polymer chain, in a linear copolymer) of two or three or more chemically different types of units. These units may be mers, in a product of addition polymerization, or residues of the condensing small molecules in a polycondensate. (2) A polymer produced by living organisms, such as cellulose, natural rubber, silk, rosin, and leather.

Biot Number *n* A dimensionless group, $h\,t/k$, important in convective heating and cooling of sheets. h = the heat-transfer coefficient at the sheet's surface, t = the sheet thickness of half-thickness, and k = the thermal conductivity of the sheet material.

Biphenol A *n* (4,4′-isopropylidenediphenol) An intermediate used in the production of epoxy, polycarbonate, and phenolic resins. The name was coined after the condensation reaction by which it may be formed – two (bis) molecules of phenol with one of acetone (A).

Biphenyl n \(▪)bī- ▪fe-nºl, -▪fē-\ n [ISV] (1923) (diphenyl, phenyl benzene) $(C_6H_5)_2$. A stable, high-boiling (256°C) liquid long used as a heat-transfer medium.

Bipolymer n A polymer derived from two species of monomer (IUPAC). The more commonly used Syn: ▶ Copolymer.

Bird Applicator n Device used for the laboratory application of coating of a prescribed thickness to test panels. It is a machined steel bar with a fixed clearance having a beveled undercut to guide the coating material. See ▶ Applicator.

Bird's-Eye n (1) Small localized areas in wood with the fibers indented and otherwise contorted to form few to many small circular or elliptical figures remotely resembling birds' eyes on the tangential surface. Common in sugar maple and used for decorative purposes; rare in other hardwood species. (2) A generic term describing a cloth woven on a dobby loom, with a geometric pattern having a center dot resembling a bird's eye. Originally birdseye was made of cotton and used as a diaper cloth because of its absorbent qualities, but now the weave is made from a variety of fibers or fiber blends for many different end uses. (3) A speckled effect on the back of a knit fabric resulting from the use of different colors on the face design. (Vincenti R (ed) (1994) Elsevier's textile dictionary. Elsevier Science and Technology Books, New York)

Birefringence \ ▪bī-ri- ▪frin-jən(t)s\ n [ISV] (1898) (1) (double refraction) The difference between any two refractive indices in a single material. When the refractive indices measured along three mutually perpendicular axes are identical, the material is said to be optically isotropic. Orientation of a polymer by drawing may alter the refractive index in the direction of draw so that it is no longer equal to that in the perpendicular directions, in which event the material is said to display birefringence. Crystalline polymers, normally birefringent, may become optically isotropic at their melting points. Studies of birefringence provide useful information regarding the shapes of molecules, degrees of orientation and crystal habits. (2) Property of anisotropic materials, those which possess different refractive indices according to the direction of vibration of light passing through a crystal; usually observed with the aid of a polarizing microscope. The degree of birefringence is expressed numerically as the difference between the highest and lowest indices of refraction, even though there may be more than two different indices of refraction. (Fox AM (2001) Optical properties of solids. Oxford University Press, UK)

Bisacki n Grade of see lac.

Bis(4-t-Butylcyclohexyl) peroxy Dicarbonate n A catalyst of the organic- peroxide family, used in reinforced plastics and vinyl polymerization. Unlike other percarbonates, it does not require refrigeration for storage or handling.

Biscuit See ▶ Preform.

Bis(ethoxyethoxyethyl) Phthalate \ ▪bis-e- ▪thäk-sē-e- ▪thäk-sē- ▪e-thəl- ▪tha- ▪lāt\ n $C_6H_4(COOC_2H_4OC_2H_4OC_2H_5)$ A good primary plasticizer for polyvinyl acetate, nitrocellulose, cellulose acetate, and many other polymers.

Bis(β-Hydroxyethyl-γ-Aminopropyltriethoxy Silane n A silane coupling agent used in reinforced epoxy resins, also in many reinforced thermoplastics such as PVC, polycarbonates, nylon, polypropylene, and polysulfones.

Bismaleimide Resins *n* Bismaleimide resins are low molecular substances (dry powders) containing imide structures already in the monomer form.

Bismarck Brown *n* Brown-colored dyestuff, which chemically is triaminoazobenzene. It dissolves in alcohol and water, to yield reddish-brown solutions. It has some application in the manufacture of mahogany stains.

Bisphenol A \ˈbis-ˈfē-ˌnōl, -ˌnól. fi-ˈ\ *n* OH-Ar-$(CH_3)_2$-Ar-OH, a diol which reacts with epichlorohydrin to form bisphenol epoxy resins. (4,4′-isopropylidenediphenol) An intermediate used in the production of epoxy, polycarbonate, and phenolic resins. The name was coined after the condensation reaction by which it may be formed – two (bis) molecules of phenol with one of acetone (A).

Bisphenol A and F *n* Dihydroxydiphenyldimethylmethane, mol wt 224.1. Insoluble in water. Used in the manufacture of phenolic and epoxy resins. Condensation product formed by reaction of two (bis) molecules of phenol with acetone. This polyhydric phenol is a standard resin intermediate, along with epichlorohydrin, in the production of epoxy resins.

Bisphenol A Polyester *n* A thermoset unsaturated polyester based on bisphenol A and fumaric acid.

Bisphenol Epoxy Resins *n* Resins based on Bisphenol A.

BISRA *n* Abbreviation for British Iron And Steel Research Association.

Bistre \ˈbis-tər\ *n* [Fr *bistre*] (ca. 1751) Brown water color pigment which is derived from the tarry soot of burned, resinous wood and beechwood. It is similar to asphaltum, in color and composition. The color varies from saffron yellow to brown-black, depending upon the source and treatment of the raw material.

Bis(Tri-n-Butyltin) Oxide *n* A liquid derived by the hydrolysis of tributyltin chloride, used to control the growth of most fungi, bacteria and marine organisms on plastics used in boat construction and in urethane foams.

Bis(Tri-n-Butyltin) Sulfosalicylate *n* An antimicrobial agent used in flexible PVC film and urethanes.

Bis(2,2,4-Trimethyl-1,3-Pentanediol) Monoisobutyrate Adipate *n* A plasticizer for cellulosic resins and polystyrene.

Bite *n* (1) The ability of an adhesive to penetrate surfaces and thereby produce an adhesive bond. (2) Syn: ▶ Nip.

Bi-tert-Butyl-p-Cresol *n* (DBPC, butylated hydroxytoluene, BHT) [C(CH$_3$)$_3$]$_2$C$_6$CH$_3$H$_2$OH *n* A white crystalline solid used as an antioxidant in polyethylene, vinyl monomers, and many other substances.

Bitter-Almond Oil, Synthetic See ▶ Benzaldehyde.

Bittiness *n* Presence of material that hinders the appearance of a smooth and uniform coating film.

Bitty *n* Said of coatings containing bits of skin or foreign matter of any type which project above the coating film and hinder the formation of a smooth and uniform film. Syn: ▶ Peppery. See also ▶ Nibs.

Bitumen \bə-ˈtyu-mən, bī-, -ˈtu-, *esp British also* ˈbit-yə-\ *n* [ME *bithumen* mineral pitch, fr. L *bitumen-*, *bitumen*] (15c) Hydrocarbon material of natural or pyrogenous origin, or combinations of both, frequently accompanied by their nonmetallic derivatives, which may be gaseous, liquid, semisolid, or solid and which is completely soluble in carbon disulfide. See ▶ Asphalt, ▶ Elaterite, ▶ Gilsonite, and ▶ Rafaelite.

Bitumens *n* A naturally occurring, almost black materials that are also obtained in mineral-oil refining, it consists of a high molecular weight hydrocarbons dispersed in oil-like material.

Bituminous Cement *n* A black substance available in solid, semisolid, or liquid states at normal temperatures; composed of mixed indeterminate hydrocarbons; appreciably soluble only in carbon disulfide or other volatile liquid hydrocarbon; especially used in sealing built-up roofing and between joints and in cracks of concrete pavements.

Bituminous Coating *n* An asphalt or tar compound used to provide a protective finish for a surface.

Bituminous Emulsion *n* A suspension of minute globules of bituminous material in water or of minute globules of water in a liquid bituminous material; used as a protective coating against weather, especially where appearance is not important.

Bituminous Paints *n* (1) Originally, the class of paints consisting essentially of natural bitumens dissolved in organic solvents: they may or may not contain softening agents, pigments, and inorganic fillers. They are usually black or dark in color. Within recent years the term "bituminous" has, by common usage, come to include bitumen-like products such as petroleum asphalt. (2) A low-cost paint containing asphalt or coal tar, a thinner, and drying oils; used to waterproof concrete and to protect piping where bleeding of the asphalt is not a problem. (Usmani AM (ed) (1997) Asphalt science and technology. Marcel Dekker)

Bituminous Varnish See ▶ Varnish.

Biuret *n* (allophanamide, carbamylurea) NH$_2$CONHCONH$_2$·H$_2$O. A white crystalline material derived from urea by heat or by reaction with an isocyanate. It is used primarily in analytical chemistry, but the biuret group is formed during some polymerization reactions, such as primary bonding in urethane elastomers. (Odian GC, (2004) Principles of polymerization, Wiley, New York)

Bivinyl See ▶ Butadiene.

Black *n* Ideally, the complete absorption of incident light; the absence of any reflection. In the practical sense, any color which is close to this ideal in a relative viewing situation, i.e., a color very low saturation and of low luminance.

Blackboard Paint *n* Rapid drying flat paint capable of permitting writing on it by chalk; usually black or green in color.

Black Body *n* The theory that atoms can exist for a duration solely in certain states, characterized by definite electronic orbits, i.e., by definite energy levels of their extranuclear electrons, and in these stationary states they do not emit radiation; the jump of an electron from an orbit to another of a smaller radius is accompanied by monochromatic radiation. A black body has an emissivity (e_B) of 1, indicating that a black body absorbs then emits all radiation at a wavelength. A black body is the opposite property of a perfect reflector of radiation at a wavelength that absorbs no radiation. (Odian GC, (2004) Principles of polymerization, Wiley, New York).

Black Boy Gum See ▶ Accroides.

Black Chalk A bluish black clay containing carbon.

Black Dammar See ▶ Black East India.

Black East India *n* Dark colored resin collected in the East Indies. It resembles the dammars in several respects, notably in solubility and free acid content. Its alternative name is "black dammar".

Black Iron Oxide *n* $FeO \cdot Fe_2O_3$ or Fe_3O_4. Pigment Black 11 (77499). Two black oxides of iron are used, namely, the natural mineral, magnetite, and the artificially produced oxide. Both types are magnetic. Neither possesses outstanding opacity. Magnetite consists of about 95% Fe_3O_4. Artificial back oxide consists of a mixture of ferric oxide (Fe_2O_3) and ferrous oxide (FeO). (Kirk-Othmer encyclopedia of chemical technology: pigments-powders (1996). Wiley, New York).

Black Japan *n* Black varnish-like material in which the higher grades of asphalt or other resinous compositions are employed, blended with suitable drying oils to produce hard, glossy black coatings which become insoluble in mineral spirits. (Weismantal GF (1981) Paint handbook. McGraw-Hill Corporation, Inc., New York).

Black *n* Another name for graphite.

Black Light *n* (Deprecated) Popular term for ultraviolet radiation without any visible radiation. It is an incorrect term because *light* is defined as *visible* radiation and ultraviolet radiation is not detected by the visual mechanism of the eye. Syn: ▶ Ultraviolet Radiation.

Black Magnetic Oxide See ▶ Black Iron Oxide.

Black Orlon *n* Orlon® (black), acrylic fiber.

Black Out Paint Opaque paint, usually black or dark in color, used during war time, on windows, skylights, etc., to prevent the passage of light to the outside of a building.

Black Varnish *n* Any varnish in which the resin component is substituted completely, or in part, by a petroleum bitumen, or natural asphaltum.

Blanc Fixe \ˈblaŋk-ˈfiks\ *n* [F] (1866) A synthetic form of barium sulfate prepared by reacting aqueous solutions containing barium ions with others containing sulfate ions and precipitating the reaction product. It is used as a special-purpose filler to impart X ray opacity and high specific gravity. Precipitated barium sulphate is used as a semi-transparent extender in printing inks.

Blanch \ˈblanch\ *v* [ME *blaunchen*, fr. MF *blanchir*, fr. OF *blanche*, feminine of *blanc*] (15c) (1) To make white. (2) As distinct from blush or bloom, a pale milky cast on a coating film.

Blank *n* (1) A piece punched or die-cut from a sheet and intended for further forming into its final shape. (2) In chemical analysis, a dummy sample or solution of the same matrix as the sample to be analyzed but containing none of the analyte sought.

Blanket *n* An unquilted bedding fabric designed primarily to provide thermal insulation.

Blanket Mark See ▶ Corrugation Mark.

Blanking (die cutting) The cutting of flat sheet stock to shape by striking it sharply with a punch while it is supported on a mating die. Punch presses are often used for the operation. An alternate method is to make the cut with a thin, sharp-edged, shaped steel blade called a ▶ Steel-rule Die. See also ▶ Die Cutting.

Blanking Die *n* A metal die used in the blanking process.

Blast Cleaning *n* Cleaning and roughening of a surface (particularly steel) by the use of natural or artificial grit or fine metal shot (usually steel), which is projected against a surface by compressed air. *Also known as Power Cleaning.*

Blast Finishing *n* The removal of flash from molded objects (and/or dulling their surfaces) by impinging media such as steel balls, crushed apricot pits, walnut shells or plastic pellets upon them with sufficient force to fracture the flash. When the material being deflashed is not sufficiently brittle at room temperature, the articles are first chilled to a temperature below their brittleness temperature. Typical blast-finishing machines consist of wheels rotating at high speeds, fed at their centers with the media, which are thrown out at high velocities against the objects.

Bleach *n* (1) Refers to the method of measuring the tinctorial strength of an ink or toner, usually accomplished by mixing a small portion of the ink (or toner) with a large amount of white base and evaluating the tinctorial strength of the ink versus a control standard. (2) A whitening agent such as an aqueous solution of sodium hypochlorite commonly used for oxidizing or whitening clothes.

Bleached Oil *n* Oil which has been refined by the acid, alkali, or mechanical process and in which the refining includes a treatment at about 88°C (190°F) with an adsorbent such as fuller's earth or activated charcoal, after which the oil is filtered.

Bleached Shellac *n* Substantially colorless product obtained by bleaching natural orange gum lac with chlorine or chlorine-containing agents, such as sodium or calcium hypochlorites. *Also known as Bone Dry Bleached Shellac.*

Bleaching *n* (1) Loss of color of a paint or varnish. This may be due to internal chemical or physical action in the paint itself, to influences from the surface on which it is applied or to weathering or contamination from the atmosphere. See also ▶ Fading and ▶ Whitening In The Grain. (2) Intentional lightening of the color of a material such as wood, vegetable oils, varnishes, etc. (3) Any of several processes to remove the natural and artificial impurities in fabrics to obtain clear whites for finished fabric or in preparation for dyeing and finishing.

Bleaching Clay *n* Clay that possesses decolorizing characteristics for use in refining of mineral, petroleum, vegetable, and animal oils from the spinneret hole involved.

Bleb \bleb\ *n* (1607) A blister or bubble on the face of a spinning jet, interrupting the extrusion of the filament.

Bleb Rate *n* The frequency of bleb formation in an extrusion operation.

Bleed (1) *n* An escape passage at the parting line of a mold, similar to an ▶ Air Vent but deeper, serving to allow material to escape or bleed out. (2) The spreading or running of a pigment color by the action of a solvent. See ▶ Bleeding.

Bleed Characteristics See ▶ Bleeding.

Bleeding *n* (1) The diffusion of colorants through a coating from a previously painted substrate due to the action of the vehicle or solvent or both. The action is dependent on the pigments; vehicles, and solvents of the systems. (2) Diffusion of a soluble colored substance from, into and through a coating from beneath, thus producing an undesirable staining or discoloration. Materials which give rise to this effect are tannins or dyes in some types of wood; wood preservatives; bituminous coatings; pigment dyestuffs; and stains. (3) Diffusion of coloring matter from the substrate; also, the discoloration arising from such diffusion. In the case of printing ink, the spreading or running of a pigment color by the action of a solvent such as water or alcohol. (4) Migration to the rubber surface of an oil, wax, or plasticizer as a film or in drops, sometimes called sweating. Also a term applied to organic pigment colors if they migrate into an adjacent stock of a different color, or when they are removable at the surface by water or other solvents. (5) The spreading or migration of an ink component into an unwanted area. The terms *migration*, *crocking*, *blooming*, and *bronzing* are sometimes used loosely to describe the same phenomenon. (6) Loss of color by a fabric or yarn when immersed in water, a solvent, or a similar liquid medium, as a result of improper dyeing or the use of dyes of poor quality. Fabrics that bleed can cause staining of white or light shade fabrics in contact with them while wet. (Martens CR (1968) Technology of paints, varnishes and lacquers. Reinhold Publishing Co., New York; Paint testing manual: physical and chemical examination of paints, varnishes, lacquers, and colors – STP 500 (1973). American Society for Testing and Materials; Paint/coatings dictionary (1978). Compiled by Definitions Committee of the Federation of Societies for Coatings Technology).

Bleeding Pigment *n* A pigment in a coating which is partially soluble in the solvent or vehicle portion of subsequent coats applied to it.

Bleedout In filament winding, the excess liquid resin that migrates to the surface of a winding.

Blemish *n* Any surface imperfection of a coating or substrate.

Blend *n* (1) A mixture, such as a mixture of solvents or a mixture of inks. (2) A yarn obtained when two or more staple fibers are combined in a textile process for producing spun yarns (e.g., at opening, carding, or drawing). (3) A fabric that contains a blended yarn (of the same fiber content) in the warp and filling. See ▶ Alloy.

Blender *n* A round, softish brush of badger hair or similar material with a blunt tip, used for blending colors and removing brush marks left by coarser brushes. *Also called Softener*. (2) A small laboratory mixer used to dispense pigment in a vehicle.

Blenders *n* Mixing of liquid/liquid and/or solids components by mixing in a vessel using paddles or other low shear rate means of mixing.

Blending *n* (1) Any process in which two or more components or ingredients are physically intermingled without significant change of the physical states of the components. (2) The bringing together of two or more polymers, using whatever means may be needed, such that the final ▶ Scale of Segregation is microscopic or finer. (3) The combining of staple fibers of different physical characteristics to assure a uniform distribution of these fibers throughout the yarn.

Blending Resin *n* (extender resin) With respect to vinyl plastisols and organosols, a blending resin is one of larger particle size and lower cost than the dispersion resins normally used, a partial replacement for the primary resin. Blending resins are sometimes sued to achieve a better balance of properties other than cost.

Blinding *n* Loss of luster of fibers after wet processing.

Blinding of Lithographic Plates *n* Loss of ink-receptivity in the plates' image areas.

Blister *n* (1) An imperfection on the surface of a plastic article caused by a pocket of air or gas beneath the surface. NOTE – A blister may be caused by insufficient adhesive; inadequate curing time, temperature or pressure; or trapped air, water, or solvent vapor. (2) A thermoformed canopy or pocket roughly hemispherical, for example an aircraft cockpit cover or a shape used in ▶ Blister Packaging.

Blister House *n* House-like structure used as an accelerated test for blister resistance of coatings. Under controlled conditions, it attempts to simulate or accelerate blistering which occurs on buildings.

Blistering *n* Formation of dome-shaped projections in paints or varnish films resulting from local loss of adhesion and lifting of the film from the underlying surface.

Blistering of Paints, Determination of Degree of *n* Test method to evaluate the degree of blistering that may develop when paint systems are subjected to conditions which will cause blistering, using photographic reference standards.

Blistering Resistance *n* The ability of a coating to resist the formation of dome-shaped, liquid- or gas-filled projections in its film resulting from local loss of adhesion and lifting of the film from the underlying surface or coating. See ▶ Blistering.

Blister Packaging *n* The enclosing of articles in thermoformed, transparent "blisters" shaped to more or less fit the contours of the articles. The preformed blisters, usually slightly oversized to provide ample room, are made of thermoplastics such as vinyl, polystyrene, or cellulosic plastics. They are placed inverted in fixtures and loaded with the articles, and then cards coated with an adhesive are applied and sealed to the flanges between and around the blisters by means of heat and pressure.

Block *n* A portion of a polymer molecule comprising many mer units that has at least one constitutional or configurational feature not present in the adjacent portions.

Block Coat See ▶ Tie Coat, ▶ Transition Primer, and ▶ Barrier Coat.

Block Copolymers *n* (1) A copolymer with chains composed of shorter homopolymeric chains that are linked together. These "blocks" can alternate regularly or randomly. Such copolymers usually have some higher properties than either of the homopolymers or their physical blends. They have distinct blocks of polymer segments such as,

AAAABBBBAAAABBBAAA

sequencing of A and B monomers. When blocks are of different monomer species, the term *block copolymer* is used. (2) An essentially linear copolymer in which there are repeated segments of different chemical structure. (Odian GC, (2004) Principles of polymerization, Wiley, New York).

Block Cutter *n* A craftsman who, by hand, hammers into a roller the brass strips and felt from which the wallpaper will be printed.

Blocked Curing Agent *n* A curing agent or hardener rendered unreactive, which can be reactivated as desired by physical or chemical means.

Block Filler *n* A pigmented coating of heavy consistency used to fill void spaces in concrete or cinder blocks, prior to the application of the top coat, in order to produce a smooth surface.

Block Flooring *n* End grain blocks glued down in decorative patterns.

Blocking *n* (1) The undesirable sticking together of two painted surfaces when pressed together under normal conditions or under specified conditions of temperature, pressure, and relative humidity. (2) Undesired adhesion between touching layers of a material, such as occurs under moderate pressure and sometimes pressure and heat, during storage of fabrication. Such agents are called ▶ Antiblocking Agents.

Blocking Point *n* Lowest temperature at which two coated surfaces in mutual contact will tick together sufficiently to injure the surfaces permanently and/or prevent easy separation. Thus, blocking point is a major quality criterion for evaluating coating-grade petroleum waxes.

Blocking Test *n* Procedure for determining the tendency of painted surfaces to stick together (block) when stacked or placed in contact with each other under a weighted load. See ▶ Block Resistance.

Block Press *n* (1) A press used to agglomerate laminate squares under heat. The squares, which have been cut from laminated sheet, are crossed to combat the anisotropy that normally occurs during laminating. (2) A press used to mold very large blocks of polystyrene foam.

Block Printing *n* A process of printing with blocks on which the unit or design stands out in relief. Pear wood is traditionally used, and a different block is needed for each color. *Also called Hard-blocking*. See ▶ Printing.

Block Resistance *n* The resistance to blocking.

Blood Compatibility See ▶ Biocompatibility.

Blood Red *n* A pigment consisting essentially of red iron oxide.

Bloom \\ˈblüm\\ *n* [ME *blome*, fr. ON *blōm*; akin to OE *blōwan* to blossom] (13c) A haziness which develops on high gloss surfaces resulting in scattering of the surface reflectance. One mechanism is by exudation of a component such as a plasticizer out of the paint film. See ▶ Blush. (2) Undesirable deposit which sometimes forms on a glossy coating, resulting in whitening or loss of gloss. (3) Similar to bleeding in that it is migration of liquids or solids to the surface of a rubber compound to cause a change of appearance in color or cloudiness at the surface. Waxes used in excess of their solubility point in rubber come to the surface as a wax bloom, as does sulfur that remains as an excess over the amount actually chemically combined with the rubber.

(4) Undesirable exudation of pentachlorophenol on the surface of wood that has been treated with a preservative solution. (5) In the plastics industry, bloom is a visible exudation on the surface of a plastic, generally caused by lubricant, plasticizer, etc. (6) The fluorescence of mineral oil; the blue or purple cast apparent when the oil is spread in a thin film. (7) The appearance of brightness of a dyed fabric when the fabric is viewed across the top while held at eye level. Syn: ▶ Fog. See also ▶ Oil and ▶ Debloomed Oil. (Wicks ZN, Jones FN, Pappas SP (1999) Organic coatings science and technology, 2nd edn. Wiley-Interscience, New York; Paint/coatings dictionary (1978). Compiled by Definitions Committee of the Federation of Societies for Coatings Technology).

Blooming See ▶ Opening.

Bloom Oil n One of the liquid decomposition products obtained from the destructive distillation of rosin. It can be regarded as a rosin oil, and has a bp of about 270°C.

Blotch See ▶ Finishing Spot.

Blotching See ▶ Mottle.

Blotch Printing See ▶ Printing.

Blowing n Porosity or sponginess occurring during cure, either deliberately through use of a gas releasing material to form sponge or expanded rubber, or inadvertently due to entrapped moisture to cause undesirable porosity. The practice of injecting a gas into a hot and flowing resin to create a part or film.

Blowing n (of Plaster) Appearance on the surface of the plaster of conical hollows (pops or blows). They are due to the presence of particles of reactive material which expand, after the plaster has set, with sufficient force to push out the plaster in front of the particles. They may occur in undercoats or finishing coats of plaster, or both.

Blowing Agent n (foaming agent) A substance that, alone or in combination with other substances, is capable of producing a cellular structure in a plastic or rubber mass. Thus, the term includes compressed gases that expand when pressure is released, soluble solids that leave pores when leached out, liquids that develop cells when they vaporize, and chemical agents that decompose or react under the influence of heat to form a gas. Liquid foaming agents include certain aliphatic and halogenated hydrocarbons, low-boiling alcohols, ethers, ketones, and aromatic hydrocarbons. The chemical blowing agents range from simple salts such as ammonium or sodium bicarbonate to complex nitrogen-releasing agents, of which azobisformamide (ABFA) is an important example.

Blown Bitumen n Generally regarded as the blown products of petroleum bitumens, which have been previously obtained as residues from the distillation of crude petroleum. The petroleum bitumens are subjected to high temperature blowing with air, which has profound effects on the properties of the material. Melting point is raised considerably, and the product becomes rubbery. For bitumens of the same melting point, the blown product has a higher penetration figure. See ▶ Bitumen and ▶ Air Blowing.

Blown Castor Oil n Product of very high viscosity and reddish-brown color, obtained by blowing raw castor oil. One of its chief applications is as a plasticizer or softener for cellulosic compositions.

Blown Film See ▶ Film Blowing.

Blown Oils n Vegetable or fish oil which has been partially oxidized by blowing with a current of air while at an elevated temperature. The characteristics of the oil such as its increased viscosity and degree of oxidation can be controlled by time, temperature, and amount of air.

Blow-up Ratio n (BUR) (1) In blow molding, the ratio of the largest diameter of a cavity into which a parison is to be blown to the outside diameter of the parison. (2) In blown-film extrusion, the ratio of the film diameter, before collapsing or gusseting, to the mean diameter of the die opening.

Blue Asbestos n (crocidolite) An iron-rich form of ▶ Asbestos, fibers of which were long used in reinforced plastics when good chemical resistance was essential.

Blue Basic Lead Sulfate (Blue Lead) n A variant of white basic lead sulfate containing 78% monobasic lead sulfate. It is not actually a blue pigment, but has a dark grey or slate color. See ▶ Basic Lead Sulfate.

Blueing Neutralizing the yellow cast of certain white pigments or paints by adding a trace of blue, thereby increasing apparent whiteness. See ▶ Blue Toner.

Blue Stone See ▶ Copper Sulfate.

Blue Toner n Pigment or dye used in small quantities to neutralize the yellow cast of certain white pigments or coatings, increasing apparent whiteness; also used to make black coatings more jet black.

Blue Verditer n Basic copper carbonate. Syn: ▶ Azurite.

Blue Vitriol See ▶ Copper Sulfate.

Bluing n A mold blemish in the form of a blue oxide film on the polished surface of a mold, caused by overheating.

Bluing Off v A term used by mold makers for the process of checking the accuracy of two mating surfaces by applying a thin coating of Prussian blue to one surface, pressing the coated surface against the other surface, and observing the areas of intimate contact where the blue color has transferred.

Blushing n (1) Film defect which appears as a milky opalescence as the film dries; can be a temporary or permanent condition. It is generally caused by rapid evaporation, moisture, or incompatibility. See also ▶ Bloom. Syn: ▶ Fog. (2) Milky opalescence which sometimes develops as a film of lacquer dries, and is due to the deposition of moisture from the air and/or precipitation of one or more of the solid constituents of the lacquer; usually confined to lacquers which dry solely by evaporation of solvent. Also applies to printing inks. (3) A white or grayish cast on high gloss paint; results from the precipitation of binder solids owing to incompatibility with water, oil, or solvent.

BMC *n* Abbreviation for ▶ Bulk Molding Compound.

BOA See ▶ Benzyl Octyl Adipate.

Board *n* (1) A heavy weight, thick sheet of paper or other fibrous substance, usually of a thickness greater than 6 mm (0.006 in.). (2) Lumber less than 2 in. (6 cm) thick and between 4 in. (10 cm) and 12 in. (30 cm) in width; a board less than 4 in. (10 cm) wide may be classified as a strip.

Board and Batten *n* A type of wall cladding for wood-frame houses: closely spaced, applied boards or sheets of plywood, the joints of which are covered by narrow wood strips.

Board and Brace *n* A type of carpentry work consisting of boards which are grooved along both edges and have thinner boards fitted between them.

Board Foot A unit of cubic content, used in measuring lumber; equal in volume to a piece 1 sq. ft and 1 in. thick, or 2,360 cm^3.

Boardy A term used to describe a fabric with a very stiff hand.

Bob-and-C \ˈbä-bən\ *n* [origin unknown] (1530) A cylindrical or slightly tapered barrel, with or without flanges, for holding slubbings, rovings, or yarns.

BOBTEX® ICS Yarn System *n* A process for producing a simulated spun yarn by embedding individual fibers in a thermoplastic or adhesive coating on a filament yarn.

Bodanyl *n* Poly(caprolactam). Manufactured by Feldmühle, Rorschach, Switzerland.

Bodied Linseed Oil See ▶ Bodied Oil.

Bodied Oil *n* In the strictest sense of the term, this is a drying oil which has been heat polymerized. The term is also used more broadly to describe drying oils which have viscosities greater than those of the oils in their natural condition, irrespective of how the increased viscosity is attained, e.g., blown oil or stand oil.

Body *n* (1) A term used loosely in the paint and adhesives industries to denote overall consistency, i.e., a combination of viscosity, density, pastiness, tackiness, etc. (2) An aspect of fabric quality, akin to ▶ Drape and ▶ Hand. (3) A general term referring to viscosity, consistency, and flow of a vehicle or ink. (4) Used also to describe the increase in viscosity by polymerization of drying oils at high temperatures. (5) A practical term widely used to give a qualitative picture of consistency. For Newtonian liquids, both is the same as viscosity.

Body Gum (#8 Varnish) In printing inks, linseed oil that has been heat polymerized to a heavy, gummy state. It is commonly used as a bodying agent.

Body Putty *n* A paste-like mixture of plant resin, often a polyester, and a filler such as talc, used to smooth and repair metal surfaces such as auto bodies.

Bodying *n* Increase in the apparent viscosity of a paint, varnish, resin, or lacquer, which occurs either deliberately during manufacture of adventitiously during storage.

Bodying Agent *n* A material added to an ink to increase its viscosity. See ▶ Thickener.

Boea *n* One of the classes of Manila copal.

Boiled Oil *n* Linseed oil, and sometimes soya oil, which has been treated in any of the following ways: (a) Heated for a short time to slightly increase viscosity and shorten drying time; (b) Heated with compounds of lead and/or cobalt and/or manganese forming fatty acid salts of the metal(s), thus shortening the drying time; (c) By adding soluble driers with or without heating. (Weismantal GF (1981) Paint handbook. McGraw-Hill Corporation, Inc., New York)

Boilers Group name for nitrocellulose and lacquer solvents.

Boil-in-Bag A type of food packaging – foil, plastic, or laminate – intend to be dropped into boiling water in order to cook the contents.

Boiling Point Temperature at which the vapor pressure of a liquid is just slightly greater than the total pressure of the surroundings. The liquid as a consequence is rapidly converted from the liquid state into a vapor.

Boiling-Point Elevation *n* A colligative property: the increase in boiling point of a solvent brought about by the presence of a solute.

For each mole-weight of solute in solvent there is a boiling point elevation and the change in B.P. is represented by,

$$\Delta T_{bp} = K_{bp} m$$

where K = 0.51 for water, and m = molality or mole-weight solute/1000 g of solvent. (Whitten KW, Davis RE, Davis E, Peck LM, Stanley GG (1994) General chemistry, Brookes/Cole, New York)

Blow Molding The process of forming hollow articles by expanding a hot plastic element against the internal surfaces of a mold is blow molding. In its most common form, the process comprises extruding a tube (*parison*) downward between the opened halves of a metal mold, closing the mold to pinch off and seal the parison at top and bottom, injecting air through a needle inserted through the parison wall, cooling the mass by contact with a chilled mold, opening the mold and removing the formed article. Improvements on this process have increased the number of products and their properties. (Strong AB (2000) Plastics materials and processing. Prentice Hall, Columbus; Ash, M, Ash I (1982–83) Encyclopedia of plastics polymers, and resins, vols. I–III, Chemical Publishing Co., New York; Pittance JC (ed) (1990) Engineering plastics and composites. SAM International, Materials Park; Carley JF (ed) (1993) Whittington's dictionary of plastics. Technomic Publishing Co., Inc., USA)

Boiling Range Range between initial and final boiling temperatures of a solvent. Many solvents have no specific boiling point but distill over a definite range of temperature; e.g. VM&P Naphtha 250–300°F.

Bole *n* Name frequently given in the arts to clay, either white or colored.

Bolster \‖bōl-stər\ *n* [ME, fr. OE; akin to OE *belg* bag] (before 12c) A spacer or filler in a mold.

Bolt *n* A roll or piece of fabric of varying length.

Boltzmann Constant *n* The number that relates the average energy of a molecule to its absolute temperature, 1.380658×10^{-23} J K^{-1}, 1.3807×10^{-16} erg K^{-1}. (Giambattista A, Richardson R, Richardson RC, Richardson B (2003) College physics. McGraw Hill Science/Engineering/Math, New York).

Boltzmann Superposition Principle *n* Strain is presumed to be a linear function of stress, so the total effect of applying several stresses is the sum of the effects of applying each one separately. (Giambattista A, Richardson R, Richardson RC, Richardson B (2003) College physics. McGraw Hill Science/Engineering/ Math, New York).

Bolus Alba \ ‖bō-ləs ‖al-bə\ See ▶ Kaolin.

BON *n* Acronym for betaoxy-naphthoic acid which can be coupled with a variety of amine compounds to form a soluble azo dye, which is then precipitated by any of several metals to form "BON" reds or maroons.

Bon-Arylamide Red *n* Any of any group of metal-free monazo pigments based on substituted 2-hydroxy-3-naphthoic acid.

Bond *n* (1) *Chemical* – Force of attraction between atoms in a molecule. (2) *Adhesive* – Adhesion between to materials, such as between an adhesive and a given surface. See also ▶ Adhere. (3) *Coatings* – Adhesion between a coating and a substrate to which it has been applied.

Bond Axis *n* A line passing through the nuclei of two bonded atoms.

Bond Coat *n* Coating used to improve the adherence of succeeding coats.

Bond Distance *n* The distance between nuclei of two bonded atoms. Also known as *Bond Length*.

BON Pigment *n* (1) Any of several brilliant reds and maroons widely used in plastics and rubbers, resistant to bleeding, migration, and crocking. The initials BON stand for β-oxynaphthoic acid, the base raw material. (2) A related azo pigment made by coupling β-hydroxynaphthoic acid to an amine and forming the barium, calcium, strontium, or manganese salt thereof. The colors range from yellowish red to deep maroon.

Bond Strength (Adhesive)*n* (1) Adhesive bond between joined substrates, the force required to break the bond divided by the bond area (Pa) (▶ Tensile-Shear Strength. The term adherence is frequently used in place of bond strength ▶ Adhesion and ▶ Bond). (2) Of fiber-reinforced laminates, the strength of the bond between fiber and matrix. (3) The degree of attraction between adjacent atoms within a molecule, usually expressed in J/mol. (Skeist I (ed) (1990) Handbook of adhesives. Van Nostrand Reinhold, New York) See ▶ Peel Adhesion.

Bonded Adhesive See ▶ Adhesive.

Bonded Fabric *n* (1) A wed of fibers held together by an adhesive medium that does not form a continuous film. (2) See ▶ Nonwoven Fabric.

Bonding *n* (1) A process for adhesive laminating two or more fabrics or fabric and a layer of plastic foam. There are two methods: the flame method used for bonding foam and the adhesive method used for bonding face and backing fabrics. (2) One of several processes of binding fibers into thin sheets, webs, or battings by means of adhesives, plastics, or cohesion (self-bonding). Several types of bonding or bonding processes are listed below. (Kadolph SJJ, Langford AL (2001) Textiles. Pearson Education, New York)

Bonding Action *n* See ▶ Bond Strength.

Bonding Equipment *n* Equipment used for welding polymer parts such as ultrasonic welding; also cementing equipment.

Bonding Orbital *n* A molecular orbital in which electrons have lower energies than they would in the unbonded atoms; an orbital characterized by a region of high electron probability density between the bonded atoms, leading to a stabilizing effect on the molecule.

(Whitten KW, Davis RE, Davis E, Peck LM, Stanley GG (1994) General chemistry, Brookes/Cole, New York).

Bonding Pair *n* A pair of electrons shared between two atoms and constituting a covalent bond model). (Whitten KW, Davis RE, Davis E, Peck LM, Stanley GG (1994) General chemistry, Brookes/Cole, New York)

Bonding, Point *n* The process of binding thermoplastic fibers into a onwoven fabric by applying heat and pressure so that a discrete pattern of fiber bonds is formed. (Kadolph SJJ, Langford AL (2001) Textiles. Pearson Education, New York).

Bonding, Print *n* A process of binding fibers into a nonwoven fabric by applying an adhesive in a discrete pattern. (Kadolph SJJ, Langford AL (2001) Textiles. Pearson Education, New York).

Bonding Resin *n* Any resin used for bonding aggregates together, e.g., holt melt epoxy, such as foundry sands, grinding wheels, abrasive papers, asbestos brake linings, and concrete masses. Plywood adhesives are sometimes called bonding resins. (Skeist I (ed) (1990) Handbook of adhesives. Van Nostrand Reinhold, New York).

Bonding, Saturation *n* A process of binding fibers into a nonwoven fabric by soaking the web with an adhesive. (Kadolph SJJ, Langford AL (2001) Textiles. Pearson Education, New York)

Bonding, Spray *n* A process of binding fibers into a nonwoven fabric involving the spray application of a fabric binder.

Bonding, Stitch *n* A bonding technique for nonwovens in which the fibers are connected by stitches sewn or knitted through the web. *Also known as Quilting*. (Kadolph SJJ, Langford AL (2001) Textiles. Pearson Education, New York).

Bonding with Binder Fibers *n* Specially engineered low-melting point fibers are blended with other fibers in a web, so that a uniformly bonded structure can be generated at low temperature by fusion of the binder fiber with adjacent fibers. (Skeist I (ed) (1990) Handbook of adhesives. Van Nostrand Reinhold, New York)

Bonds, Chemical *n* The electronic linkages between atoms such as -C-H which can be covalent, ionic and others. (Goldberg DE (2003) Fundamentals of chemistry. McGraw-Hill Science/Engineering/Math, New York).

Bone Black *n* Made by charring animal bones in closed retorts. Bone black is blue-black in color and is fairly smooth in texture. It contains carbon, calcium phosphate, and calcium carbonate. (Kirk-Othmer encyclopedia of chemical technology: pigments-powders (1996). Wiley, New York).

Bone Brown *n* Pigment related to bone black. Both are obtained from bones by calcinations, but with bone brown the temperature of calcinations is much lower and less carbonization occurs. the pigment is not widely used.

Bone Dry Bleached Shellac *n* A bleached, light-colored orange gum shellac. See ▶ Bleached Shellac.

Bone Glue *n* Impure gelatin prepared from bones. See ▶ Gelatin and ▶ Glue. (Skeist I (ed) (1990) Handbook of adhesives. Van Nostrand Reinhold, New York).

Bone Oil *n* Fatty oil obtained by the dry distillation of bones.

Bone White *n* Pigment made by acclimating animal bones. Composed chiefly of tricalcium phosphate. Calcium carbonate and minor constituents make up the rest. Bone white is a grayish white and slightly gritty powder. (Kirk-Othmer encyclopedia of chemical technology: pigments-powders (1996). Wiley, New York).

Bookbinders' Varnish *n* Rapid drying varnish based on shellac, sandarac or gum mastic, or mixtures of these resins. (Kirk-Othmer encyclopedia of chemical technology: pigments-powders (1996). Wiley, New York).

Booster Ram *n* A hydraulic ram used as an auxiliary to main ram of a molding press.

BOP Abbreviation for ▶ Butyl Octyl Phthalate. See ▶ Butyl Ethylhexyl Phthalate.

BOPP Abbreviation for ▶ Biaxially Oriented Polypropylene.

Boralloy *n* Boronitride., manufactured by Union Carbide, USA.

Borax \ˈbōr-ˌaks, ˈbór-, -əks\ *n* [ME *boras*, fr. MF, fr. ML *borac-*, *borax*, fr. Arabic *būraq*, fr. Persian *būrah*] (14c) $Na_2B_4O_7 \cdot 10H_2O$. Natural hydrated sodium borate, found in salt lakes and alkali soils. Also, the commercial name for sodium borate. It is used in some fire-retardant paints, and to dissolve casein in distemper.

Border *n* A narrow strip around an edge. A border wallcovering is used for trimming, or as a frieze, generally just under the ceiling. See ▶ Soffit. (Harris CM (2005) Dictionary of architecture and construction, McGraw-Hill Co.)

Boric-Acid Ester *n* One of a family of flame retardants for plastics, etc. and plasticizers. Examples are the trimethyl, tri-n-butyl, tricyclohexyl, and tri-*p*-cresyl borates. (Kidder RC (1994) Handbook of fire retardant coatings and fire testing services. CRC Press, Boca Raton).

Borneol \ˈbór-nē-ˌól, -ˌōl\ *n* [ISV, fr. *Borneo*] (1876) $C_{10}H_{18}O$. Constituent of pine oil. A terpenic alcohol. It is a solid at ordinary temperature, with a mp of 203°C (397.4°F), bp of 212°C (413.6°F), sp gr of 1.011. It is sometimes described as Bornyl alcohol or borneo camphor. Solubility characteristics are good, and it has some application as a plasticizer.

Bornyl Acetate *n* A solvent and plasticizer for nitrocellulose.

Boron-Epoxy Composite *n* A composite in which ▶ Boron Fibers are embedded in an epoxy matrix. Modulus is about 200 GPa, tensile strength about 1.6 GPa, rising with fiber content. The combination of high specific modulus and high specific strength has made these composites attractive for aerospace vehicles in spite of their high cost. (Harper CA (ed) (2002) Handbook of plastics, elastomers and composites, 4th edn. McGraw-Hill, New York).

Boron Fiber *n* An advanced reinforcing fiber produced by passing 10-μm, resistively heated tungsten wire through an atmosphere of boron trichloride and hydrogen. Hydrogen reduces the BCl^3 and the boron deposits on the wire to make a filament from 120 to 140 μm in diameter. Density is low, 2.4–2.6 g/cm^3. Boron fiber is extremely strong and stiff with strength near 3.1 GPa (450 kpsi) and modulus near 400 GPa (58 Mpsi). The very high cost of these fibers (ca $1/g) has limited their use. (Harper CA (ed) (2002) Handbook of plastics, elastomers and composites, 4th edn. McGraw-Hill, New York).

Boron Hydride See ▶ Diborane.

B——H

Boron Resins *n* Esters derived from boric acids and polyhydric alcohols. Characterized by solubility in water.

Boron Trifluoride Etherate *n* A metal halide $BF_3 \cdot O(C_2H_5)_2$ which serves as a cationic initiator for the polymerization of monomers.

Boss *n* (1) A protuberance provided on an article to add strength, facilitate alignment during assembly, or for attaching the article to another part. (2) That part of a drafting roll of largest diameter where the fibers are gripped. It may be an integral part of the roll, as in steel rolls, or it may have a covering of leather, cork, etc. In the former case, the boss is fluted.

Boston Round *n* A family of variously sized, blown bottles either glass or plastic, of circular-cylindrical shape with a short curved shoulder and length-to-diameter ratio in the body of about 1.7:1.

Boston Stone, The *n* Originally a paint mill imported around 1700 by a painter, Thomas Child. Later used as a landmark or central point from which distances from Boston were measured. It is the earliest known implement of the paint industry in America. The round stone was rolled back and forth in the trough of a larger flat stone underneath, in which oil and pigment were dispersed. The Boston Stone is the official insignia of the Federation of Societies for Coatings Technology. (Paint/coatings dictionary (1978). Federation of Societies for Coatings Technology, Philadelphia).

Botany Bay Resin See ▶ Accroides.

Bottom *n* (1) Ship's bottom. (2) Hard deposit in paint cans.

Bottom-Drying *n* Drying of a film from the bottom towards the top of the film; e.g., lead naphthenate is used as a bottom drier.

Bottoms See ▶ Tailings.

Bouclé \bu-▪klā\ *n* [F *bouclé*] 1895) (1) A fabric woven or knit with bouclé yarns. (2) An uneven yarn of three piles one of which forms loops at intervals.

Bouclé Yarn *n* A novelty yarn with loops which give fabrics a rough appearance. Some bouclé yarns have cotton cores with other fibers wound around them. Bouclé yarns may be made from wool, cotton, silk, linen, manufactured fibers, or combinations of fibers. (Vincenti R (ed) (1994) Elsevier's textile dictionary. Elsevier Science and Technology Books, New York).

Bouguer's *n* **(Lambert's) Law** Describes the proportionality between the absorption of radiant energy by a nonscattering material and its optical path length $-LogT_i/1 = K$, where T_i is the internal transmittance, 1 is the path length, and K is the unit absorption coefficient. See ▶ Beer-Bougure Law. (Giambattista A, Richardson R, Richardson RC, Richardson B (2003) College physics. McGraw Hill Science/Engineering/Math, New York).

Bounce Back *n* A spray rebound.

Boundary Surface *n* A surface of constant electron probability density, Ψ^2.

Bourrelet A double-knit fabric with raised loops running horizontally across the surface of the cloth giving a rippled or corded effect.

Bow *n* (1) Distortion of a board in which the face is convex or concave longitudinally. (2) The greatest distance, measured parallel to the selvages, between a filling yarn and a straight line drawn between the points at which this yarn meets the selvages. Bow may be expressed directly in inches or as a percentage of the width of the fabric at that point.

Bow Window See ▶ Bay.

Boxed Heart *n* Term used when the pith falls entirely within the four faces anywhere in the length of a piece of wood.

Boxing *v* Combining of two or more separate batches to one uniform batch.

Box Loom *n* A loom using two or more shuttles for weaving fabrics with filling yarns that differ in fiber type, color, twist, level, or yarn size. The box motion is automatic, changing from one shuttle to another. Examples of fabrics made on box looms are crepes and ginghams.

Box Mark *n* A fine line parallel to the filling caused by shuttle damage to a group of filling yarns.

Boyer-Beaman Rule *n* A rule of thumb stating that the ratio of polymer's glass-transition temperature to its melting temperature (T_g/T_m), both in Kelvin, usually lies between 0.5 and 0.7. For symmetrical polymers, such as polyethylene, it is close to 0.5; for unsymmetrical ones, such as polystyrene and polyisoprene, it is near 0.7.

Boyle's Law \bóilz-\ [Robert *Boyle*] *np* (ca. 1860) The part of the ideal-gas law, due to Robert Boyle, stating that the volume of a gas is inversely proportional to its pressure. This can also be stated as, at a constant temperature the volume of a given quantity of any gas varies inversely as the pressure to which the gas is subjected. For a perfect gas, changing from pressure p and volume v to pressure p' and volume v' without change of temperature, $pv = p' v'$. (Atkins PW, De Paula J (2001) Physical chemistry, 7th edn.WH Freeman Co., New York).

Bozzo See ▶ Abbozzo.

BPF *n* Abbreviation for British Plastics Federation.

BR *n* Poly(butadiene) Abbreviation for Butadiene Rubber (▶ British Standards Institution). See ▶ Polybutadiene.

Br *n* Chemical symbol for the element bromine.

Barbender Plastograph *n* A rheometer for measuring shear-strain with temperature relationships. (Strong AB (2000) Plastics materials and processing. Prentice Hall, Columbus).

Braid *n* (1) A narrow textile band, often used as trimming or binding, formed by plaiting several strands of yarn. The fabric is formed by interfacing the yarns diagonally to the production axis of the material. (2) In aerospace textiles, a system of three or more yarns which are interlaced in such a way that no two yarns are twisted around each other.

Braid Angle *n* The acute angle measured from the axis of a fabric or rope to a braiding yarn.

Braided Fabric *n* A narrow fabric made by crossing a number of strands diagonally so that each strand passed alternatively over or under one or more of the other strands. They are frequently used in shoelaces and suspenders.

Braiding *n* The intertwining of three or more strands to make a cord. The strand form a regular diagonal pattern down the length of the cord.

Brake Lining Medium *n* A medium for bonding together the chosen filling agent for the brake lining, e.g., asbestos.

Branched *n* Refers to side chains attached to the main chain in molecular structure of polymers, as opposed to linear arrangement.

Branching *n* (1) The growth of a new polymer chain from an active site on an established chain, in a direction different from that of the original chain. Branching occurs as a result of chain-transfer processes or from the polymerization of difunctional monomers, and is an important factor influencing polymer properties. (2) Two-dimensional polymers having relatively short sidechains branching from the main backbone. (Odian GC, (2004) Principles of polymerization, Wiley, New York).

Brashness *n* Condition of wood characterized by low resistance to shock and by an abrupt failure across the grain without splintering.

Brayer *n* A small hand roller for distributing ink.

Brazil Wax *n* A name sometimes applied to a carnauba wax.

Brazil-Wood *n* A natural red dye from the wood of *Caesalpinia braziliensis*.

Break The break in an oil is the flocculent material or foots which separate on long standing or upon heating.

Breakdown *n* To soften or plasticize rubber by working it on a rubber mill or in an internal mixer. Same as to mill or masticate.

Breakdown Voltage *n* The voltage required, under specific conditions, to cause the failure of an insulating varnish or other insulating material. See ▶ Dielectric Breakdown Voltage and ▶ Dielectric Strength.

Breakfree Oil *n* Oil which has been refined mechanically by heating to about 88°C (190°F), adding an adsorbent such as fuller's earth or activated charcoal, agitating, and filtering. *Also known as Nonbreak Oil*.

Break Factor *n* A measure of yarn strength calculated as: (1) the product of breaking strength times indirect yarn number, or (2) the product of breaking strength times the reciprocal of the direct yarn number.

Breaking *v* Separation of an emulsion or latex into two phases.

Breaking Elongation See ▶ Elongation.

Breaking Extension *n* Elongation necessary to cause rupture of a test specimen; the tensile strain at the moment of rupture. See ▶ Elongation.

Breaking Length *n* A measure of the breaking strength of a yarn; the calculated length of a specimen whose weight is equal to its breaking load. The breaking length expressed in kilometers is numerically equal to the breaking tenacity expressed in grams-force per tex. (Kadolph SJJ, Langford AL (2001) Textiles. Pearson Education, New York).

Breaking Ratio See ▶ Break Factor.

Breaking Strength (Load) *n* (1) The maximum resultant internal force that resists rupture in a tension test. The expression "breaking strength" is not used for compression tests, bursting tests, or tear resistance tests in textiles. (2) The load (or force) required to break or rupture a specimen in a tensile test made according to a specified standard procedure. ( ) Also see ▶ Ultimate Strength.

Breaking Tenacity *n* The tensile stress at rupture of a specimen (fiber, filament, yarn, cord, or similar structure) expressed as newton per tex, grams-force per tex, or gram-force per denier. The breaking tenacity is calculated from the breaking load and linear density of the unstrained specimen, or obtained directly from tensile testing machines which can be suitably adjusted to indicate tenacity instead of breaking load for specimens of known linear density. Breaking tenacity expressed in grams-force per tex is numerically equal to breaking length expressed in kilometers. (Shah V (1998) Handbook of plastics testing technology. Wiley, New York).

Break Spinning *n* A direct spinning process for converting manufactured fiber tows to spun yarn that incorporates prestretching and tow breaking with subsequent drafting and spinning in one operation.

Break-Out See ▶ Smash.

Break Test (Oils) *n* Break in drying oils is a measure of the materials rendered insoluble in carbon tetrachloride under the conditions of test by a prescribed method.

Breathable Film *n* A film that is at least somewhat permeable to gases due to the presence of open cells throughout its mass or to minute perforations. (Shah V (1998) Handbook of plastics testing technology. Wiley, New York).

Brewster's Law *n* The tangent of the polarizing angle for a substance is equal to the index of refraction. The polarizing angle is that angle of incidence of which the reflected polarized ray is at right angles to the refracted ray. If n is the index of refraction and θ the polarizing angle, $n = \tan \theta$. (Moller KD (2003) Optics. Springer, New York).

Bridging *n* (1) In the flow by gravity of powders or pellets in feed hoppers and the throats of extruders and injection molders, the stoppage of flow caused by the formation of an arch across the flow path.

Bridle *n* Series of rolls that keeps the tension on the strip of a continuous line.

Bright *n* The term applied to fibers whose luster has not been reduced by physical or chemical means; the opposite of dull or matte. (Vincenti R (ed) (1994) Elsevier's textile dictionary. Elsevier Science and Technology Books, New York).

Brightener, Optical *n* fluorescent dye, or pigment which absorbs UV radiation and re-emits light of a violet or bluish hue. Used to increase the luminance factor and to remove the yellowish of white or off-white materials. (Bart J (2005) Additives in polymers: industrial analysis and applications. Wiley, New York).

Brighteners *n* Additives for providing optical brilliance to a plastic part. (Bart J (2005) Additives in polymers: industrial analysis and applications. Wiley, New York).

Brightening Agents *n* (optical brighteners, fluorescent bleaches, and optical whiteners) Brightening agents are chemicals used primarily in fibers but also to some extent in molded and extruded products, to overcome yellow casts and to enhance clarity or brightness. In contrast to *bluing agents*, which act by removing yellow light, the optical brighteners absorb ultraviolet rays and convert their energy into visible blue-violet light. Thus, they cannot be used ion compounds containing UV-absorbing agents. Optical brighteners are used in PVC sheet and film, fluorescent lighting fixtures, vinyl flooring, nylon fishing line, polyethylene bottles, etc. A few examples of optical brighteners are coumarins, naphthotriazolyl stilbenes, benzoxazolyl-, benzimidazolyl-, naphthylimide-, and diaminostilbene disulfonates. (Bart J (2005) Additives in polymers: industrial analysis and applications. Wiley, New York). See ▶ Brightener, ▶ Optical.

Brightfield Illumination *n* The usual form of microscope illumination with the image of the sample on a bright, evenly lighted field; or in vertical illumination, light reflected through the objective by means of prism or semireflecting plane glass. (Loveland RP (1981) Photomicrography, Krieger Publishing Co., New York).

Brightness *n* (1) Brightness is measured by the flux emitted per unit emissive area as projected on a plane normal to the line of sight. The unit of brightness is that of a perfectly diffusing surface giving out one lumen per square centimeter of projected surface and is called the *lambert*. The millilambert (0.001 lambert) is a more convenient unit. *Candle per square centimeter* is the brightness of a surface which has, ion the direction considered, a luminous intensity of one candle per cm². (2) (Deprecated) The dimension of color that is referred to an achromatic scale, ranging from black to white, also called lightness or luminous reflectance or transmittance. Because of confusion with saturation, the use of this term should be discouraged. (McDonald (ed) (1997) Colour physics for industry, 2nd edn. Roderick, Society of Dyers and Colourists, West Yorkshire).

Brightness Flop *n* Change in lightness with direction of observation.

Brilliance *n* A subjective term referring to: (1) Clarity or freedom from visible suspended matter or opalescence in clear coatings. (2) Cleanness or lack of a muddy or dirty tone in pigmented coatings. (3) The combined effect of lightness and strength, or purity of tone of printing inks. (Wicks ZN, Jones FN, Pappas SP (1999) Organic coatings science and technology, 2nd edn. Wiley-Interscience, New York)

Brilliant Ultramarine Synthetic ultramarine blue. See ▶ Ultramarine Blue.

Brine \brīn\ *n* [ME, fr. OE *bryne*; akin to MD *brīne* brine] (before 12c) Concentrated aqueous salt solution. The term can be applied to solutions of any salt, but it is commonly applied to solutions of sodium chloride.

Brinell Hardness \brə-nel-\ *n* [Johann A. *Brinell* † Swedish engineer] (1915) The hardness of a material as determined by pressing a hardened steel or carbide ball, 10 mm in diameter, into the specimen under a fixed load. The Brinell number is the load in kilograms divided by the area in mm² of the spherical impression formed by the ball. For nonferrous materials, the prescribed load is 500 kg applied for 30 seconds. The Brinell test is used mostly for ductile metals, for which the Brinell number is related to yield strength in a simple way. Also known as *Brinell Hardness Number*. (Shah V (1998) Handbook of plastics testing technology. Wiley, New York).

Bring Forward *n* In repainting, to repair local defective areas with the appropriate paints so as to bring them into conformity with the surrounding areas before applying the finishing coats.

Brinkman Number *n* A dimensionless group relevant for heat transfer in flowing viscous liquids such as polymer melts. It is defined by $N_{Br} = \mu \cdot V^2 / (k \cdot \Delta T)$, in which μ = viscosity, V = velocity, k = thermal conductivity, and ΔT = the difference in temperature between the stream and the confining wall. The number represents the ratio of the rates of heat generation and heat conduction. (Shenoy AV (1996) Thermoplastics melt rheology and processing. Marcel Dekker, New York).

Bristle \bri-səl\ *n* [ME *bristil*, fr. *brust* bristle, fr. OE *byrst*; akin to OHGr *burst* bristle] (14c) (1) A generic term for a short, stiff, coarse fiber. It is frequently hog hair. (2) Any grade of nylon monofilament used in toothbrushes, hairbrushes, paintbrushes (*tapered bristle*), etc.

British Gum See Dextrin.

British Standards Institution A national organization (corresponding to the American National Standards Institute and the American Society for Testing and Materials) which establishes and publishes standard specifications and codes of practice.

British Thermal Unit *n* (Btu) Before the introduction of the SI system, the Btu was variously defined as the quantity of heat required to raise the temperature of one pound (avoirdupois) of water 1°F, either at or near 39°F (the temperature of maximum density), at 50°F or 60°F, or averaged from 32 to 212°F ("mean Btu"). The Btu is defined as 1,055.056 J.

Brittle \bri-t³l\ *adj* [ME *britil*; akin to OE *brēotan* to break. ON *brjōta*] (14c) Easily broken when bent rapidly or scratched as with a knife blade or the finger nail; the opposite of tough.

Brittle Fracture An abrupt breaking of a material in which there is very little or no elongation or distortion of the part before failure. In a tensile test, the stress-strain graph is nearly linear to the breaking point. (Grellman W, Seidler S (eds.) (2001) Deformation and fracture behavior of polymers. Springer, New York).

Brittle Temperature *n* Temperature at which a material transforms from being ductile to being brittle, i.e., the critical normal stress for fracture is reached before the critical stress for plastic deformation.

Brittleness *n* The lack of flexibility and the tendency to crack when deformed or struck.

Brittleness Temperature The temperature at which a plastic or elastomer breaks in cantilever-bending impact under specific conditions. The brittleness temperature is related to that of the ▶ Glass Transition, for those plastics with T_gs below room temperature, such as flexible PVC. (Wickson EJ (ed) (1993) Handbook of polyvinyl chloride formulating. Wiley, New York).

Brittle Point The temperature at which a polymer no longer exhibits viscoelastic properties. (Shah V (1998) Handbook of plastics testing technology. Wiley, New York).

Brittle Point, Brittle Temperature Lowest temperature at which a material withstands an impact test under standardized conditions.

Brittle Strength *n* The strength of a polymer at the brittle point or temperature.

Broadcast \bród-\ *adj* (1767) To apply or sprinkle solid particles on an uncured coating surface.

Broadcloth \-ˈklóth\ *n* (15c) (1) Originally, a silk shirting fabric so named because it was woven in widths exceeding the usual 29 in. (2) A tightly woven, lustrous cotton or polyester/cotton blend fabric in a plain weave with a crosswise rib. It resembles poplin, but the rib is finer, and broadcloth always has more picks that poplin. The finest qualities are made with combed pima or Egyptian cotton. (3) A smooth, rich-looking, woolen fabric with a napped face and a twill back. Better grades have a glossy, velvety hand. (Vincenti R (ed) (1994) Elsevier's textile dictionary. Elsevier Science and Technology Books, New York).

Broadgoods A fabrics-industry term for woven materials, including glass fabrics that are over 46 cm in width.

Broadloom \-ˈlüm\ *adj* (1925) A term that refers to carpets woven in widths from 54 in. to 18 ft, as distinguished from narrow loom widths of 27–36 in. (Vincenti R (ed) (1994) Elsevier's textile dictionary. Elsevier Science and Technology Books, New York).

Brocade \brō-ˈkād\ *n* [S *brocado*, fr. Catalan *brocat*, fr. I *broccato*, fr. *broccare* to spur, brocade, fr. *brocco* small nail, fr. L *broccus* projecting] (1588) (1) A rich, Jacquard-woven fabric with an all over interwoven design of raised figures or flowers. The pattern is emphasized by contrasting surfaces or colors and often has gold or silver threads running through it. The background may be either a satin or a twill weave. (2) A term describing a cut-pile carpet having a surface texture created by mixing twisted and straight standing pile yarns. (Vincenti R (ed) (1994) Elsevier's textile dictionary. Elsevier Science and Technology Books, New York).

Brocatelle \ˈbrä-kə-ˈtel\ *n* [Fr, fr. It *broccatello*] (1669) A fabric similar to brocade with a satin or twill figure in high relief on a plain or satin background. (Vincenti R (ed) (1994) Elsevier's textile dictionary. Elsevier Science and Technology Books, New York).

Broken Color *n* General multicolored effect brought about by the automatic merging of wet paints of various colors or by manipulation which produce apparently accidental effects. (Paint/coatings dictionary (1978). Federation of Societies for Coatings Technology, Philadelphia).

Broken-Color Work See ▶ Antiquing.

Broken End *n* A broken, untied warp thread in a fabric. There are numerous causes, such as slubs, knots, improper shuttle alignment, shuttle hitting the warp shed, excessive warp tension, faulty sizing, and rough reeds, heddles, dropwires, and shuttles.

Broken Pick *n* A broken filling thread in a fabric. Usual caused include too much shuttle tension, weak yarn, or filling coming into contact with a sharp surface.

Broken Selvage See ▶ Cut Selvage.

Broken White *n* Off-white in which the initial pure whiteness of the principal pigment has been very slightly "broken" or modified in a yellow or brown direction.

Bronze \ˈbränz\ *n* {*often attributive*} [F, fr. I *bronzo*] (1739) An appearance characteristic of some printed films in which the apparent color of the print depends upon the angles of viewing and illumination. (Leach RH, Pierce RJ, Hickman EP, Mackenzie MJ, Smith HG (eds) (1993) Printing ink manual, 5th edn. Blueprint, New York)

Bronze Blue *n* One of the names applied to the complex ferric ferrocyanide or iron blues. (Leach RH, Pierce RJ, Hickman EP, Mackenzie MJ, Smith HG (eds) (1993) Printing ink manual, 5th edn. Blueprint, New York)

Bronze Gold Powders *n* Metallic powders made from alloys of copper. They range in color from bonze gold to green yellow. (Solomon DH, Hawthorne DG (1991) Chemistry of pigments and fillers. Krieger Publishing Co., New York).

Bronze Paint *n* Paint incorporating bronze gold powder in a bronzing liquid. (Solomon DH, Hawthorne DG (1991) Chemistry of pigments and fillers. Krieger Publishing Co., New York)

Bronze Paste *n* Paste consisting of fine metallic flakes in a volatile medium. The metal may be aluminum, copper, or alloys of copper. (Solomon DH, Hawthorne DG (1991) Chemistry of pigments and fillers. Krieger Publishing Co., New York).

Bronze Pigment *n* Simulated bronze- or gold-colored pigments made by staining aluminum flakes with brown or yellow colorants. (Solomon DH, Hawthorne DG (1991) Chemistry of pigments and fillers. Krieger Publishing Co., New York).

Bronzing *v* (1) A subjective, descriptive, appearance term applied to the metal-like reflectance which sometimes appears at the surface of nonmetallic colored materials. It is perceived at the specular angle, by observing the image of a white light source, for example, and is characterized by a distinct hue of different dominant wavelength than the hue of the paint film itself. The origin of the selective specular reflectance observed is generally considered to be reflectance from very small particle size pigments partially separated from surrounding vehicle at or near the surface. (2) Characteristic metallic luster shown by certain highly colored pigments in full strength, e.g., certain iron and phthalocyanine blues. May also occur with bleeding type pigments in making finishes. (3) Application of imitation gold or other metals either in powder form or in leaves. (4) Printing with a sticky size and dusting same with finely powdered metal particles or flakes to give the appearance of metallic printing. *Also* *known as Bronze Dusting.* (Leach RH, Pierce RJ, Hickman EP, Mackenzie MJ, Smith HG (eds) (1993) Printing ink manual, 5th edn. Blueprint, New York).

Brookfield Viscometer *n* The Brookfield "Synchrolectric" viscometer is the most widely used instrument for measuring the viscosities of plastisols and other liquids, both Newtonian and nonNewtonian, (Goodwin JW, Goodwin J, Hughes RW (2000) Rheology for chemists. Royal Society of Chemistry, Cambridge). About a dozen models are available to accommodate subranges of viscosity in the overall range from 0.01 to 1000 Pa·s. It is portable and can be hand held. A synchronous motor provides 4 or 8 spindle speeds by shifting gears. Near the tip- of the spindle, and concentric with it, is a horizontal disk whose drag torque in the liquid is detected by a torsion spring; a pointer indicates viscosity. By taking readings at different rotational speeds, one can estimate the pseudoplasticity of the liquid. For accurate work, the spindle guard should be removed and the diameter of the vessel containing the liquid should be at least five times that of the disk. (Pierce PE (1969) Rheology of coatings. J Paint Tech., 41(533): 383–395).

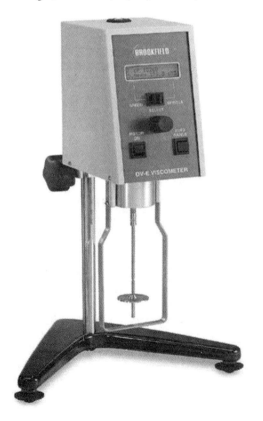

Brookite \bru-‖kīt\ *n* [*Henry J. Brooke* † 1859 English mineralogist] (1825) Rutile form of titanium dioxide characterized by its orthorhombic form. Its color varies from black to reddish brown. It is not used directly as a pigment.

Brown Coat *n* The coat of roughly finished plaster beneath the finish coat; in three-coat work, the second coat of plaster, applied over a scratch coat and covered by the finish coat; in two-coat work, the base-coat plaster applied over lath or masonry; may contain a greater proportion of aggregate than the scratch coat. *Also known as Floating Coat.*

Browne Heat Test Heat polymerization test designed to determine the purity of tung and similar oils. The result is reported as the number of minutes required to cause gelation at the specific temperature (282°C). (Paint: pigment, drying oils, polymers, resins, naval stores, cellulosics esters, and ink vehicles, vol. 3. (2001) American Society for Testing and Material)

Brown Hard Varnish *n* Spirit varnish prepared from shellac dissolved in industrial alcohol. Cheaper grades may consist of shellac-rosin mixtures, spirit manila, spirit manila-rosin mixtures, or alcohol soluble rosin-modified phenolic resin.

Brownian Motion \braú-nē-ən-\ *n* [*Robert Brown* † 1858 Scottish botanist] (ca. 1889) Refers to the continual movement of extremely small particles suspended in a fluid. The accepted explanation is that movement of the particles is due to bombardment by molecules of the fluid. The average length of path *L* followed by a particle in any given direction at a given time *t* is directly proportional to the square root of the absolute temperature *T* and the time *t*; it is also inversely proportional to the square root of the viscosity η of the medium and to the diameter *d* of the particle (Avogadro's number N and the gas constant R enter in as constants):

$$L = \frac{2R}{N}\sqrt{\frac{T \cdot t}{\eta d 3\pi}}$$

(Whitten KW, Davis RE, Davis E, Peck LM, Stanley GG (1994) General chemistry, Brookes/Cole, New York)

Brownian Movement *n* A continuous agitation of particles in a colloidal solution caused by unbalanced impacts with molecules of the surrounding medium. The motion may be observed with a microscope when a strong beam of light is caused to traverse the solution across the line of sight. (Whitten KW, Davis RE, Davis E, Peck LM, Stanley GG (1994) General chemistry, Brookes/Cole, New York).

Brown Rot *n* (1894) See ▶ Dry Rot.

Brunswick Black \\▎brənz-(▁)wik-\\ Solution consisting of gilsonite, petroleum pitch, or similar material in spirit or aromatic hydrocarbons.

Brush \\▎brəsh\\ *n* [ME *brusshe*, fr. MF *broisse*, fr. OF *broce*] (14c) A tool composed of bristles set into a handle; most often used to apply coatings. Bristles may be synthetic or natural. (Weismantal GF (1981) Paint handbook. McGraw-Hill Corporation, Inc., New York).

Brushability *n* The ability or ease with which a coating can be brushed.

Brush (or Roller) Cleaner *n* A combination of chemicals in which a brush or rollers may be soaked in order to permit cleaning with soap and water. (Weismantal GF (1981) Paint handbook. McGraw-Hill Corporation, Inc., New York).

Brush Graining *n* An imitation effect of wood grain; produced by drawing a clean dry brush through a dark liquid stain, applied over a dry, light base coat.

Brushing Consistency *n* Rheological conditions at which a coating can be properly applied to a substrate with a brush. Brushing is used on sweaters, scarves, knit underwear, wool broadcloths, etc.

Brush Mark *n* A small ridge or valley produced in a paint film by the combining action of the bristles of a brush.

Brushout The application of paint on a small surface for testing.

BS *n* Abbreviation for British Standard Code of Practice.

BSI *n* Abbreviation for ▶ British Standards Institution.

Btu (BTU) See ▶ British Thermal Unit.

Bubble \\▎bə-bəl\\ *n* {*often attributive*} [ME *bobel*] (14c) (1) In blown-film extrusion, the expanding tube moving from the die to the collapsing rolls at the top of the tower. (2) A void within a molding or an extrusion.

Bubble Coating *n* Coating wherein the light scattering characteristics are derived partially or wholly from small bubbles or microvoids in a transparent binder. The air-filled or vapor-saturated voids having diameters in the range of the wavelength of light, are trapped in a water soluble medium such as soya protein. This type of coating has a low specific weight and high lightness. See ▶ Microvoids.

Bubble Forming *n* A variant of ▶ Sheet Thermoforming, in which the plastic sheet is clamped in a frame suspended above a mold, heated, expanded into a blister shape with air pressure, then molded to its final shape by means of a descending plug applied to the blister and forcing it downward into the mold. (Weismantal GF (1981) Paint handbook. McGraw-Hill Corporation, Inc., New York).

Bubbler *n* A device inserted into a mold force, cavity, or core that delivers **Bubble Tube Viscometer** *n* One of the bubble viscometers which is used by comparing the speed of a rising air bubble in a liquid of unknown viscosity with that of a liquid of known viscosity, both liquids being in corked glass tubes of identical size, and at the same temperature. Syn: ▶ Air Bubble Viscometer. (Patton TC (1979) Paint flow and pigment dispersion: a rheological approach to coating and ink technology. Wiley, New York; Paint and coating testing manual (Gardner-Sward Handbook) (1995) MNL 17, 14th edn. ASTM, Conshohocken).

Bubbling *n* Film defect, temporary or permanent, in which bubbles of air or solvent vapor, or both, are present in the applied film.

Bucket Spinning See ▶ Pot Spinning.

Buckling *v* (14c) (1) A crimping of the fibers in a composite material that may occur in glass-reinforced thermosets due to shrinkage of the resin during cure. (2) The principal mode of failure of axially loaded, slender structural members such as columns and panels.

Buckram \\▎bə-krəm\\ *n* [ME *buckeram*, fr. OF *boquerant*, from OP *bocaran*] (15c) A scrim fabric with a stiff finish, often used as interlining.

Buffer \\▎bə-fər\\ *n* {*often attributive*} [*buff*, v., to react like a soft body when struck] (1835) A solution which contains moderate or high concentrations of a Brønsted-Lowery conjugate acid-base pair; a solution whose pH does not change greatly in response to added acids or bases.

Buffer Coat *n* A finishing material applied over an old coating to protect it from solvent action of the new finish.

Buffing *n* A surface-finishing method used with plastics and other materials in which the object is rubbed with cloths or cloth wheels that may contain fine, mild abrasives of waxes or both. Also see ▶ Tumbling.

Buffing Compounds *n* Polishing grit for developing a smooth surface.

Buhr Mill *n* Stone mill in which a viscous pigment-medium paste is dispersed between the flat surfaces of two large stones, one of which rotates.

Buhrstone \\▎bər-▁stōn\\ *n* (1690) Type of porous flint, formed in France, used for millstones.

Build *n* Real or apparent thickness, fullness, or depth of a dried film

Building *n* A more or less enclosed and permanent structure for housing, commerce, industry, etc., distinguished from mobile structures and those not intended for occupancy.

Building Block *n* (1846) A rectangular masonry unit, other than a brick, made of brunt clay, cement, concrete, glass, gypsum, or any other material suitable for use in building construction. (Harris CM (2005) Dictionary of architecture and construction, McGraw-Hill Co.)

Building Board *n* Any sheet of building material, often faced with paper or vinyl; suitable for use as a finished surface on walls, ceilings, etc. (Harris CM (2005) Dictionary of architecture and construction, McGraw-Hill Co.)

Building Coats *n* Coats used to build up a surface or surfaces before rubbing.

Building Code *n* A collection of rules and regulations adopted by authorities having appropriate jurisdiction to control the design and construction of buildings, alteration, repair, quality of materials, use and occupancy, and related factors of buildings within their jurisdiction; contains minimum architectural, structural, and mechanical standards for sanitation, public health, welfare, safety, and the provision of light and air.

Building Paper *n* A heavy, relatively cheap, durable paper, such as asphalt paper, used in building construction, especially in frame construction, to improve thermal insulation and weather protection and to act as a vapor barrier. Special types are: sheathing paper, used between sheathing and siding; floor lining paper, used between rough and finish floors. (Harris CM (2005) Dictionary of architecture and construction, McGraw-Hill Co.)

Buildup *n* A term applied to substantivity of dye for a textile material. It refers to the ability of a dye to produce deep shades.

Built-up Roofing *n* A continuous roof covering made up of laminations or plies of saturated or coated roofing felts, alternated with layers of asphalt or coal tar pitch and surfaced with a layer of gravel or slag in a heavy coat of asphalt or coal tar pitch or finished with a cap sheet; generally used on flat or low-pitched roofs. (Harris CM (2005) Dictionary of architecture and construction, McGraw-Hill Co.)

Bulk Density *n* The density of a particulate material (granules, powder, flakes, chopped fiber, etc) expressed as the ratio of weight to total volume, voids included.

Bulk Development *n* Any of various relaxation treatments to produce maximum bulk in textured or latent crimp yarns or in fabrics made therefrom. The essential conditions are heat, lubrication, movement, and the absence of tension. Bulk development may be accomplished during wet processing or may be a separate operation such as hot-air tumbling, steam-injection tumbling, or dry cleaning.

Bulk Factor *n* The ratio of the volume of a given mass of plastic particles to the volume of the same mass of material after molding or forming. The bulk factor is also equal to the ratio of the density after molding or forming to the apparent density (▶ Bulk Density) of the material as received.

Bulking *n* In the process of formulating coatings, the step wherein ingredient weights are converted to their volume equivalents.

Bulking Value *n* Reciprocal of apparent density. Solid volume of a unit weight of material, usually expressed as gallons per pound or l/kg. For practical purposes this is 0.120 divided by the specific gravity (ASTM). Syn: ▶ Specific Volume. *Also called Specific Volume.*

Bulk Material *n* A material or product in large quantity such as a drum or sack.

Bulk Modulus *n* The modulus of volume elasticity, i.e., the resistance of a solid or liquid to change in volume with change in pressure, at constant temperature. The thermodynamic definition is:

$$B = -v(\partial P/\partial v)_T = -(\partial P/\partial \ln v)_T$$

in which v = specific volume, P = pressure, and T = the specific temperature. This may be approximated over an interval of pressure as $0.5 \cdot (v_2 + v_1) \cdot (P_2 - P_1)/(v_1 - v_2)$. With plastics, B gradually diminishes as temperature rises and increase with rising pressure. (Mark JE (ed) (1996) Physical properties of polymers handbook. Springer, New York)

Bulk Molding Compound *n* (BMC) See ▶ Premix.

Bulk Polymerization *n* The polymerization of a monomer (mass polymerization) in the absence of any medium other than a catalyst or accelerator. The monomer is usually a liquid, but the term also applies to the polymerization of glass and solids in the absence of solvents or any other dispersing medium. Polystyrene, polymethyl methacrylate, low-density polyethylene, and styrene-acrylonitrile copolymers are examples of polymers most frequently produced by bulk polymerization. Acrylic monomers may be simultaneously polymerized and formed into products by conducting the polymerization in molds such as those for rods and sheets. (Odian GC, (2004) Principles of polymerization, Wiley, New York; Mark JE (ed) (1996) Physical properties of polymers handbook. Springer, New York)

Bulk Specific Gravity *n* The specific gravity of a porous solid when the volume of the solid used in the calculation includes both the permeable and impermeable voids. Compare ▶ Specific Gravity and ▶ Bulk Density.

Bulk Yarn *n* Yarn of glass (or other) fiber in bulk form, as opposed to roving, mat, or woven forms.

Bumping *n* Sudden, explosive boiling following the superheating of a liquid.

Buna N \\ˈbyü-nə, ˈbü-\\ Copolymer from butadiene and acrylonitrile, manufactured by Hüls, Germany. See ▶ Acrylonitrile-Butadiene Copolymer.

Buna Rubber *n* A brand of synthetic rubber made by polymerizing or copolymerizing butadiene with another material, as acrylonitrile, styrene, or sodium.

Buna-S or SS *n* (GR-S) A synthetic elastomer produced by the copolymerization of butadiene and styrene. Manufactured by Hüls, Germany. *Also known as Styrene-Butadiene Rubber.*

Bunghold Oil See ▶ Boiled Oil.

Bunghole \\ˈbəŋ-ˌhōl\\ *n* (1571) Orifice, usually in a steel drum, a barrel, or a keg, which is stopped by a "bung" or large stopper.

Bunting \\ˈbən-tiŋ\\ *n* [perhaps fr. E dialect bunt (to sift)] (1711) A soft, flimsy, loose-textured, plain weave cloth most frequently used in flags. Bunting was originally made from cotton or worsted yarns, but today's flags are made primarily from nylon or acrylic fibers.

BUR *n* Abbreviation for ▶ Blow-up Ratio.

Burgundy Pitch See ▶ Pitch, ▶ Burgundy.

Burl \\ˈbər(-ə)l\\ *n* [ME *burle*, fr. (ass.) MF *bourle* tuft of wool, fr. (ass.) VL *burrula*. dimin. of LL *burra* shaggy cloth] (15c) Swirl or twist in the grain of wood that usually occurs near a know, but does not contain a knot.

Burlap \\ˈbər-ˌlap\\ *n* [origin unknown] (ca. 1696) A coarse, heavy, plain weave fabric constructed from singles yarn of jute. Used for bags, upholstery lining, in curtains and draperies. See ▶ Jute.

Burling *v* (1) The process of removing loose threads and knots from fabrics with a type of tweezers called a burling iron. (2) The process of correcting loose tufts and replacing missing tufts following carpet construction.

Burn Mark *n* Any visual sign of burning or charring at a particular spot on a part. In injection molding with poorly vented molds, burn marks can be caused by severe compression-heating of the air trapped in the mold cavity and consequent ignition of the molten plastic in contact with the hot air pocket (*dieseling*).

Burned *v* Showing evidence of excessive heating during processing or use of a plastic, as evidenced by blistering, discoloration, charring, or distortion.

Burned Finish *n* Wood finish in which the hard portion of the grain stands out in relief, the effect being produced by using a blowtorch and a stiff-bristled brush. *Also known as Fiery Finish.*

Burning-In *v* Process of repairing scratches and other damaged places in a finish, by melting compounds (stick shellac) into the defect with a heated tool.

Burning Off *v* Removal of paint by a process in which the paint is softened by heat, e.g., from a flame, and then scraped off while still soft.

Burning Rates *n* (1) The oxidation in air of a polymer with time. (2) The speed at which a fabric burns. It can be expressed as the amount of fabric affected per unit time, in terms of distance or area traveled by the flame, afterglow, or char. See ▶ Flammability, ▶ Oxygen-Index Flamm-Ability Test, ▶ Self-Extinguishing.

Burnishing *adj, n* [ME, fr. MF *bruniss-*, stem of *brunir*, literally, to make brown, fr. *brun*] (14c) Shiny or lustrous spots on a paint surface caused by rubbing or polishing

Burnish Resistance *n* Resistance of a coating to an increase in gloss or sheen due to polishing or rubbing.

Burn-Out Printing See ▶ Printing.

Burnt Lime See ▶ Calcium Oxide.

Burnt Sienna See ▶ Iron Oxide.

Burnt Umber *n* Naturally occurring ferric oxide containing manganese dioxide and varying amounts of silica and alumina. Characteristic color is a rich, dark brown. Ferric oxide range, 45–65%; manganese dioxide range, 3–20%. Produced by calcining raw umber. See ▶ Umber.

Burr A device that assists in loop formation on circular-knitting machines equipped with spring needles.

Bursting Strength *n* Of rigid plastic tubing, the internal liquid pressure required to cause rupture of a test specimen. Tubes with internal diameters between 3.2 and 152 mm (1/8–6 in.) may be tested and diameter and wall thickness must be reported. The term has much the same meaning for filament-wound pressure vessels. (2) The force required to rupture a fabric by distending it with a force applied at right angles to the plane of the fabric under specified condition. Bursting strength is a measure widely used for knit fabrics, nonwoven fabrics, and felts where the constructions do not lend themselves to tensile tests. The two basic types of bursting tests are the inflated diaphragm method and the ball-bust method. (Shah V (1998) Handbook of plastics testing technology. Wiley, New York) (3) Bursting strength of a material, such as plastic film, is the minimum force per unit area or pressure required to produce rupture. The pressure is applied with a ram or a diaphragm at a controlled rated to a specified area of the material held rigidly and initially flat but free to bulge under the increasing pressure. (Sepe MP

(1998) Dynamic mechanical analysis. Plastics Design Library, Norwich).

Bush Kauri Copal *n* Special grade of the Kauri copal, which has fossilized on trees at the point of exudation and above ground level.

Butacite Poly(vinyl butyral). Manufactured by DuPont, USA.

Butcher's Linen *n* A plain weave, stiff fabric with thick and thin yarns in both the warp and the filling. The fabric was originally made of linen but is now duplicated in 100% polyester or a variety of blends such as polyester/rayon or polyester/cotton.

Butadiene \ ˈbyü-tə-ˌdī-ˌēn, -ˌdī-ˈ\ *n* [ISV *butane* + *di-* + *-ene*] (1900) Buta-1,3-diene, 1,3-butadiene, erythrene, vinylethylene, bivinyl, divinyl. $CH_2=CHCH=CH_2$. A gas, insoluble in water but soluble in alcohol and ether, obtained from cracking of petroleum, from coal-tar benzene, or from acetylene. It is widely used in the formation of copolymers with styrene, acrylonitrile, vinyl chloride and other monomers, imparting flexibility to the products made from them. Its homopolymer is a synthetic rubber. As noted it is a synthetic chemical compound, used principally in the manufacture of synthetic rubber, nylon, and latex paints.

Butadiene

Butadiene-Acrylonitrile Copolymer (NBR) See ▶ Acrylonitrile-Butadiene Copolymer.

Butadiene Rubber (BR) See ▶ Polybutadiene.

Butadiene Styrene Latex A synthetic latex similar to synthetic rubber; used for latex paints. *Also known as Styrene Butadiene.*

Butadiene-Styrene Thermoplastics See ▶ Styrene-Butadiene Ther-Moplastic.

Butaldehyde \ ˈbət ˈal-də-ˌhīd\ See ▶ Butyraldehyde.

Butanal See ▶ Butyraldehyde.

1,4-Butanedicarboxylic Acid See ▶ Adipic Acid.

1,3- or 1,4-Butanediol See ▶ 1,3- or 1,4-Butylene Glycol.

1,2,4-Butanetriol *n* A nearly colorless liquid, an intermediate for alkyd resins, and a plasticizer for cellulosics.

1,2,4-Butanetriol

Butanoic Acid See ▶ Butyric Acid.

1-Butanol \ ˈbyü-tᵊn-ˌȯl\ See ▶ *n*-Butyl Alcohol.

Butene \ ˈbyü-ˌtēn\ *n* [ISV *butyl* + *-ene*] (1885) Any of the monounsaturated C_4 hydrocarbons listed below:

IUPAC NAME	Alternative Names
1-butene	α-butylene
cis-2-butene	*cis*-β-butylene
trans-2-butene	*trans*-β-butylene

The term butanes refers to the first three compounds above as a group. The butylenes are used as monomers for rubbery homopolymers and copolymers with styrene, acrylics other olefins, and vinyls. They are also used in adhesives for many plastics, and in making plasticizers. (Carley JF (ed) (1993) Whittington's dictionary of plastics. Technomic Publishing Co., Inc.)

2-Butene-1,4-diol *n* A nearly colorless, odorless liquid, an intermediate for alkyd resins, plasticizers, nylon, and a cross-linking agent for resins.

Buton *n* Cross-linkable plastic produced at high polymerization temperatures from butadiene and styrene. Manufactured by Esso, Great Britain.

2-Butoxyethanol See ▶ Ethylene Glycol Monobutyl Ether.

2-Butoxyethyl Pelargonate *n* A plasticizer for polystyrene, vinyl chloride polymers and copolymers, and cellulosics.

Butoxyethyl Stearate *n* Octadecanoic acid 2-butoxy ethylester. A high-boiling plasticizer for nitrocellulose, polystyrene, ethyl cellulose, and polyvinyl acetate.

Butt *n* To met without overlapping.

Butt Fusion *n* A method of joining pipe, sheet or other forms of a thermoplastic resin wherein the ends of the two pieces to be joined are heated into the lower end of the polymer's melting range, then rapidly pressed together and allowed to cool, forming a homogeneous bond. ASTM D 2657 describes a recommended practice for butt-joining polyolefin pipe by heat fusion.

Butt Joint *n* (1823) A joint made by fastening and/or bonding two surfaces that are perpendicular to the main surfaces of the parts being joined.

Button Lac Lac refined in the form of buttons, which are usually about 3 in. (7.5 cm) in diameter and ¼ in. (6 mm) in thickness.

Butt Veneer *n* Veneer having a strong curly figure, caused by roots coming into the trunk at all angles.

Butyl \ˈbyü-t³l\ *n* [ISV *butyric* + *-yl*] (1869) (1) The radical C_4H_9-, occurring only in combination. (2) Abbreviation used by British Standards Institution for ▶ Butyl Rubber.

Butyl Acetate *n* $CH_3COOC_4H_9$. A pleasantly aromatic solvent of moderate strength for ethyl cellulose, cellulose nitrate, vinyls, polymethyl methacrylate, polystyrene, coumarone-indene resins, and certain alkyds and phenolics. See ▶ N-Butyl Acetate.

n-Butyl Acetate *n* $CH_3COOC_4H_9$. Limpid, colorless liquid with fruity odor. Prepared by heating and distillation, after contact of n-butyl alcohol with acetic acid in the presence of a catalyst such as sulfuric acid. Solvent used in production of lacquers, natural gums and synthetic resins.

n-Butyl acetate

Butyl Acetoxystearate *n* $CH_3(CH_2)_5CH(CH_3COO)(CH_2)_{10}COOC_4H_9$. A plasticizer similar to ▶ Butyl Acetyl Ricinoleate, but with the double bond saturated. It is compatible with cellulosic and vinyl resins.

Butyl Acetyl Ricinoleate *n* $CH_3(CH_2)_5CH(CH_3COO)CH_2CH=CH(CH_2)_7-COOC_4H_9$. A yellow, oily liquid derived from castor oil, butyl alcohol and acetic anhydride, used as a plasticizer, compatible with cellulosics and vinyls.

Butyl Alcohol *n* (ca. 1871) C_4H_9OH Any of four flammable alcohols used in organic synthesis and as solvents. See ▶ N-Butyl Alcohol.

n-Butyl Alcohol *n* Colorless liquid, with vinous, irritating odor which causes coughing. Solvent used for resins, coatings, shellac, and butylated melamine resins. Syn: ▶ 1-Butanol and ▶ Propyl Carbinol.

n-Butyl Acrylate *n* $CH_2 = CHCOOC_4H_9$. Colorless liquid. A monomer used in the manufacture of synthetic resins and as a solvent for cellulose esters. Sp gr, 0.898; bp, 145°C.

Butylated Hydroxytoluene *n* (di-*tert*-*p*-cresol, BHT) A white, crystalline solid, the most widely used antioxidant for plastics such as ABS and LDPE. It has been approved by the FDA for use in foods and food-packaging materials.

Butylated Resin *n* A resin containing the butyl radical, C_4H_9-.

Butyl Benzenesulfonamide *n* (*N*-*n*-butyl benzenesulfonamide) $C_6H_5SO_2$-NHC_4H_9. A plasticizer for some synthetic resins and an intermediate for resin manufacturer.

Butyl Benzoate *n* (*n*-butyl benzoate) A plasticizer and solvent for cellulosics.

Butyl Benzyl Phthalate *n* (BBP) (C_4H_9OOC)C_6H_4($COOCH_2C_6H_5$). A clear, oily liquid used as a plasticizer for cellulosic and vinyl resins. It imparts good stain resistance, low volatility at calendaring and extruding temperatures, low oil extraction and good heat and light stability.

Butyl Benzyl Sebacate *n* $C_4H_9OOC(CH_2)_8COOCH_2C_6H_5$. An ester-type plasticizer with a light straw color. It combines the desirable properties of dibenzyl sebacate and dibutyl sebacate.

Butyl Borate See ▶ Tributyl Borate.

Butyl Carbitol *n* Diethylene glycol monobutyl ether, butyldiglycol, 2-[2-(1-butoxy)ethoxy] ethanol. Practically odorless oil, solvent, water miscibile liquid used as a coalescent in latex paints; bp, 230.4°C; mp, −68.1°C; sp gr, 0.9536.

Butyl Carbitol Acetate *n* $CH_3COO(CH_2CH_2O)_2C_4H_9$. Diethylene glycol Monobutyl ether acetate.

Butyl Cellosolve Ethylene glycol monobutyl ether, 2-butoxyethanol. Liquid soluble in 20 parts water; soluble in most organic solvents. It is used as a coalescent in latex paints; bp, 171–172°C; sp gr, 0.901.

Butyl Cellosolve Acetate *n* Ethylene glycol monobutyl ether acetate.

Butyl Citrate *n* Colorless or pale yellow, stable, odorless, nonvolatile liquid. Plasticizer, anti-foam agent, solvent for cellulose nitrate. *Also known as Tributyl Citrate.* See ▶ Citrates.

Butyl Cyclohexyl Phthalate *n* A plasticizer for PVC, other vinyls, cellulosics, and polystyrene.

Butyl Decyl Phthalate *n* A plasticizer for polystyrene, PVC, and vinyl chloride-acetate copolymers.

Butyl Diglycol Carbonate *n* [diethyleneglycol bis(n-butylcarbonate)] A colorless liquid of low volatility, used as a plasticizer with many resins.

Butylene See ▶ Butene.

1,3-Butylene Dimethacrylate A polymerizable monomer used in PVC and rubber systems to obtain rigid or semi rigid products from materials that are normally flexible. The monomer acts as a plasicizer at room temperature and crosslinks at processing temperatures.

1,3-Butylene Glycol *n* (1,3-butanediol) $CH_3CHOHCH_2CH_2OH$. A colorless liquid made by catalytic hydrogenation of aldol (3-hydroxy-n-butyradehyde). Its most important use is as an intermediate in the manufacture of polyester plasticizers.

1,4-Butylene Glycol *n* (1,4-butanediol, tetramethylene glycol) A stable, hygroscopic, colorless liquid used in the production of polyesters by reaction with dibasic acids, and in the production of polyurethanes by reaction with diisocyanates.

1,3-Butylene Glycol Adipate Polyester *n* (Santicizer® 334F) A polymeric plasticizer for PVC.

Butyl Epoxy Stearate *n* A plasticizer for PVC, imparting low-temperature flexibility.

Butyl Ethylhexyl Phthalate *n* (butyl octyl phthalate, BOP) A mixed ester of butanol and 2-ethylhexanol, widely used as a primary plasticizer for PVC compounds and plastisols, in which it performs like dioctyl phthalate in most respects. It is also compatible with vinyl chloride-acetate copolymers, cellulose nitrate, ethyl cellulose, polystyrene, chlorinated rubber, and at lower concentrations, with polymethyl methacrylate.

Butyl Formate *n* A solvent for several resins, including cellulose acetate.

Butyl Isodecyl Phthalate *n* (decyl butyl phthalate) (C_4H_9OOC)C_6H_4-($COOC_{10}H_{21}$). A plasticizer for PVC and polystyrene.

Butyl Isohexyl Phthalate *n* (C_4H_9OOC)C_6H_4(COO C_6H_{13}). A plasticizer for cellulosics, acrylic resins, polystyrene, PVC, and other vinyl resins.

Butyl Lactate *n* $CH_3CHOHCOOC_4H_9$. Slow evaporating liquid which is miscible with most of the organic solvents. It is a solvent for nitrocellulose, ethyl cellulose, oils, dyes, natural gums, and synthetic resins.

Butyl Octadecanoate See ▶ Butyl Stearate.
Butyl Octyl Phthalate See ▶ Butyl Ethlyhexyl Phthalate.
Butyl Oleate *n* $CH_3(CH_2)_7CH=CH(CH_2)_7COOC_4H_9$. A solvent, plasticizer and lubricant, used mainly with neoprene and other synthetic rubbers, chlorinated rubber, and ethyl cellulose; also a mold lubricant.

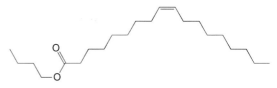

Butyl Phthalate See ▶ Phthalate Esters.
Butyl Phthalyl Butyl Glycollate *n* A plasticizer with good light stability, used mainly with PVC and polystyrene, but compatible with most other thermoplastics. It has been approved by the FDA for contact with foods.

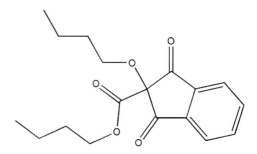

Butyl Ricinoleate *n* Yellow to colorless oleaginous liquid derived from castor oil fatty acid (ricinoleic acid) and butyl alcohol. Used as a plasticizer or lubricant.

Butyl Rubber *n* (1940) A synthetic elastomer produced by copolymerizing isobutylene with a small amount (ca 2%) of isoprene or butadiene. It has good resistance to hear, oxygen, and ozone, and low gas permeability. Thus, it is widely used in inner tubes and to line tubeless tires. Butyl rubber is a vinyl polymer, and is very similar to polyethylene and polypropylene in structure, except that every other carbon is substituted with two methyl groups. It is made from the monomer isobutylene, by cationic vinyl polymerization. It can also go by the name of *Polyisobutylene*. (1) Generic name for vulcanizable elastic copolymers of isobutylene and small amounts of diolefins. (2) Mixture of isobutylene, 98% and butadiene or isoprene, 2%. (3) Poly (isobutylene) with 2% isoprene. Manufactured by Bayer, Germany.

Butyl Stearate *n* (butyl octadecaonoate) A mold lubricant and plasticizer, compatible with natural and synthetic rubbers, chlorinated rubber, and ethyl cellulose. It can be used in vinyls in very low concentrations as a nontoxic, secondary plasticizer and lubricant. In the production of polystyrene, butyl stearate is added to the emulsion polymerization to impart good flow properties to the resin.

Butyl (Tetra) Titanate *n* Ti(OC$_4$H$_9$)$_4$. Used for adhesion promoter, catalyst, and in heat-resistant paints.

Butyraldehyde \byü-tə-ral-də-hīd\ *n* [ISV] (ca. 1885) (butaldehyde, n-butanal, n-butyl aldehyde, butyric aldehyde) An aldehyde sometimes used in place of formaldehyde in the production of resins. Butyraldehyde reacts with polyvinyl alcohol to form polyvinyl butyral. It is used in the manufacture of rubber and synthetic resins. Sp gr, 0.817; bp, 75.7°C. Syn: ▶ Butaldehyde, ▶ Butyl Aldehyde, and ▶ Butyric Aldehyde.

Butyrate \byü-tə-rāt\ *n* (1873) (1) A salt or ester of ▶ Butyric Acid. (2) The industry nickname for ▶ Cellulose Acetate Butyrate (CAB).

Butyrates *n* Name applied to esters of butyric acid, such as cellulose butyrate, ethyl butyrate, etc.

Butyric Acid *n* (n-butyric acid, butanoic acid, ethylacetic acid, propylformic acid) CH$_3$(CH$_2$)$_2$COOH. A water soluble liquid with a strong butter-rancid odor, used in the production of ▶ Cellulose Acetate Butyrate. Derivatives of butyric acid are used in the production of plasticizers for cellulosic plastics.

Butyric Alcohol See ▶ *n*-Butyl Alcohol.

γ-Butyrolactone *n* (butyrolactone) A hygroscopic, colorless liquid obtained by the dehydrogenation of 1,4-butanediol with the structure: It is a solvent for cellulosics, epoxy resins, and vinyl copolymers.

C

C *n* \sē\ (1) Abbreviation for SI prefix, ► Centi-. (2) Abbreviation for ► Cubic.

C *n* (1) Chemical symbol for the element carbon. (2) Abbreviation for ► Celsius or ► Centigrade. (3) Symbol for electrical capacitance.

ca *adj* Circa; about, approximate.

Ca *n* Chemical symbol for the element calcium.

CA See ► Cellulose Acetate.

C or Δc *n* Abbreviations for ► Chromaticity or ► Chromaticity Difference. See ► Receptivity and ► Color Difference Equations.

C 23 *n* Ethylene/propylene copolymer. Manufactured by Montecatini, Italy.

CAB *n* Cellulose acetobutyrate.

CAB See ► Cellulose Acetate Butyrate.

Cabinet *n* A basic part of the manufactured-fiber spinning machine where, in dry spinning, the filaments become solidified by solvent evaporation and, in melt spinning, the filaments are solidified by cooling.

Cabinet Finish *n* A varnished or polished hardwood interior finish as distinguished from a painted softwood finish.

Cabled Yarn *n* A yarn formed by twisting together two or more plied yarns.

Cable Stitch *n* A knit effect produced by crossing a group of stitches over a neighboring stitch group.

Cable Twist *n* A construction of thread, yarn, cord, or rope in which each successive twist is in the direction opposite the preceding twists; i.e., and S/Z/S or Z/S/Z construction.

Cable Varnishes and Lacquers *n* Employed for impregnating the cloth or similar covering of cables. Normally, oil varnishes are used.

Cacahuananche Oil *n* This oil is obtained from the nuts of the tree, *Licania arborea*. So far as the usual laboratory tests are concerned, this oil and Brazilian oiticica oil are much alike. The raw oil becomes lard-like on aging but may be permanently liquefied by heat. The raw and lightly heat-treated oil wrinkles as it dries, similarly to oiticica and tung oils.

CAD/CAM *n* Acronym for Computer-Aided Design and ► Computer-Aided Manufacturing.

Cadmium \ˈkad-mē-əm\ *n* [NL, fr. L *cadmia* zinc oxide, fr. Gk *kadmeia*, fr. feminine of *kadmeios* Theban, from *Kadmos*] (1822) A bluish white malleable ductile toxic bivalent metallic element used in protective platings and in bearing metals.

Cadmium Ethylhexanoate *n* A metallic soap used as a stabilizer for vinyls, especially to avoid plate-out in calendaring compounds.

Cadmium Green *n* Pigment Green 18. Composed of 93–94% hydrated chromium, oxide, the balance being cadmium yellow.

Cadmium Lithopone *n* Pigment Red 108 (77202). Range of yellow, orange, and red colors consisting of either cadmium sulfide or cadmium red reduced on barium sulfate. They are usually produced by coprecipitation followed by calcinations. See ► Cadmium Red.

Cadmium/Mercury Lithopones *n* Orange, red, and maroon pigments consisting of calcined co-precipitations of cadmium sulfide and mercury sulfide, with barium sulfate.

Cadmium/Mercury Sulfides *n* CdS·HgS. Pigment Orange 23 (77201). A series of cadmium/mercury sulfides of mixed crystal composition. Colors vary according to CdS/HgS ratio from deep orange to maroon. Known for high heat stability, solvent insolubility, good chemical resistance, and excellent lightfastness. CdS/HgS ratio, 89.1/10.9. Density 4.8–5.2 g/cm^3 (40–43 lb/gal); O.A., 33.2; particle size, 0.1–1.0 μm. Syn: ► Mercadium®, ► Mercury Cadmium Cadmium Red/Orange and ► Cadium Orange.

Cadmium Orange *n* Mixture or coprecipitate of cadmium sulfide and cadmium selenide used as a pigment. See ► Cadmium Sulfide.

Cadmium Plating *n* Electrodeposition of cadmium on iron wire, steels articles, etc., to make them relatively rust-proof. This surface treatment makes the substrate more suitable for subsequent painting.

Cadmium Pigment *n* Any inorganic pigment based on cadmium sulfide or cadmium sulfoselenide, used widely in PVC, polystyrene, and polyolefins. Included are cadmium-maroon, -orange, -red, and -yellow. The cadmium pigments have good resistance to heat (up to 500°C) and to alkalis, and do not bleed. Light stability is good in solid colors, but may be poor when used for tints with white pigments. Resistance to acids is poor.

Cadmium Red *n* Pigment Red 108 (77202). Pigments of Calcined coprecipitations of compounds of cadmium, sulfur, and selenium; a brilliant red pigment, opaque, with good staining power, fast to light, unaffected by exposure to sulfur fumes; considerable resistance to heat. Density, 4.9 g/cm^3 (41 lb/gal); O.A., 20 lb/100 lb.

Cadmium Ricinoleate *n* [CH$_3$(CH$_2$)$_5$CHOHCH$_2$CH=CH(CH$_2$)$_7$COO]$_2$–Cd. A white powder derived from castor oil, used as a heat stabilizer for vinyl chloride polymers and copolymers.

Cadmium Selenide See ▶ Cadmium Red and ▶ Cadmium Orange.

$$Cd^{++} \quad Se^{--}$$

Cadmium Stearate *n* (C$_{17}$H$_{35}$COO)$_2$Cd. A heat- and light-stabilizer, used when good clarity in transparent compositions is desired.

Cadmium Sulfide *n* (ca. 1893) (Cd/Zn)S or CdS Pigment Yellow 37 (77199). Class of yellow, orange, and red pigments that include pure cadmium sulfides and blends of cadmium sulfides with ZnS and CdSe and used with BaSO$_4$ to form lithopone type pigments. They show excellent heat and good alkali resistance especially for baking finishes. Density, 4–64.7 g/cm^3 (38–39 lb/gal); O.A., 18–22 lb/100 lbs. Syn: ▶ Cadmium Yellow and ▶ Cadmium Orange.

$$Cd^{++} \quad S^{--}$$

Cadmium Yellow See ▶ Cadmium Sulfide.

Cage Effect *n* Free radicals that exist very close together in a solvent or monomer molecule called a "cage."

Caisson \ˈkā-ˌsän, -sən, *British also* kə-ˈsůn\ *n* [Fr, fr. MF, fr. OP, fr. *caissa* chest, fr. L *capsa*] (ca. 1702) Recessed compartment in a vault or ceiling; recessed panel in a ceiling or soffit.

Cajeputene See ▶ Dipentine.

Cake, Press *n* The thick cake of wet pigment that is withdrawn from the filter press.

Caking *n* (1) Settling of pigment particles of a paint into a compact mass which is not easily redispersed by stirring. (2) In printing inks, caking is the collecting of pigment upon plates, rollers or blankets caused primarily by the inability of the vehicle to hold the pigment in suspension.

Calcareous \kal-ˈkar-ē-əs, -ˈker-\ *adj* [L *calcarius* of lime, from *calc-*, *calx*] (1677) Any material containing calcium or calcium compounds.

Calcicoater \ˈkal-sik ˈkotər\ *n* (1) Vegetable oil product that has been reacted with lime to form a heavy gel-like soap, which has almost infinite tolerance for mineral spirits; used in flat wall paints, primers, and sealers. (2) Pigmented flat paints using such heavy gel-like soap vehicles.

Calcimine \ˈkal-sə-ˌmīn\ *n* [alter. of *kalsomine*] (ca. 1859) Also spelled "kalsomine." Essentially, chalk and glue ready to mix with water. Used as a decoration for interior surfaces. It will not withstand washing. In Britain, it is referred to as powdered distemper.

Calcination \ˌkal-sə-ˈnā-shən\ *n* (1) Process of heating or roasting a material to a high temperature, but below its fusing point, to cause it to lose moisture or other volatile material or to be oxidized or reduced. Originally a heat process for the production of lime (CaO) from limestone. (2) The process of subjecting a sorptive mineral to prolonged heating at fairly high temperature, resulting in the removal of water, and an increase in the hardness, physical stability and absorbent properties of the material.

Calcined Clay *n* China clay (kaolin) that has been heated until the combined water is removed and the plastic character of the clay is destroyed. This produces an air-solid interface within the particle which increases hiding in he resulting coating. See ▶ Clay.

Calcite \ˈkal-ˌsīt\ *n* (1849) Naturally occurring form of calcium carbonate. It is an essential material of limestone, marble, and chalk.

Calcium Acetate *n* (vinegar salts, gray acetate, lime acetate, brown acetate) $(CH_3COO)_2Ca \cdot H_2O$. A stabilizer.

Calcium Carbide *n* (ca. 1888) A usually dark gray crystalline compound CaC_2 used for the generation of acetylene and for making calcium cyanamide.

Calcium Carbonate *n* (1873) (aragonite, calcite, chalk, limestone, lithographic stone, marble marl, travertine, whiting) $CaCO_3$. Grades of calcium carbonate suitable as fillers for plastics are obtained from naturally occurring deposits as well as by chemical precipitation. The natural types are prepared by dry grinding, yielding particles usually over 20 μm, used in stiff products such as floor tiles; or by wet grinding, yielding particles under 16 μm, used in flexible products. The chemically precipitated types range from 0.05 to 11 μm in size, and are most often used in plastisols and highly flexible products. Both the wet-ground and precipitated types are available with coating such as resins, fatty acids, and calcium stearate. These coated grades have low oil absorption, of particular value in compounding plastisols. The calcium stearate coatings provide improved electrical properties, heat stability and lubricity during processing, which is beneficial in extrusion compounds.

Calcium Carbonate Fillers *n* Fine powder of calcium carbonate (white) to fill spaces in a polymer or coating.

Calcium Carbonate, Natural *n* $CaCO_3CaMg(CO_3)_2$. Pigment White 18 (77220). White extender pigment derived from natural chalk, limestone, or dolomite, consisting of calcium carbonate with up to about 44% magnesium carbonate. Density, 2.71 g/cm³ (22.6 lb/gal); O.A., 6–15; particle size, 1.5–12 μm. Syn: ▶ Calcite, ▶ Limestone, ▶ Whiting, ▶ Marble Flour, ▶ Chalk, ▶ Ground Oyster Shells, ▶ Iceland Spar and ▶ Spanish White.

Calcium Carbonate, Synthetic *n* $CaCO_3$. Pigment White 18 (77220). Calcium carbonate manufactured by a precipitation process in order to obtain a finer or more uniform particle size range. Four commercial processes are known. Density, 2.65 g/cm³ (22.07 lb/gal); O.A., 28–58; particle size, 0.6–30 μm. Syn: ▶ Precipitated Calcium Carbonate.

Calcium Chloride *n* (1885) $CaCl_2$. Salt used in the manufacture of some lakes and toners from acid dyestuffs, fireproof paints, sizing compounds, wood preservatives, snow melter, and anti-freeze.

Calcium Glycerophosphate *n* (calcium glycerinophosphate) $CaC_3H_7O_2–PO_4$. A white, crystalline powder, odorless and nearly tasteless, used as a stabilizer for plastics.

Calcium Hydroxide *n* (ca. 1889) $Ca(OH)_2$. A white crystalline strong alkali that is used in mortar and plaster. *Also known as Slaked Lime.*

Calcium Hypochlorite *n* (ca. 1889) CaCl$_2$O$_2$ A white power used as a bleaching agent and disinfectant.

Calcium Linoleate *n* White amorphous powder soluble in alcohol and ether; insoluble in water. Used for waterproofing compounds, emulsifying agents, and as a stabilizer for oleoresinous paints.

Calcium Lithol Red See ▶ Lithol Red.
Calcium Metasilicate See ▶ Calcium Silicate, Natural.
Calcium Naphthenate *n* Calcium salt of naphthenic acids. Used as an auxiliary drier, dispersing and stabilizing aid.
Calcium Octoate *n* Calcium salt of 2-ethyl hexoic acid. Used as auxiliary drier, dispersing, and stabilizing aid.

Calcium Oxide *n* (ca. 1885) (calx, lime, quicklime, burnt lime) CaO. A white powder with affinity for water, with which it combines to form calcium hydroxide. It has been used to remove traces of water in vinyl plastisols. Also known as Quicklime.

Ca^{++} O^{--}
Calcium oxide

Calcium Phosphate *n* (1869) (calcium orthophosphate, tricalcium phosphate, tricalcic phosphate, tertiary calcium phosphate) Ca$_3$(PO$_4$)$_2$. A stabilizer.

Calcium Phosphate, Dibasic *n* (calcium biphosphate, acid calcium phosphate, primary calcium phosphate) CaH$_4$(PO$_4$)$_2$·H$_2$O. A stabilizer.
Calcium Phosphate, Monobasic (Dicalcium orthophosphate, bicalcic phosphate, secondary calcium phosphate) CaHPO$_4$ or CaHPO$_4$·H$_2$O. A stabilizer.
Calcium Resinate *n* Calcium salt of rosin used as auxiliary drier and dispersing and stabilizing aid. Commonly known as ▶ Limed Rosin, which is really rosin, the acidity of which has been substantially neutralized.
Calcium Ricinoleate *n* [CH$_3$(CH$_2$)$_5$CHOHCH$_2$CH=CH (CH$_2$)$_7$COO]$_2$. A white powder derived from castor oil, used as a nontoxic stabilizer for PVC.

Calcium Silicate *n* (1888) (wollastonite) CaSiO$_3$. A naturally occurring mineral found in metamorphic rocks, used as a reinforcing filler in low-density polyethylene, polyester and other thermosetting molding compounds. It imparts smooth molded surfaces and low water absorption.

Calcium Silicate, Natural *n* CaSiO$_3$. It has white color and acicular particles. It is characterized by high flatting action, combined with low oil absorption, and is used as an extender in paints, ceramics and plastics. Density, 2.9 g/cm^3 (24.2 lb/gal); O.A., 25–30; particle size, 7 μm (fine particle grade); hardness (mohs), 4.5. Syn: ▶ Wollastonite and ▶ Calcium Metasilicate.

Calcium Silicate, Synthetic *n* CaSiO$_3$·nH$_2$O. Extender pigment with some dry hiding opacity. Density, 2.26 g/cm^3 (18.8 lb/gal); O.A., 280; particle size, 10–12 μm.

Calcium Stearate *n* (C$_{17}$H$_{35}$COO)$_2$. (1) A nontoxic stabilizer and lubricant. It is not often used alone because of its early color development, but is used in combination with zinc and magnesium derivatives and epoxides in manufacturing other nontoxic stabilizers. (2) Calcium salt of stearic acid. A metallic soap used in paints as wetting aid for pigments, flatting agents, and additive for improving sanding sealers.

Calcium Sulfate *n* (ca. 1885) (anhydrite) CaSO$_4$·2H$_2$O. (1) A filler and white pigment. Notwithstanding its slight solubility in water, it is used to some extent in distempers and water paints, It forms the base of certain lakes and other pigments. The hydrated forms are known as *gypsum*, *terra alba*, and Plaster of Paris.

Calcium Sulfate, Anhydrous *n* CaSO$_4$. Manufactured by calcining natural gypsum (CaSO$_4$·2H$_2$O) or as byproducts from other precipitation processes. Notwithstanding its slight water solubility, it is used in distempers and water paints. It also forms the base of certain lakes and other pigments. Plaster of Paris is partially calcined gypsum. Density, 2.96 g/cm^3 (24.7 lb/gal); O.A., 20; particle size, 1.8–2.2 μm. Syn: ▶ Anhydrite, ▶ Dead Burned Calcium Sulfate, and ▶ Dehydrated Gypsum.

Calcium Sulfate Hemihydrate *n* CaSO$_4$·½H$_2$O. See ▶ Gypsum (Calcined) and ▶ Plaster of Paris.

Calcium Tallate *n* Calcium salt of tall oil acids used as an auxiliary drier and dispersing and stabilizing aid.

Calcium Thiocyanate *n* (calcium sulfocyanate, calcium rhodanate) Ca–(SCN)$_2$·3H$_2$O. In water solution, a solubilizer for acrylic and cellulosic resins.

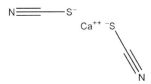

Calcium-Zinc Stabilizer *n* Any of a family of stabilizers based on compounds and mixtures of compounds of calcium and zinc. Their effectiveness is limited, but they are among the few that have been approved by the FDA for materials to be contacted by foods.

Calendar *n* (1) To produce or process sheets of material by pressing between a series of revolving heated rolls. (2) The machine performing the process of ▶ Calendering.

Calender Coating *n* A roller coating on a film or other material. The process of applying plastics to substrates such as paper or fabric by passing both the substrate and a plastic film through calender rolls. See ▶ Coating.

Calendered Cloth *n* Basis of varnished cloths for electrical insulation. The process of calendering cloths involves the passing of moist cloth between heavy rollers, resulting in a smooth or glazed appearance.

Calendered Film & Sheet *n* Rolled film and sheet.

Calendered Papers Wallpapers with hard coatings.

Calendering *n* The process of forming thermoplastics sheeting and films by passing the material through a series of rigid, heated rolls and rolling to press out voids. Four rolls are typical. The gap between the last pair of heated rolls determines the thickness of the sheet. Subsequent chilled rolls cool the sheet. The plastic compound is usually premixed and plasticated on separate equipment, then fed continuously into the nip of the first pair of calender rolls.

Calendering Rolls *n* (1) The main cylinders on a calender. (2) Smooth or fluted rolls used on carious fiber-processing machines such as pickers and cards to compress the lap or sliver as it passes between them.

Calender with Rubber *n* A machine for impregnating fabric.

Calibrate \ˈka-lə-ˌbrāt\ *vt* (ca. 1864) (1) To determine the exact relationship between the indicated or recorded readings of a measuring instrument or method and the true values of the quantities measured, over the instrument's range. (2) (mainly European usage) To bring the dimensions of an extruded tube or pipe within their specified ranges.

Calibration *n* (1) The process of ascertaining the errors of a measuring technique or instrument by comparing its readings with the corresponding values of known standards or the readings of a more accurate technique or instrument. Calibration standards for the whole gamut of measurements are available from the (U.S.) National Institute of Standards and Technology (NIST, formerly the National Bureau of Standards), Gaithersbury, MD 20899. "They can also furnish standard materials, some of them polymeric. See ▶ NIST special publication 260, *Standard Reference Materials Catalog*. NIST also calibrates instruments on request; see special publication 250, *NIST Calibration Services Users' Guide*. Both publications are available from the U.S. government Printing Office in Washington, DC. (2) See ▶ Calibrate (2).

Calico \ka-li-ˌkō\ *n* [*Calicut*, India] (1578) A plain, closely woven, inexpensive cloth, usually cotton or a cotton/manufactured fiber blend, characteristically having figured patterns on a white or contrasting background. Calico is typically used for aprons, dresses, and quilts.

Caliper \ˈka-lə-pər\ *n* [alteration of *caliber*] (1588) (1) The thickness of film or sheet, typically stated in thousandths of an inch (mils). (2) Any of several types of precise instruments used to measure thickness, as *micrometer caliper*.

Calk \ˈkók, ˈkó-kər\ *n, vt* [prob. alter. of *calkin*, fr. ME *kakun*, fr. MD or ONF *calcain* heel, fr. L *calcaneum*, fr. *calc-, calx* heel] (1587) See ▶ Caulk.

Calking See ▶ Caulking Compound.

Calorie \ˈka-lə-rē, ˈkal-rē\ *n* [F *calorie*, fr. L *calor* heat, fr. *calēre*] (1866) (small calorie) A deprecated, small unit of heat energy: the amount of heat required to raise the temperature of one gram of water, at or near 4°C, 1°C. The ASTM "Standard for Metric Practice" (E 380) lists five slightly different calories, all nearly equal to 4.185 J. In future, most of these will fade away, leaving only the "calorie (International Table)," equal to 4.186800 J. The term calorie is also used loosely, especially in the nutrition field, to mean 1,000 calories, the kilocalorie, or "large calorie."

Calorimeter *n* A instrument for measuring the heat liberated (or absorbed) during chemical reactions or physical changes of state.

Calorimetry *n* The process of measuring quantities of absorbed or evolved heat, often used to determine specific heat.

Calutron \ˈkal-yə-ˌträn\ *n* [*Ca*lifornia *U*niversity cyclo*tron*] (1954) An apparatus operating on the principle of the mass spectrograph and used for separating U^{235} from U^{238}.

Calvary Twill *n* A pronounced, raised cord on a 63-degree ▶ Twill weave characterizes this rugged cloth usually made from wool or wool blend yarns.

CAM *n* A rotating or sliding piece or projection used to impart timed or periodic motion to other parts of a machine. It is used chiefly as a controlling or timing element in machines rather than as part of a power transmission mechanism. Cams are particularly important in both knitting and weaving machinery. Acronym for ▶ Computer-Aided Manufacturing.

Camber \ˈkam-bər\ *v* [F *cambrer*, fr. MF *camber*, fr. L *camur*] (1627) Strip under tension, as on a coating line that has a tendency to deviate from the horizontal or to slope to one side.

Cambogia See ▶ Gamboge.

Cambric \ˈkām-brik\ *n* [obs. Flemish *Kameryk*] (1530) A soft, white, closely woven, cotton or cotton blend fabric that has been calendered on the right side to give it a slight gloss. Cambric is used extensively for handkerchiefs.

Camel Hair *n* A term inaccurately applied to a fine haired brush, the hair of which is obtained from squirrels native to Russia and Siberia.

Camphol See ▶ Oxanilide.

Camphor \ˈkam(p)-fər\ *n* [ME *caumfre*, fr. AF, fr. ML *camphora*, fr. Arabic *kāfūr*, fr. Malay *kapur*] (*d*-2-camphanone, 2-keto-1,7,7-trimethylnorcamphane) $C_{10}H_{16}O$. A bridged-ring, naturally occurring ketone with a mp of 175°C and a bp of 204°C with the structural formula:

It is a colorless, aromatic, crystalline material originally derived from camphor oil but now mostly synthesized from pinene. It is used as a plasticizer for celluloid, cellulose nitrate, and lacquers. In 1870, the Hyatt brothers were awarded a U.S. patent for a horn-like compound of camphor and cellulose nitrate, which they called celluloid. The event marked the start of the U.S. plastics industry.

Camphoric Acid n $C_8H_{14}(COOH)_2$. A dibasic-acid plasticizer for cellulose nitrate, derived from CAMPHOR by oxidizing the =C=O and adjacent =CH$_2$ groups and opening the right-side ring.

Can n (1) A cylindrical container, about 3 ft high and 10–12 in. in diameter, that is used to collect sliver delivered by a card, drawing frame, etc. (2) See ▶ Drying Cylinders.

CAN n Abbreviation for Cellulose Acetate Nitrate.

Canada Balsam n Oleoresin which exudes naturally from *Pinus balsamea*, the Canadian balsam pine. It is essentially a resin dissolved in an essential oil. When freshly exuded it is a viscous liquid, but it hardens on exposure. Its chief use is for cementing lenses and other glass objects, because its refractive index, 1.53, is near that for glass. It is also used in the manufacture of fine lacquers.

Canada Turpentine See ▶ Canada Balsam.

Candela \kan-ˈdē-lə, -ˈde-, -ˈdā-; ˈkan-də-lə\ n [L, candle] (1949) (cd) One of the six basic units of the SI system, the unit of luminous intensity, defined as the luminous intensity normal to the surface of 1/60 of a square centimeter of a black body at the temperature of freezing platinum (1,772°C) under a pressure of 101.325 kPa. In older literature, the same abbreviation has been used for both *candlepower* and *candle*, which have different values.

Candelilla Wax n Natural wax obtained from the stems of a plant which grows in Mexico and Texas. It varies in color from pale yellow to a yellow or greenish-brown. It is one of the harder waxes, although not so hard as carnauba wax. Used in varnishes, electric insulation compositions, waterproofing and insect proofing, and paint removers.

Candle n (new unit) $\frac{1}{60}$ of the intensity of 1 cm^2 of a blackbody radiator at the temperature of solidification of platinum (2,042 K).

Candle n (unit of luminous intensity) Candle power is a measure of intensity of a source of light as compared with a standard candle. See ▶ Foot-Candle.

Candlenut Oil n Oil derived from the nuts of the candlenut tree (*Aleurites moluccana*) of tropical Asia and Polynesia.

Candlewick Fabric n An unbleached muslin base fabric used to produce a chenille-like fabric by applying candlewick (heavy-plied yarn) loops and cutting the loops to give a fuzzy effect.

Cangle Filter n A small filter interposed between the spinning pump and spinning jet to effect final filtration of the spinning solution prior to extrusion.

Cangle Water Temperature n The temperature of the water surrounding the candle filter or within the heating jacket during fiber extrusion.

Cannon-Fenske Viscometer n Capillary viscometer (calibrated) used for measuring relative flow time for liquids, useful for determining relative, specific and inherent viscosities of polymer solutions and extrapolation of intrinsic viscosities. It is convenient to clean and calibrate, but lacks correction due to different volume levels leading to weight due to gravity in the viscometer.

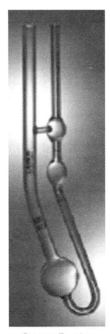

Cannon-Fenske Viscometer

Canopy Ceiling n A decoration composed of a ceiling paper, or a sidewall paper such as a strip, in which the strips are cut in triangles and hung so that the apex will terminate in the center of the ceiling to produce a striking domed effect.

Can Stability See ▶ Shelf Life and ▶ Storage Stability.

Cantilever-Beam Stiffness n A method of determining stiffness of plastics by measuring the force and angle of bend of a cantilever beam made of the specimen material. The ASTM test is D 747. See also ▶ Flexural Modulus.

Canton Flannel *n* A heavy cotton or cotton blend material with a twilled face and a napped back. The fabric's strength, warmth, and absorbance make it ideal for interlinings and sleeping garments.

Canvas \\kan-vəs\ *n, v, vt* [ME *canevas*, fr. ONF. fr. (assumed VL *cannabaceus*, fr. L *cannabis*] (13c) A coarse cloth made from cotton, hemp, or flax. It may be used for artist's canvas or a picture painted on canvas. See ▶ Duck.

Canvas Board *n* A paper board with primed canvas fastened to one face.

Caoutchouc \ˈkaú-ˌchúk, -ˌchúk, -ˌchǘ\ *n* [F, fr. obs. Sp *cauchuc* (now *caucho*), fr. Quechua *kawchu*] (1775) An early name for pure natural (raw) rubbers, still in use in the French literature.

CAP See ▶ Cellulose Acetate Propionate.

Capacitance \kə-ˈpa-sə-tən(t)s\ *n* [capacity] (1893) (C, electric capacity) The property of a system of two conducting surfaces (plates or foils, typically) separated by a nonconductor (*dielectric*) that permits storage of electric charge in proportion to the voltage difference between the conductors. The SI unit is the farad (F). A capacitor storing one coulomb of charge at a potential difference of one volt has a capacitance of one farad. Plastics are much used as capacitor dielectrics. Capacitance is measured by the charge which must be communicated to a body to raise its potential one unit. Electrostatic unit capacitance is that which requires one electrostatic unit of charge to raise the potential one electrostatic unit. The farad = $0 \times 1,011$ electrostatic units. A capacitance of one farad requires one coulomb of electricity to raise its potential one volt. Dimensions, [ε *l*]; [$\mu^{-1} l^{-1} t^2$]. A conductor charged with a quantity Q to a potential V has a capacitance,

$$C = \frac{Q}{V}$$

Capacitance of a spherical conductor of radius *r*,

$$C = Kr$$

Capacitance of two concentric spheres of radii *r* and *r'*

$$C = K\frac{rr'}{r - r'}$$

Capacitance of a parallel plate condenser, the area of whose plates is *A* and the distance between them *d*,

$$C = \frac{KA}{4\pi d}$$

Capacitances will be given in electrostatic units if the dimensions of condensers are substituted in centimeters. *K* is the dielectric constant of the medium.

Capillarity \ˌka-pə-ˈlar-ə-tē\ *n* (1830) Force of attraction between like and dissimilar substances. It is exhibited, for example, by the rise of a liquid up a capillary tube, or by the wetting of solids by liquids.

Capillary Action *n* A phenomenon associated with surface tension and contact angle. Examples are the rise of liquids in capillary tubes and the action of blotting paper and wicks.

Capillary Constant *n* Also known as Specific Cohesion,

$$a^2 = \frac{2T}{(d_1 - d_2)q} = hr$$

where *T* is surface tension, d_1 and d_2, the densities of the two fluids, *g* the acceleration due to gravity, *h* the height of the rise in a capillary tube of radius r. See ▶ Surface Tension.

Capillary Viscometers sed for concentrated solutions or polymer melts and described just above, the other, described here, used for measuring dilute-solution viscosities. The most widely used of the latter types employ a glass capillary tube and means for timing the flow of a measured volume of the solution (e.g., polymer in solvent) through the tube under the force of gravity. This time is then compared with the time taken for the same volume of pure solvent, or of another liquid of known viscosity, to flow through the same capillary. Relevant tests are from ASTM (www.astm.org). See also ▶ Dilute-Solution Viscosity and ▶ Viscometer.

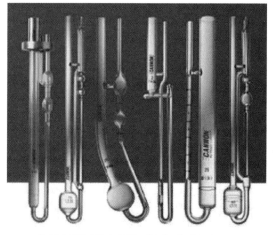

Capillary Viscometers, Cannon instrument company

Capric Acid *n* \ˈka-prik-\ *n* (decanoic acid, decylic acid, decoic acid) $CH_3(CH_2)_8COOH$. A plasticizer and an intermediate for resins.

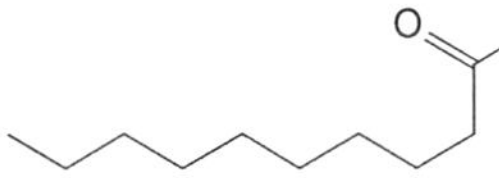

Caprolactam \ˌka-prō-ˈlak-ˌtam\ *n* [*capro*ic acid + *lact*one + *am*ide] (1944) (ε-caprolactam, hexanoic acid-ε-amino lactam) (1) A cyclic amid having the structure shown below. When the ring is opened, caprolactam is polymerizable to a nylon resin known as NYLON 6 or *polycaprolactam*. It is also used as a cross-linking agent for polyurethanes, and as a plasticizer. In the late 1960s it was found that caprolactam could be rotationally cast by heating the solid monomer to its melting point (ca. 71°C) and introducing the molten monomer into a mold along with a catalyst, then heating and rotating the mold in the usual manner. The liquid gradually thickens and gels against the mold in the manner of a plastisol, and conversion to nylon 6 is accomplished in a few minutes. (2) A white, hygroscopic, crystalline solid or leaflets.

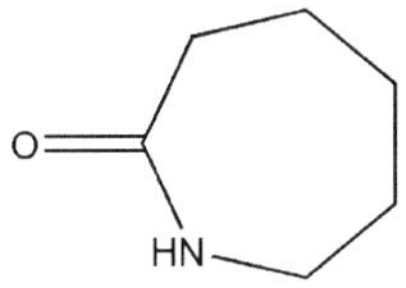

Cap Spinning *n* A system of spinning employing a stationary, highly polished metal cap just large enough to fit over the take-up bobbin, which revolves at a high rate of speed. The cap controls the build and imparts sufficient tension to the yarn for winding. The yarn is twisted and wound onto packages simultaneously.

Caprylic Acid \kə-ˈpri-lik\ *n* [ISV *capryl*] (1845) (octanoic acid, octoic acid, octylic acid, caprilic acid) $CH_3(CH_2)_6COOH$. A plasticizer and organic intermediate.

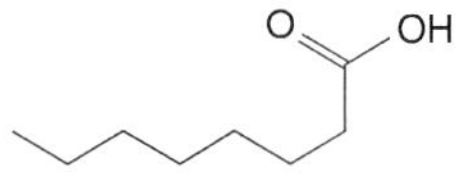

Caproic Acid \kə-ˈprō-ik-\ *n* $C_6H_{12}O_2$. Hexanoic acid. Natural unsaturated fatty acid found in oils or made synthetically and used in pharmaceuticals and flavors.

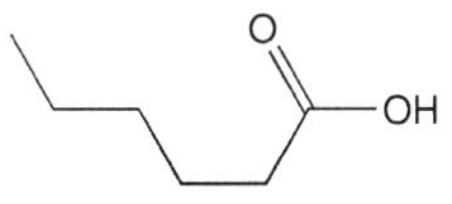

Capstan \ˈkap-stən, -ˌstan\ *n* [ME, prob. fr. MF *cabestant*] (14c) In the plastics industry, a drum or pulley that controls the speed of a filament, wire, or web between production stages.

Captive Production *n* Production of materials or components by a manufacturer for its own use or for later incorporation in its products. Compare ▶ Custom Molder.

CAR Carbon fiber.

Carbamide \ˈkär-bə-ˌmīd, kär-ˈba-məd\ *n* [ISV *carb-* + *amide*] (1865). See ▶ Urea.

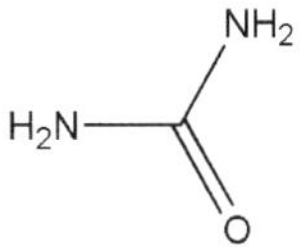

Carbamide Phosphoric Acid *n* (urea phosphoric acid) $CO(NH_2)_2 \cdot H_3PO_4$. A catalyst for acid-setting resins.

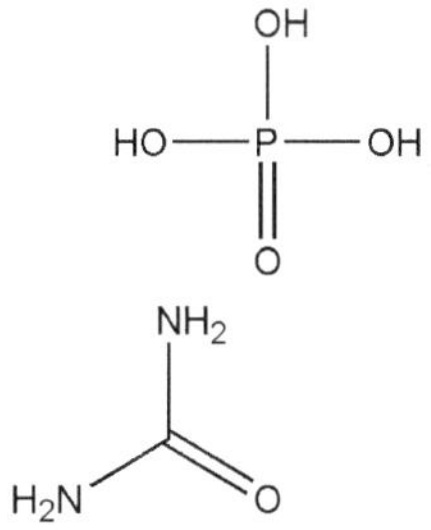

Carbamyl Urea See ▶ Biuret.

Carbanion \kär-ˈba-ˌnī-ən, -ˌnī-ˌän\ *n* (1933) An organic ion carrying a negative charge (-) on a carbon atom.

Carbazole \ˈkär-bə-ˌzōl\ *n* [ISV *carb-* + *az-* + *-ole*] (1887) (dibenzopyrrole, diphenylenimine) Pigment Violet 23 (51319). Extracted from coal tar, unsaturated compound.

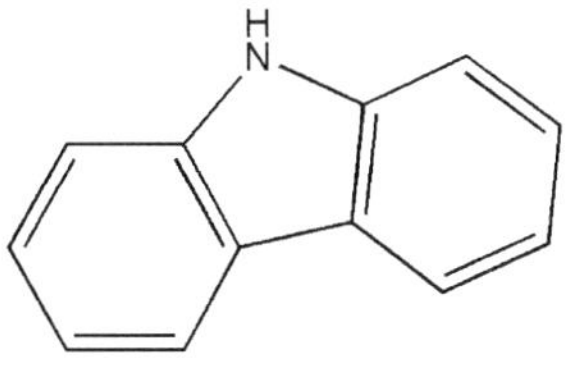

These pigments are manufactured in a variety of shades from red-shade to blue-shade violets. The latter appear more popular for toning (reddening) phthalocyanine blues. These pigments have excellent fastness properties even at high degrees of dilution but exhibit poorer color retention in metallized (aluminum flake) finishes than in light TiO_2 pastels. They are produced in both toner and lake (alumina hydrate) forms. They are used in the manufacture of ▶ Poly(N-Vinylcarbazole).

Carbenes *n* Constituents of bituminous type products which are insoluble in the chlorinated hydrocarbons, carbon tetrachloride and chloroform, but soluble in carbon disulfide.

Carbenium Ions *n* Stable carbenium ions, e.g., triphenylmethyl and tropylium salts, and their merit is that can initiate the polymerization of certain olefins by direct addition.

Carbinol \ˈkär-bə-ˌnȯl, -ˌnōl\ *n* [ISV, fr. obs. Gr *Karbin* methyl, fr. Gr *karb-*] (ca. 1885) (1) –CH$_2$OH. Monovalent primary alcohol radical. It may be part of an aliphatic or an aromatic alcohol. For example, ethyl alcohol (C$_2$H$_5$OH) can be described as methyl carbinol, or benzyl alcohol (C$_6$H$_5$CH$_2$OH) as phenyl carbinol. (2) It is sometimes used as a Syn: ▶ Methanol.

Carbitol [2,(2-ethoxyethoxy)ethanol, ethyl cellosolve] C$_2$H$_5$OCH$_2$–CH$_2$OCH$_2$CH$_2$OH and also written as C$_2$H$_5$O(CH$_2$)$_2$O(CH$_2$)$_2$OH. This is a trade name of an active, water-soluble solvent, the monoethyl ether of diethylene glycol. Properties: bp, 195°C; sp gr, 0.990/20°C; flp, 96°C (205°F); vp, less than 0.1 mmHg/20°C. If the ethyl (C$_2$H$_5$) group is substituted by methyl (CH$_3$), butyl (C$_4$H$_9$) groups, etc., the corresponding methyl and butyl carbitols are obtained.

Carbodiimide Polymers *n* Self addition of isocyanates with the aid of organic phosphine and arsine oxides as catalysts form yields of polymers referred to as *carbodiimides*; example, *n*RN=C=NR.

Carbohydrate \-ˌdrāt, -drət\ *n* (ca. 1869) An aldehyde or ketone which is a polyol, or a polymer of these.

Carbohydrates *n* Compounds containing carbon and the elements of water, and thus their only elements are carbon, hydrogen, and oxygen. They are all variations of the same basic formula: C$_m$(H$_2$O)$_n$. The carbohydrates include the mono-, di-, tri, tetra-, and poly-saccharides, as well as the conjugated saccharides.

Carbolic Acid \kär-ˈbä-lik\ *n* [ISV *carb-* + L *oleum* oil] (ca. 1859) A Syn: ▶ PHENOL.

Carbon \ˈkär-bən\ *n*. [F *carbone*, fr. L *carbon-*, *carbo* ember, charcoal] (1789) (1) A nonmetallic tetravalent element, atomic no. 6, atomic wt. 12.011; the major bioelement. It has two natural isotopes, ^{12}C and ^{13}C (the former, set at 12.00000, being the standard for all molecular weights), and two artificial, radioactive isotopes of interest, ^{11}C and ^{14}C. The element occurs in three pure forms (diamond, graphite, and in the fullerines), in amorphous form (in charcoal, coke, and soot), and in the atmosphere as CO$_2$. Its compounds are found in all living tissues, and the study of its vast number of compounds constitutes most of organic chemistry. (2) A nonmetallic chiefly tetravalent element found native (as in the diamond and graphite) or as a constituent of coal, petroleum, and asphalt, of limestone and other carbonates, atomic number 6, atomic weight 12.011 (longest living isotope) (Smith MB, March J (2001) Advanced organic chemistry, 5th edn. Wiley, New York).

Carbon 14 *n* (radiocarbon) Radioactive carbon of mass number 14, naturally occurring bur usually made by irradiating calcium nitrate. It has been used as a beta-ray source in gages for measuring the thickness of plastics films (Smith MB, March J (2001) Advanced organic chemistry, 5th edn. Wiley, New York).

Carbon Black *n* (ca. 1889) Pigment Black 7 (77266) Finely divided carbon formed by any one of the following processes: (1) Incomplete combustion of natural gas in burners under moving channel irons (channel carbon black). (2) Incomplete combustion of natural gas and petroleum in large, closed furnaces (furnace carbon black) (3) Decomposition of gasses in large converters filled with hot refractory brick checkerwork. These carbon blacks vary in particle size and some of them may be surface treated. A generic term for the family of colloidal carbons. More specifically, carbon black is made by the partial combustion and/or thermal cracking of natural gas, oil. Or another hydrocarbon. *Acetylene black* is the carbon black derived from burning acetylene. *Animal black* is derived from bones of animals. *Channel blacks* are made by impinging gas flames against steel plates or channel irons (hence the name), from which the deposit is scraped at internals. *Furnace black* is the term sometimes applied to carbon blacks made in a refractory-lined furnace. *Lamp black*, the properties of which are markedly different from other carbon blacks, is made by burning heavy oils or other carbonaceous materials in closed systems equipped with settling chambers for collecting the soot. *Thermal black* is produced by passing natural gas through a heated brick checker work where it thermally cracks to form a relatively coarse carbon black. Carbon blacks are widely used as fillers and pigments in PVC, phenolics, polyolefins, and several other resins, also imparting resistance to ultraviolet rays. In polyethylene, carbon black acts as a crosslinking agent, in rubbers, as a reinforcement (Donnet J-B, Wang M-J (1993) Carbon black. Marcel Dekker, New York; Pierson HO (1994) Handbook of carbon, graphite, diamond and fullenes. Noyes Data Corporation/Noyes Corporation, New York). (4) A black colloidal carbon filler made by the partial combustion or thermal cracking of natural gas, oil, or another hydrocarbon. There are several types of carbon black depending on the starting material and the method of manufacture. Each type of carbon black comes in several grades. Carbon black is widely used as a filler and pigment in rubbers and plastics. It reinforces, increases the resistance to UV light and reduces static charging (Sepe MP (1998) Dynamic mechanical analysis. Plastics Design Library, Norwich).

Carbon Blacks & Graphite *n* Carbon black is the amorphous form and graphite is the crystalline form of carbon (Pierson HO (1994) Handbook of carbon, graphite, diamond and fullenes. Noyes Data Corporation/Noyes Corporation, New York; Kirk-Othmer encyclopedia of chemical technology: pigments-powders (1996). Wiley, New York; Pierson HO (1994) Handbook of carbon, graphite, diamond and fullenes. Noyes Data Corporation/Noyes Corporation, New York).

Carbon Dating *n* (1951) Carbon dating is a variety of radioactive dating which is applicable only to matter that was one living and presumed to be in equilibrium with the atmosphere, taking carbon dioxide from the air for photosynthesis. Cosmic ray protons bombard nuclei n the upper atmosphere, producing neutrons that in turn bombard nitrogen, the radioactive isotope carbon-14 (^{14}C). The ^{14}C isotope combines with oxygen to form carbon dioxide and is incorporated into the cycle of living things. The ^{14}C isotope forms as at a rate that is constant, so that by measuring the radioactive emissions from once-living matter and comparing its activity with the equilibrium level of living things, a measurement of the time elapsed can be made (i.e., the activity of a sample can be directly compared to the equilibrium activity of living matter and the age calculated) (Lowe JJ (1997) Radiocarbon dating. Wiley, New York). The half-life of ^{14}C is about 5,730 years by the emission of an electron of energy 0.016 MeV. This changes the number of the nucleus to 7, producing a nucleus of ^{14}N. The low activity limit of the ^{14}C limites age determinations to the order of 50,000 years by counting techniques, and extended to perhaps 1,000,000 years by accelerator techniques (Higham T, Ramsey B, Owen C (2004) Radiocarbon and archaeology. Oxford University School of Archaeology, UK). Fossil fuels have no ^{14}C content, and atmospheric testing of nuclear weapons in the 1950s and 1960s increased the ^{14}C content of the atmosphere (Levin HL (2005) The Earth through time. Wiley, New York).

Carbon Dioxide *n* (1873) CO_2. A gas commonly found in the atmosphere. In a solid state (below freezing temperature) it is called dry ice. *Also known as Carbonic Acid Gas* (Levin HL (2005) The earth through time. Wiley, New York).

$$O = C = O$$

Carbon Disulfide *n* (1869) CS_2 A colorless flammable poisonous liquid used as a solvent for rubber and as an insect fumigant. *Also called Carbon Bisulfide.*

$$S = C = S$$

Carbon Fiber *n* (1960) (1) A high-tensile fiber or whisker made by heating rayon or polyacrylonitrile fibers or petroleum residues to appropriate temperatures. Fibers may be 7–8 μ in diameter and are more that 90% carbonized. (2) Any fiber consisting mainly of elemental carbon and increasingly used in reinforced-plastics products. They may be prepared by growing single crystals in a carbon electric arc under high-pressure inert gas; by growth from a vapor state via pyrolysis of a hydrocarbon gas; or by pyrolysis of organic fibers, the most widely used method. Polyacrylonitrile and rayon fibers are most commonly used as starting materials. The terms "carbon fibers" and "graphite fibers" are used somewhat interchangeably. However, PAN-based carbon fibers are 93–95% C by elemental analysis, whereas graphic fibers are usually 99+% C. The difference is due mainly to the temperature of formation. 1,315°C for fibers formed from PAN, while the high-modulus graphite fibers are graphitized at 3,450°C. The higher the graphite content, the higher the elastic modulus but the lower the strength. Properties hold to very high temperatures in inert atmospheres. (Properties transverse to the fiber length are much lower than along the length.) In recent years, carbon fibers have become the leading reinforcement for high-performance composites. The less expensive carbon fibers produced from pitch have broadened the markets for these reinforcements. Strength and modulus range considerably depending on the supplier and grade, from 1.7 to 3.5 GPa (250–500 kpsi) for strength and from 230 to 830 GPA (34–120 Mpsi) for modulus (Chung DD (1994) Carbon fiber compositers. Elsevier Science and Technology Books, New York; Pierson HO (1994) Handbook of carbon, graphite, diamond and fullenes. Noyes Data Corporation/Noyes Corporation, New York).

Carbon 14 *n* (1936) A heavy radioactive isotope of carbon of mass number 14 used in tracer studies and in dating archaeological and geological materials (Lowe JJ (1997) Radiocarbon dating. Wiley, New York).

Carbon Monoxide *n* (1873) A colorless odorless very toxic gas CO that burns to carbon dioxide with a blue flame and is formed as a product of the incomplete combustion of carbon.

Carbon Steel *n* (1903) A strong hard steel that derives its physical properties from the presence of carbon.

Carbon Tetrachloride *n* (1866) (tetrachloromethane) CCl_4. A clear, dense pungent-smelling liquid similar to that of chloroform, a powerful solvent for many resins and miscible with most other organic solvents, and a starting compound in the synthesis of ▶ Nylon 7. It has a sp gr of 1.629 and is nonflammable. Solvent action on oils and similar products and soft resins is good. It is frequently used to reduce fire hazards. Very toxic. (No flash or fire point.) Its use as a solvent has now been severely curtailed because of its high toxicity and known carcinogenicity. Syn: ▶ Tetrachloro-Methane, ▶ Perchloromethane, ▶ Tetrachloride.

$$\begin{array}{c} Cl \\ | \\ Cl - \!\!\!-\!\!\!- Cl \\ | \\ Cl \end{array}$$

Carbon 13 *n* (1939) An isotope of carbon of mass number 13 that constitutes about 1/70 of natural carbon and is used as a tracer in spectroscopy utilizing nuclear magnetic resonance (Breitmaier E, Voelter W (1986) Carbon-13 nmr spectroscopy. Wiley, New York).

Carbon 12 *n* (1946) An isotope of carbon of mass number 12 that is the most abundant carbon isotope and is used as a standard for measurements of atomic weight (Goldberg DE (2003) Fundamentals of chemistry. McGraw-Hill, New York).

Carbonaceous \kär-bə-ˈnā-shəs\ *adj* (1791) Matter containing carbon. See ▶ Organic (1).

Carbon-Arc Lamp *n* A type of fading lamp which utilizes an arc between two carbon electrodes as the source of radiation.

Carbonate White Lead *n* $Pb_3O_8C_2H_2$. Pigment White 1 (77597). Basic carbonate of lead, the composition of which approximates to the formula $Pb(OH)_2 \cdot 2PbCO_3$. It is the normal white lead of industry and is often given the above description to differentiate it from white lead sulfate which is also used as a pigment (Gooch JW (1993) Lead based paint handbook. Plenum Press, New York. It is marketed in several grades designated according to the method of manufacture, e.g., stack, chamber, electrolytic or precipitated. The latter is further subdivided into high, medium, and low stain varieties. Syn: ▶ Basic Carbonate of Lead, ▶ Basic Lead Carbonate, ▶ Cremnitz White, ▶ Kremnitz White, and ▶ White Lead (Gooch JW (1993) Lead based paint handbook. Plenum Press, New York).

Carbon-Carbon Composite *n* A combination of carbon or graphite fibers in a carbon or graphite matrix, produced by impregnating a carbon- or graphite-fiber cloth or mat structure with a carbonizable binder such as pitch.

Carbon-Fiber-Reinforced Plastics *n* (CRP) Plastics, either thermosetting or thermoplastic – most commonly epoxies or high-performance resins – that contain carbon or graphite fibers (Chung DD (1994) Carbon fiber compositers. Elsevier Science and Technology Books, New York).

Carbonific *n* Chemical compound which, upon decomposition, produces a mass of carbon which frequently occupies a volume much greater than the original unburned materials. Carbonifics function to produce the insulating, relatively incombustible properties of intumescent coatings.

Carbonium \kär-ˈbō-nē-əm-\ *n* [*carb-* + *-onium*] (1942) A positively charged organic ion such as H_3C, having one less electron than a corresponding free radical and behaving chemically as if the positive charge were localized on the carbon atom.

Carbonium Ions *n* Positively charged organic molecules in which every atom has an octet of electrons which makes it more stable, e.g., $RCH=O^+R$.

Carbonium-Ion Polymerization *n* Cationic $-C^+$ initiated polymerization reaction.

Carbonization *n* Process of degrading organic matter to elemental carbon. The term is used chiefly in describing the decomposition of varnish films and the like, when subjected to elevated temperatures. Carbonization of films on stoving is always associated with pronounced darkening.

Carbonizing *n* A chemical process for eliminating cellulosic material from wool or other animal fibers. The material is reacted with sulfuric acid or hydrogen chloride gas followed by heating. When the material is dry, the carbonized cellulose material is dust-like and can be removed.

Carbonyl (carbonyl group) \ˈkär-bə-ˌnil, -ˌnēl\ *n* (1869) An organic functional group $-C=O$, found only in combination, as in aldehydes, ketones and organic acids.

$$\cdot HC = O$$

Carbonyl Addition-Elimination *n* The single most important type of reaction mechanism which has been applied to the preparation of step-growth polymers is the "addition-elimination reaction" of the carbonyl double bond of carboxylic acids and carboxylic acid derivatives; included in this general type of reaction are esterification amidation and anhydride formation from carboxylic acids, esters, amides, anhydrides and acid halides.

Carbonyl Addition-Substitution *n* Step-growth polymerization schemes based on carbonyl addition-substitution reactions are almost entirely concerned with reaction of aldehydes; step-growth polymers prepared by this type of reaction include polyacetals, phenolformaldehyde polymers, ureaformaldehyde polymers, and melamineformaldehyde polymers.

Carbonyl Group *n* The bivalent radical, –C=O, especially in aldehydes or ketones.

Carbonyl Value *n* Keto group –C=O, is sometimes referred to as the carbonyl radical. Thus the amount of keto groups in a compound is its carbonyl value.

Carborane *n* (dicarbadodecaborane) $C_2B_{10}H_{12}$. An icosahedral cage compound containing mostly boron, with two active hydrogens on the carbon atoms, and available in several isomers. It is polymerizable, but only the silicone copolymers have been manufactured.

Carborundum See ▶ Silicon Carbide.

Carbowax *n* Poly(ethylene glycol), manufactured by Union Carbide, U.S. (now Dow Chemical).

Carboxy Nitroso Rubber *n* (CNR) A fluorocarbon elastomer, synthesized as a terpolymer from tetrafluoroethylene trifluoronitrosomethane, and nitro-soperfluorobutyric acid. CNR has unique resistance to strong oxidizers and is nonflammable in pure oxygen, hence has found applications in the aerospace field. The gum can be processed on standard rubber-mixing equipment for molding, or dissolved for application by spraying, dipping or brushing.

Carboxyl End Group *n* The chain-terminating (–COOH) group found in polyamide and polyester polymers.

Carboxyl Group *n* The radical –COOH, characteristic of most organic acids.

Carboxylate Ion *n* The anion of a carboxylic acid.

Carboxylic *n* Term for the –COOH group, the radical occurring in organic acids.

Carboxylic Acid *n* Organic acid which possesses one or more carboxyl groups. The simplest member of the series is formic acid, H·COOH and has the general formula.

$$R-\underset{\underset{\text{Carboxylic acid}}{}}{\overset{\overset{O}{\|}}{C}}-OH$$

Carboxymethyl Cellulose (CMC) \kär-▮bäk-sē-▮me-thəl, -▮sel-yə-▮lōs, -▮lōz\ *n* (1947) The common name for a cellulose ether of glycolic acid. It is an acid ether derivative of cellulose formed by the reaction of alkali cellulose with chloroacetic acid. It is usually marketed as a water-soluble sodium salt, more properly called sodium carboxymethyl cellulose. The sodium salt of this compound is commonly used as a stabilizer or an emulsifier. In the early literature, it is sometimes called cellulose glycolate or cellulose glycolic acid.

Carburizing *n* A process for case-hardening steels in which the objects to be hardened are heated with carbonates and charcoal in the absence of air for bout 24 h at 840–950°C, then quenched in oil. The depth of hardening is about 1.7 mm and the surface hardness is from 50 to 55 on the Rockwell C scale.

Carcinogen \kär-▮si-nə-jən, ▮kär-s°n-ə-▮jen\ *n* (1853) Any material that has been tested and found to cause cancer in laboratory animals or that, through statistical studies, is correlated with the incidence of cancers in humans (Merriam-Webster's collegiate dictionary (2004) 11th edn., Merriam-Webster, Springfield). An example from the plastics industry is vinyl chloride monomer which is believed to have caused human liver cancer. PVC polymer, on the other hand, in noncarcinogenic (Wickson EJ (ed) (1993) Handbook of polyvinyl chloride formulating. Wiley, New York). A list of known carcinogens is available from OSHA (U.S. Office of Health and Safety Administration).

Card Choking See ▶ Cylinder Loading.

Card Clothing *n* The material used to cover the working surfaces of the card, i.e., cylinder and rolls or flats. The clothing consists of either wire teeth set in a foundation fabric or rubber, or narrow serrated metal flutes which are spirally arranged around the roll. The metallic wire has the appearance of band-saw blade.

Card Conversion Efficiency *n* The efficiency of the carding process, expressed as a percentage obtained from ratio of sliver output to staple input.

Carded Yarn *n* A cotton yarn that has been carded but not combed. Carded yarns contain a wider range of fiber lengths and, as a result, are not as uniform or as strong as combed yarns. They are considerably cheaper and are used in medium and course counts.

Cardigan \▮kär-di-gən\ *n* [James Thomas Brudenell, 7th Earl of *Cardigan* † 1868 English soldier] (1868) (1) A modification of the rib-knitting stitch to allow tucking on one (half cardigan) or both(full cardigan) sets of needles. (2) A sweater that buttons down the front (Merriam-Webster's collegiate dictionary (2004) 11th edn. Merriam-Webster, Springfield).

Carding *n* A process in the manufacture of spun yarns whereby the staple is opened, cleaned, aligned, and formed into a continuous, untwisted strand called a sliver.

Cardol *n* One of the original constituents of cashew nutshell liquid, occurring to the extent of about 10%.

It is a dihydroxy phenol, containing a side chain with two double bonds.

Care Label *n* The label that gives directions for cleaning, ironing, and otherwise maintaining a fabric of fiber product.

Carene *n* Terpene hydrocarbon, which is a constituent of certain turpentines. Carene has been reported in German, Indian, Russian, and Finnish turpentines.

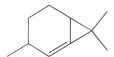

Cariflex *n* Block copolymer of styrene/butadiene/styrene. Manufactured by Shell, The Netherlands.

Carlona *n* Poly(ethylene), manufactured by Shell, The Netherlands.

Carlona Pt *n* Poly(propylene), manufactured by Shell, The Netherlands.

Carmine \\ˈkär-mən, -ˌmīn\\ *n* [F *carmine*, fr. ML *carminium*, irreg. fr. Arabic *qirmiz* kermes + L *minium* cinnabar] (1712) Aluminum lake of a pigment from cochineal.

Carmine Lake *n* Natural Red 4 (75470). Barium lake of the dyestuff produced by coupling 7-amino-1-naphthalenesulfonic acid with R-sale (2-naphthol-3:6-disulfonic acid).

Carmine Vermillion See ▶ Mercuric Sulfide.

Carnauba Wax \\kär-ˈnó-bə, ˌkär-nə-ˈü-bə\\ *n* (1854) Extremely hard wax obtained from the leaves of a Brazilian tree. It appears on the market in several grades, such as bleached, yellow, fatty grey, and chalky grey. The wax has the following approximate constants: mp, 84°C, sp gr, 0.998; acid value, 2; saponification value, 80; iodine value, 13. It is a constituent of wax polishes, and has some application in matt, overprinting, and baking varnishes. Its solubility is of a very low order in most varnish solvents and other constituents. Syn: ▶ Brazil Wax.

Carnaubic Acid *n* $C_{23}H_{47}COOH$. One of the constituent acids for carnauba wax; monocarboxylic acid. MP, 72°C.

Carnaubyl Alcohol *n* $C_{24}H_{49}OH$. Alcoholic constituent of carnauba wax. Mp, 69°C.

Carnot Cycle *n* A sequence of operations forming the working cycle of an ideal heat engine of maximum thermal efficiency. It consists of isothermal expansion, adiabatic expansion, isothermal compression, and adiabatic compression to the initial state.

Carotene (Carotin) *n* \\ˈkar-ə-ˌtēn\\ *n* [ISV, fr. LL *carota* carrot] (1861) $C_{40}H_{56}$. Hydrocarbon found in linseed and other vegetable oils where it is believed to act as an anti-oxidant.

Carothers Equation *n* In step-growth polymerization, the Carothers Equation (Carothers' Equation) gives the degree of polymerization, X_n, for a given fractional monomer conversion, p (Odain, 2004). For the case of *linear polymers - two monomers in equimolar quantities*: The simplest case refers to the formation of a strictly linear polymer by the reaction (usually by condensation) of two monomers in equimolar quantities. An example is the synthesis of nylon-6,6 whose formula is $[-NH-CH_2)_6-NH-CO-(CH_2)_4-CO-]_n$ from one mole of hexamethylenediamine, $H_2N(CH_2)_6NH_2$, and one mole of adipic acid, $HOOC-(CH_2)_4-COOH$. For this case (Cowie, 1991; Rudin, 1982).

$$\bar{X}_n = \frac{1}{1-p}$$

In this equation, X_n is the number-average value of the degree of polymerization, equal to the average number of monomer units in a polymer molecule. For the example of nylon-6,6 where $X_n = 2n$ (n diamine units and n diacid units); p is the extent of reaction (or conversion to polymer), defined by $p = (N_0 - N)/N_0$, where N_0 is the number of molecules present initially and N is the number of unreacted molecules at time t. This equation shows that a high monomer conversion is required to achieve a high degree of polymerization. For example, a monomer conversion, p, of 98% is required for $X_n = 50$, and $p = 99\%$ is required for $X_n = 100$. Linear polymers: one monomer in excess: If one monomer is present in the stoichiometric excess, then the equation becomes (Allock, 2003).

$$\bar{X}_n = \frac{1+r}{1+r-2rp}$$

where r is the stoichiometric ratio of reactants, the excess reactant is conventionally the denominator so that $r < 1$. If neither monomer is in excess, then $r = 1$ and the equation reduces to the equimolar case above. The effect of the excess reactant is to reduce the degree of polymerization for a given value of p. In the limit of complete conversion of the limiting reagent monomer, $p \to 1$ and

$$\bar{X}_n = \frac{1+r}{1-r}$$

Thus for a 1% excess of one monomer, r = 0.99 and the limiting degree of polymerization is 199, compared to infinity for the equimolar case. An excess of one reactant can be used to control the degree of polymerization.

For the case of *branched polymers: multifunctional monomers*: The functionality of a monomer molecule is the number of functional groups which participate in the polymerization. Monomers with functionality greater than two will introduce branching into a polymer, and the degree of polymerization will depend on the average functionality f_{av} per monomer unit. For a system containing N_0 molecules initially and equivalent numbers of two functional groups A and B, the total number of functional groups is $N_0 f_{av}$ and

$$f_{av} = \frac{\Sigma N_i \cdot f_i}{\Sigma N_i}$$

And the modified Carothers Equation is (Carothers, 1936) is

$$x_n = \frac{2}{2 - p f_{av}}$$

where p equals to

$$\frac{2(N_0 - N)}{N_0 \cdot f_{av}}$$

Related to the Carothers equation are the following equations for the simplest case of linear polymers formed from two monomers in equimolar quantities:

$$\bar{X}_w = \frac{1 + p}{1 - p}$$

$$M_n = M_0 \frac{1}{1 - p}$$

$$\bar{M}_w = M_0 \frac{1 + p}{1 - p}$$

$$PDI = \frac{\bar{M}_w}{\bar{M}_n} = 1 + p$$

Where: X_w is the weight average degree of polymerization, M_n is the number average molecular weight, M_w is the weight average molecular weight, M_o is the molecular weight of the repeating monomer unit; and PDI is the polydispersity index. The last equation shows that the maximum value of the PDI is 2, which occurs at a monomer conversion of 100% (or p = 1). This is true for step-growth polymerization of linear polymers. For chain-growth polymerization or for branched polymers, the PDI can be much higher. In practice the average length of the polymer chain is limited by such things as the purity of the reactants, the absence of any side reactions (i.e., high yield), and the viscosity of the medium. See ▶ *Gel Point*. References: Odian G, (2004) Principles of polymer science. Wiley, New York; Cowie JMG (1991) Polymers: chemistry & physics of modern materials, 2nd edition Blackie; Rudin A, (1982) The elements of polymer science and engineering. Academic, New York; Allcock HR, Lampe FW, Mark JE (2000) Contemporary polymer chemistry, 3rd edn. Pearson, New York; Carothers W (1936) Polymers and polyfunctionality. Trans of the Faraday Soc 32:39–49.

Carpet Backing *n* A primary backing through which the carpet tufts are inserted is always required for tufted carpets. The backing is usually made of woven jute or nonwoven manufactured fiber fabrics. A secondary backing, again made of jute or manufactured fibers, is normally added at the latex backcoating stage. Carpet backings are an important end use for nonwoven fabrics.

Carpet Underlay *n* A separate fabric which is used to provide cushioning for carpet. Carpet underlays are made of hair and jute, sponge rubber, bonded urethane or foamed urethane.

Carpets *n* Heavy functional and ornamental floor coverings consisting of pile yarns or fibers and a backing system. They may be tufted or woven. (Also see ▶ Tufted Carpet.)

Carrageen \ˈkar-ə-ˌgēn\ *n* [*Carragheen*, near Waterford, Irland] (1829) (sometimes spelled "Carrageenan" or "Carrageenin"). Hydrocolloid obtained from a group of sea plants related to *Chondrus crispas*, colloquially called Irish moss or moss. It is a complex carbohydrate made up of galactose, dextrose, and levecose residue. Useful for mucilaginous and gelatizing qualities. *Also known as Irish Moss, Irish Gum, Pearl Moss, Pig Wrack, and Rock Salt Moss.*

Carrier *n* (1) A product added to a dyebath to promote the dyeing of hydrophobic manufactured fibers and characterized by affinity for, and ability to swell, the fiber. (2) A moving holder for a package of yarn used on a braiding machine. (3) A term sometimes used to describe the tube or bobbin on which yarn is wound (Tortora PG (ed) (1997) Fairchild's dictionary of textiles. Fairchild Books, New York).

Carrierless Dyeing Variants *n* Polymers that have been modified to increase their dyeability. Fibers and fabrics made from these polymers can be dyed at the boil without the use of carriers.

Carter Lead *n* White lead manufactured in the U.S.A. by what is known as the Carter process. This is somewhat similar to the chamber process except that the metallic lead is used in a finely divided state instead of sheets.

The process is therefore quicker, but the resulting product is very similar to chamber white lead.

Cartoon n (1) Preliminary sketch or detailed drawing for a painting, mural, etc., either to scale or actual size. The design is transferred from the paper to the working surface. (2) Term also used to describe a comic drawing or an animated film based on a succession of comic, grotesque drawings.

Cartridge Heater n A rod-shaped electrical heating element, consisting of a metal outer shell, sealed within which is a Nichrome-wire coil embedded in a thermally stable, electrically insulating powder, such as magnesium oxide. Cartridge heaters come in a wide range of physical sizes and wattages and are used in heating and controlling the temperatures of dies and molds, injection nozzles, hot stampers, etc.

Carvone n Ketone derived from the terpene dipentine.

Cascade Coating n A process for applying epoxy and other thermosets to objects such as electrical resistors and capacitors, in which finely powdered resin is poured over the preheated object. The article is usually rotated as the powder is applied.

Cascade Control n (piggy-back control) In automatic control, a system in which the output of one unit is the input of the next, the goal being to obtain closer control of an important, final variable by controlling more sensitive, linked variables. A cascade system may be of the open-loop or closed-loop type.

Case Hardening n Any of several processes by which the working surfaces of steel tools and molds are hardened after being machined in their softer, original states. It can also be described as the surface hardening without thorough drying of the film. See ▶ Carburizing, Flame Hardening, Nitriding.

Casein n (1) A protein usually obtained from milk. Used to make sizings, adhesive solutions, and coatings. Used as a binder in aqueous dispersions of pigments. (2) The protein substance occurring in milk and cheese. It can be obtained by treating skim milk with a dilute acid, but the type used mainly for plastics (rennet casein or paracasein) is made by treating warm skim milk with a rennet extract. See also ▶ Casein Plastic (Skeist I (ed) (1977) Handbook of adhesives. Reinhold Publishing Co., New York; Dainth J (2004) Dictionary of chemistry. Oxford University Press, New York).

Casein Finish See ▶ Casein Paint and ▶ Distemper.

Casein Paint n Coatings in which casein replaces the ordinary drying oils, or is used as an emulsifying agent in emulsion paints. Both types may be thinned with water.

Casein Plastic n A family of thermosetting plastics derived from ▶ Casein, used widely in the early years of the plastics industry but less important now. Casein plastics have poor water resistance and dimensional stability, which limits their applications.

Caseinates n Metallic salts derived from casein.

Casement A window sash that opens on hinges at the vertical edge. See ▶ Casement Window.

Casement Cloth n A general term applied to lightweight, sheer fabrics used for curtains and for screening purposes and as a backing for heavy drapery fabrics of the decorative type. This type of fabric is sometimes made in small fancy weaves for dresswear.

Casement Window n Window hinged at the side to swing in or out.

Cashew Net Shell Oil n Natural oil from the shells of the nuts of the species *Anacardium occidentale*, containing a high proportion of cardol – a substituted phenol. By reaction of this oil with formaldehyde, a resin is produced which gives films with good chemical resistance.

Cashew Nut Shell Liquid Resin n Resin derived from the liquid obtained from the shells of the nuts of the species, *Anacardium occidentale*, which grows chiefly in India and South America. The naturally occurring liquid is a mixture of a dihydroxy phenol cardol and anacardic acid. This acid readily decarboxylates on heating to yield a monohydroxy phenol with an unsaturated side chain, anacardol.

Cashmere n The extremely soft hair of the Cashmere goat. Cashmere is often blended with sheep's wool in fabrics.

Casing n A term coined by Bell Telephone Laboratories, an acronym for the process of **C**rosslinking by **A**ctivated **S**pecies of **IN**ert **G**ases developed to impart printability and adhesive receptivity to polymers such as PTFE and polyethylene. In this process, articles are exposed to a flow activated inert gases in a glow-discharge tube, forming a shell of highly crosslinked molecules having high ▶ Surface Energy on the article surfaces.

Casing Knife n In paper handing, a knife used to trim wallpaper around casings, at moldings, baseboards, etc.

Cassel Brown See ▶ Vandyke Brown.

Cassel Earth See ▶ Vandyke Brown.

Cassel Yellow n $PbCl_2 \cdot 7PbO$. Lead oxychloride pigment.

Casson Equation n Rheology expression used to relate share rate, viscosity, and at infinite shear rate for dispesrsions (pigment coatings, etc.). Casson's equation is:

$$\tau^{0.5} = \tau_0^{0.5} + \eta_\infty^{0.5} \gamma^{0.5}$$

where τ = shear stress, γ = shear rate and η_∞ = viscosity at of the liquid dispersion at infinite shear rate, a condition where all structural viscosity due to pigment flocculation or other source has been completely eliminated due to the shearing forces. This equation holds well for for many pigmented systems such as printing inks, plasticizer/iigment dispersions, and coatings. Other values for n may be more experimentally accurate, but 0.5 is a good starting point (Patton TC (1979) Paint flow and pigment dispersion: a rheological approach to coating and ink technology. Wiley, New York).

Cassone Painting n Pictures used to decorate the sides of an Italian marriage chest (cassone). The most common subjects were scriptural, chivalrous, mythological, and heraldic.

Cast n (1) A tinge; a subjective term used in combination with a directional indicator to describe a slight deviation from the norm in color or appearance. The term is most often applied to color, and refers to a small difference in hue. For example, four pure hues are generally recognized psychologically, red, green, blue, and yellow. Each is unique and contains no quality of the others. If a blue hue seems to contain some element of green that blue is said to have a green "cast." (2) The term may also be applied to appearance aspects other than color. For example, a material which is normally expected to be transparent (nonscattering) but which is slightly cloudy (due to a small amount of light scattering) is said to have a "milky cast" or a "cloudy cast." See ▶ Casting (McDonald, Roderick (1997) Colour physics for industry, 2nd edn. Society of Dyers and Colourists, West Yorkshire; Skeist I (ed) (1977) Handbook of adhesives. Reinhold Publishing Co., New York).

Cast Coating See ▶ Coating.

Cast Embossing n A process of casting films against an embossed temporary carrier. Vinyl plastisols, organosols, solutions, or lattices are used as film formers, which may be backed up with layers of foam or fabric. The temporary carrier is often paper, embossed with the desired pattern and treated so as to be easily stripped from the fused film or laminate.

Cast Film n Film produced by pouring or spreading a solution, hot melt, or dispersion of plastic material onto a temporary carrier – typically a polished metal roll, hardening the material by suitable means, and stripping the solidified film from the surface. Cellulosic, polystyrene, and vinyl films are often produced in this manner.

Cast Film Extruders n Equipment used for preparing thin polymer films by use of one or two free-rolling casting and polishing rollers.

Cast Film Extrusion See ▶ Chill Roll Extrusion.

Castanha de Cotia Kernel Oil n A drying oil from a tree indigenous to Brazil. It is believed to contain a substantial proportion of eleostearic acid, or acid of similar type. Its rate of polymerization is much slower than that of tung oil. Its iodine value is 154.

Cast-Coated Paper n A paper or board, the coating of which is allowed to harden or set while in contact with a finished casting surface. Cast-coated papers have a high gloss.

Caster Oil (Dehydrated) n Castor oil from which some chemically combined water has been removed to improve its drying properties.

Cast-in Heater n A type of heater used on cylindrical surfaces, such as extruded barrels, in which a rod-type heating element is bent to a semi-cylindrical shape and cast within an aluminum channel shape whose inner surface has the cylinder's radius. Pairs of such cast-in elements are then strapped tightly to the barrel with Monel bands and high-strength bolts, leaving the heater terminals exposed for electrical connections. Copper tubing may also be cast into the same elements, permitting circulation of water or, better, an involatile heat-transfer liquid, for barrel cooling.

Cast-in Lining See ▶ Bimetallic Cylinder.

Casting n Manufacturing process dating back to at least 4000 B.C. for inexpensively producing large or complex parts. The process of forming solid or hollow articles from fluid plastic mixtures or resins by pouring or injecting the fluid into a mold or against a substrate with little or no pressure, followed by solidification and removal of the formed object. See also ▶ Cast Embossing, ▶ Centrifugal-, Film-, Slush-, Solid-, Rotational-, ▶ Embedding, ▶ Encapsulation, ▶ Potting and ▶ Drawdown. (2, n) The finished product of a casting operation.

Casting Plaster See ▶ Gypsum.

Casting Resin n Resin which can be cast and hardened in a mold to form a shaped article.

Casting Syrup n (casting resin) Liquid monomers or partially polymerized polymers, usually containing catalysts or curing agents, capable of polymerizing to the solid state after they have been cast in molds. The materials most generally used as the acrylics, styrenes, polyesters, epoxies, silicones, and nylons. Also called *Potting Syrups* when used for encapsulating articles such as electrical components or assemblies.

Castor Oil *n* (ricinus oil) A pale-yellowish oil derived from the seeds of the castor bean, *Ricinus communis*, and consisting essentially or ricinolein. Its principal characteristics are light color, relatively high specific gravity and viscosity, and its solubility is alcohol. It differs from other oils in that its composition is mostly hydroxy fatty acids. It is essentially a nondrying oil, but it may be converted to a drying oil by "chemical dehydration" by which a hydroxy group and an adjacent hydrogen atom are removed as water to form a drying oil fatty acid ester with two double bonds, one of them being conjugated. This dehydration yields what commonly is known as dehydrated castor oil. In its original undehydrated form, castor oil is well known for its use in resins and as a plasticizer for cellulose ester lacquers. It is an important starting material for plasticizers, certain nylons, and alkyd resins; and an ingredient in certain urethane forms (Merck Index (2001) 13th ed. Merck and Company, Inc., Whitehouse Station; Paint: pigment, drying oils, polymers, resins, naval stores, cellulosics esters, and ink vehicles (2001), vol 3. American Society for Testing and Material).

CAT *n* (1) Abbreviation for the ISCC Color-Matching Amplitude Test. This test was devised by the Subcommittee for Problem 10 of the Inter-Society Color Council as a measure of discrimination ability for small differences in saturation. It is not a test for defective color vision. Results depend on experience as well as on inborn ability. It is sold by the Federation of Societies for Coatings Technology, 1315 Walnut Street, Philadelphia, PA 19107. (2) Abbreviation for ▶ Catalyst.

Cat Eyes *n* Undissolved globules of rubber in a cement generally made from only rubber and solvent. Expression also used for PVC sheet and paint films. etc. *Also called Fish Eyes.* See ▶ Pinhole.

Catalyst \ˈka-tᵊl-əst\ *n* (1902) A substance that causes or accelerates a chemical reaction when added to the reactants in a minor amount, and that is not consumed in the reaction. A negative catalyst (Inhibitor, Retarder) decreases the rate of reaction or prevents it altogether. See also ▶ Hardener, ▶ Inhibitor, ▶ Acid Catalysts, ▶ Accelerator, ▶ Curing Agent, and ▶ Initiator.

Catalysts and Promoters *n* Substances whose presence increases the rate of a chemical reaction. They are added in a small quantity as compared to the amounts of primary reactants, and do not become a component part of the chain; they are referred to as an initiator. In some cases the catalyst functions by being consumed and regenerated; in other cases the catalyst seems not to enter the reaction and functions by virtue of surface characteristics of some kind. They are added to doped solvents to produce polymerization at room temperature or a temperature blow the softening point of the thermoplastic. A negative catalyst (inhibitor, retarder) slows down a chemical reaction.

Catalytic Agent *n* A substance which by its mere presence alters the velocity of a reaction, and may be recovered unaltered in nature or amount at the end of the reaction.

Catalytic Curing *n* Mechanism by which a coating is crosslinked by the action of a catalyst as opposed to oxidation, etc. Examples of such systems are two-part (pot) and epoxies and polyurethanes.

Cataphoresis \ˌka-tə-fə-ˈrē-səs\ *n* [NL] (1889). See ▶ Electrophoresis.

Catch-Up In lithography, the printing of non-image areas of a plate and is overcome generally by increasing the amount of fountain solution applied to it. It may occur as a run is started before the dampening adjustment is correctly set.

Caterpillar *n* (caterpillar puller) A device used downstream of the extruder in extrusion of pipe and profiles, consisting of two driven and counter-rotating belts, having an elongated oval shape about 0.8 m long, with pads attached to the outsides of the belts. One of the belts is elevatable to adjust the clearance between them so as to firmly grip the extrudate being pulled away from the cooling tank, yet not so strongly as to deform it. Belt speed is adjustable over a wide range to accommodate different rates of extrusion.

Cathode \ˈka-ˌthōd\ *n* [Gr *kathodos* way down, fr. *kata-* + *hodos* way] (1834) (1) In an electrolytic cell through which current is being forced by an external emf, the cathode is the negative electrode, giving up electrons to cations in the electrolyte. In a cell or battery *delivering current*, the cathode is the positive terminal. Also used in connection with cathodic protection.. (2) In an electro-chemical cell, the electrode at which reduction occurs. (3) In a vacuum tube the cathode is the electrode from which electrons are emitted. The negatively charged electrode in a gas-discharge tube. See ▶ Anode.

Cathode Ray *n* (1880) The stream of electrons emanating from the cathode in a discharge tube.

Cathode Ray Tube *n* (1905) A vacuum tube in which a beam of electrons is projected on a fluorescent screen to produce a luminous spot at a point on the screen determined by the effect of the electron beam of a variable magnetic field within the tube.

Cathode Sputtering See ▶ Vacuum Metallizing.

Cathodic Corrosion *n* Corrosion due to the development of alkalinity by a reaction at a cathode – sometimes experienced with aluminum and lead. Alkalinity from this source may also effect paints vulnerable to alkali attack. See ▶ Cell and ▶ Electrolytic.

Cathodic Polarization *n* Portion of the reduction in the initial potential of a corrosion cell that occurs at the cathode.

Cathodic Protection *n* Reduction or elimination of corrosion of a metal achieved by making current flow to it from a solution by connecting it to the negative pole of some source of current. The source of the protective current may be a sacrificial metal, such as zinc, magnesium, or aluminum. The current may also be derived from a rectifier, generator, or battery applied through an appropriate anode which may be consumed by the applied current, as in the case of steel, or remain substantially unaffected by the current, as in the case of graphite or platinum (Uhlig HH (2000) Corrosion and corrosion control. Wiley, New York).

Cation \ˈkat-ˌī-ən\ *n* [Gk *kation*, neut., prp. of *katienai* to go down cap, fr. *kata*-cata- + *ienai* to go] (1834) An atom, molecule or radical, usually in aqueous solution, that has lost an electron and has become positively charged. It can also be described as a positively charged atom or radical, which moves to the negative electrode or cathode during electrolysis. Metallic ions, such as iron and copper, etc., are cations.

Cationic \ˈkat-(ˌ)ī-ˈä-nik\ *adj* (ca. 1920) Pertaining to any positively charged atom, radical, or molecule; or to any compound or mixture containing positively charged groups.

Cation-Exchange Resin See ▶ Ion-Exchange Resin.

Cationic Detergent *n* A detergent that produces positively charged colloidal ions in solution.

Cationic Dyeable Variants *n* Polymers modified chemically to make them receptive to cationic dyes.

Cationic Dyes See ▶ Dyes, ▶ Basic Dyes.

Cationic Polymerization *n* Process in which the active end of the growing polymer molecule is a positive ion. If the ion is a carbonium ion, it is referred to as carbonium ion polymerization. A Polymerization Reaction With A Positive Or Cationic Initiator, I.E., A Cationic Initiator. See ▶ Ionic Polymerization.

Cationic Surfactant *n* Surfactant which gives a positively charged ion in aqueous solution.

Cauchy's Dispersion Formula *n*,

$$n = A + \frac{B}{\lambda^2} + \frac{C}{\lambda^4} + \ldots$$

An empirical expression giving an approximate relation between the refractive index *n* of a medium and the wavelength λ of the light; *A*, *B*, and *C* being constants for a given medium.

Cauchy's Equation *n* An empirical relationship between the refractive index and wavelength of light for a particular transparent material. It is named for the mathematician Augustin Louis Cauchy, who defined it in 1836.

The most general form of Cauchy's equation (Jenkins, 1981) is $n(\lambda) = A + \frac{B}{\lambda^2} + \frac{C}{\lambda^4} \ldots$, where n is the refractive index, λ is the wavelength, A, B, C, etc., are coefficients that can be determined for a material by fitting the equation to measured refractive indices at known wavelengths. The coefficients are usually quoted for λ as the vacuum wavelength in micrometers.

Usually, it is sufficient to use a two-term form of the equation: $n(\lambda) = A + \frac{B}{\lambda^2}$, where the coefficients *A* and *B* are determined specifically for this form of the equation. A table of coefficients for the common optical materials is shown below:

Material	A	B (μm^2)
Fused silica	1.4580	0.00354
Borosilicate glass BK7	1.5046	0.00420
Hard crown glass K5	1.5220	0.00459
Barium crown glass BaK4	1.5690	0.00531
Barium flint glass BaF10	1.6700	0.00743

The theory of light-matter interaction on which Cauchy based this equation was later found to be incorrect. In particular, the equation is only valid for regions of normal dispersion in the visible wavelength region. In the infrared, the equation becomes inaccurate, and it cannot represent regions of anomalous dispersion. Despite this, its mathematical simplicity makes it useful in some applications.

The Sellmeier equation is a later development of Cauchy's work that handles anomalously dispersive regions, and more accurately models a material's refractive index across the ultraviolet, visible, and infrared spectrum.

Caul *n* A sheet of metal, wood, or other material used in laminating to apply and equalize pressure. See ▶ Platen.

Caulk (Calk) *n* To fill voids with plaster or semiplastic materials; to fill crevices in the adherend surface with adhesive materials; to provide a seal against moisture or solvent intrusion.

Caulking See ▶ Caulking Compound.

Caulking Cartridge *n* An expendable container made of plastic, fiberboard or metal, filled with caulking compound, for use in a caulking gun. A common type is 2 in. (5 cm) in diameter, approximately 8 in. (20 cm) long, and fitted with a plastic nozzle.

Caulking Compound *n* A soft, plastic, putty-like material, consisting of pigment and vehicle, used for sealing joints in buildings and other structures where normal structural movement may occur, or for preventing leakage. Caulking compound retains its plasticity for an extended period after application. It is usually available in two consistencies: "gun grade," for use with a caulking gun, and "knife grade," for application with a putty knife; extruded preformed shapes are also available.

Caulking Gun *n* A device for applying caulking compound by extrusion. In a hand gun, the required pressure is supplied mechanically by hand; in a pressure gun, the pressure required usually is greater and is supplied pneumatically.

Cause-and-Effect Diagram *n* (fishbone diagram) A graphical way of analyzing a process, based on ideas and experiences of workers and engineers concerning the materials, machines, and methods of the process, in order to identify possible causes of product defects.

Caustic *n* A strong chemical base.

Caustic Potash *n* (KOH) Archaic name for potassium hydroxide. A strong base.

Caustic Soda *n* (NaOH). Archaic name for sodium hydroxide. A strong base.

Cave Painting *n* The art produced in the form of paintings on the walls of caves from the beginning of the Old Stone Age to the end of the New Stone Age, around 3,000 B.C. The entire period covers a span of some 200,000 years. The Paleolithic cave paints at Altamira (Spain) and those at Lascaux (France) are believed to date between 40,000 and 10,000 B.C. See ▶ Prehistoric Art.

Cavity \ ˈka-və-tē\ *n* [MF *cavité*, fr. LL *cavitas*, fr. L *cavus*] (1541) A depression, or sometimes the set of matching or associated depressions, in a plastics molds that forms the outer surfaces of the cast or molded article(s). The cavity may surround a CORE, the portion of the mold that forms the inner surfaces of a hollow article.

Cavity Side *n* (British) The side of an injection mold that is adjacent to the nozzle.

Cavity-Retainer Plate *n* A plate in a mold that holds the cavities and forces. Such plates are at the mold parting line and usually contain the guide pins and bushings. *Also called Force-Retainer Plate.*

Cavity-Side Part *n* (U.S.) The stationary part of an injection mold.

Cavity-Transfer Mixer *n* A two-piece device installed at the end of an extruder screw to accomplish both distributive and dispersive mixing. The *stator* is a barrel extension into whose inside surface is machined an array of many hemispherical cavities. The *rotor* is a screw extension whose exterior is similarly contoured. The lands of rotor and stator have the usual close clearance of screw and barrel. As the melt steam passes through, it is smeared between the lands and is repeatedly cut into small globs and recombined, passing from rotor to stator, stator to rotor, until it emerges.

CBA *n* In the plastics-foam industry, abbreviation for Chemical Blowing Agent.

cd *n* SI abbreviation for Candela.

Cd *n* Chemical symbol for the element cadmium.

CDP *n* Abbreviation for ▶ Cresyl Diphenyl Phosphate.

Cedar *n* A durable softwood generally noted for decay resistance; includes western red cedar, incense cedar, and eastern red cedar.

Cedar Nut Oil *n Pinus cembra,* from the seeds of which this oil is derived, grows prolifically in several parts of the world. The main constituent acids are linoleic, linolenic, and oleic acids, and the oil possesses useful drying properties. Iodine values up to 160 have been reported.

Ceiling *n* The overhead surface of a room, usually a covering or decorative treatment used to conceal the floor above or the roof.

Ceiling Temperature *n* The temperature above which polymerization will not occur, symbolized T_c.

Celanese Acetate See ▶ Acetate Fiber.

Celcon *n* Poly(formaldehyde) (from trioxane with some ethylene oxide). Manufactured by Celanese, U.S.

Celestial Blue Pigment *n* Iron blue precipitated on barytes. Usually only a few percent of blue is used.

Celestite *n* Mineral which is chiefly strontium sulfate. When ground, it has some application as a filler, being used as an alternative to barytes. The ground mineral is also known as *strontium white.*

Cell *n* (1) A small etched depression in a gravure cylinder that carries the ink. (2) In the cellular-plastics industry, a single void produced by a blowing agent, by mechanically entrained gas, or by the evaporation of a volatile constituent. When the void is completely surrounded by polymer, the cell is said to e *closed. A completely open* cell has no wall membranes but is part of a three-dimensional network of connected fibers or rods.

Cell Collapse *n* A defeat in foamed plastics characterized by slumping and cratered surfaces, with the internal cells resembling a stack of leaflets when viewed in cross section under a microscope. The condition is caused by tearing of the cell walls, weakened by plasticization or other mechanism.

Cell, Electroylic *n* Source of electrical current that is responsible for corrosion, consisting of an anode and a cathode immersed in an electrolyte and electrically bonded together. The anode and cathode may be separate metals or dissimilar areas on the same metal. They will develop a difference in potential that causes current to flow and corrosion at the anode when the electrodes are in electrical contact with each other.

Cellidor *n* Thermoplast based on Cellit. Manufactured by Bayer, Germany.

Cellit Cellulose acetate or acetobutyrate. Manufactured by Bayer, Germany.

Celloidin *n* (celluidine, photoxylin) A form of cellulose nitrate made by precipitation from an ether-alcohol solution of collodion cotton. See ▶ Cellulose Nitrate and ▶ Collodion.

Cellon *n* Cellulose acetate. Manufactured by Dynamit Nobel, Germany.

Cellophane *n* (1) Regenerated cellulose film, chemically similar to RAYON, made by mixing cellulose xanthate with dilute sodium hydroxide solution to form a viscose, then extruding the viscose into an acid bath for regeneration. Cellophane is coated on one or both sides to render it moisture proof and capable of being sealed with heat or solvent. (The term *rayon* is used when the regenerated material is in fibrous form.) Cellophane is widely used for packaging, most often with coatings of other polymers to overcome its tendency to absorb moisture and to improve the film's heat-sealability. Trade name for viscose film. (2) Hydrate cellulose from pump. Manufactured by Kalle, Germany.

Cellosolve *n* $C_2H_5OCH_2CH_2OH$. Proprietary name for the monoethyl ether of ethylene glycol, ethoxyethanol. It is also the generic name for a comprehensive series of ethers or similar type such as methyl cellosolve, butyl cellosolve, etc. A relatively slow drying flexographic ink solvent frequently used as a retarder. Cellosolve is water miscible. Properties: bp, 135°C; sp gr, 0.931/20°C; flp, 42°C (107°F); vp, 4 mmHg/20°C. See ▶ Ethylene Glycol Monoethyl Ether.

Cellosolve® Acetate *n* $CH^3COOCH_2CH_2OCH_2CH_3$. Medium high boiling solvent (ethylene glycol monoethyl ether acetate). Properties: bp, 156°C (313°F); sp gr, 0.975/20°C; flp, about 53°C (128°F); and vp, 1.5 mmHg/20°C.

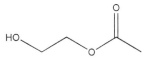

Cellular Mortar See ▶ Syntactic Foam.

Cellular Plastic *n* (expanded plastic, foamed plastic) A plastic with numerous cells of gas distributed throughout its mass. The terms *cellular-*, *expanded-*, and *foamed plastic* are used synonymously. A cellular plastic may be produced by (1) incorporating a blowing agent that decomposes to liberate a gas; (2) mechanically whipping in a gas or vaporizable liquid; (3) by adding a water-soluble salt or a solvent-extractable agent to the mix prior to forming, then leaching out the agent after forming to leave voids; or (4) other techniques described under ▶ Epoxy Foam, ▶ Polystyrene Foam, ▶ Syntactic Foam, ▶ Urethane Foam. Cellular plastics range in density from some slightly less than that of the parent resin to less than 0.01 g/cm³. The cells may be open or closed, depending on the process and density. See also ▶ Structural Foam.

Cellular Striation *n* In a cellular plastic, a layer of cells differing in size or nature from the majority of cells in the same mass.

Cellular Vinyls *n* Vinyls containing occulted gas in bubbles or cells. Cellular vinyls are used to form polymers with very low densities, down to below 0.1 g cm^{-3}, due to the high gas volume fraction, and hence can have exceptionally low thermal conductivities.

Cellulase *n* Enzyme that attacks and breaks down cellulosic substrates such as wood, cellulosic thickeners in paint, etc.

Celluloid *n* (1) An old trade name, now generic, for Cellulose Nitrate compounded with camphor and ethanol. The ethanol is removed after processing by heating, leaving behind the camphor which toughens the compound. Originally the trade name and now the common name of a synthetic plastic made by mixing pyroxylin, or cellulose nitrate, with pigments and fillers in a solution of camphor in alcohol. (2) Cellulose nitrate, plasticized with camphor. Manufactured by Dynamit Nobel, Germany.

Celsius See ▶ Temperature.

Cellulose \ˈsel-yə-ˌlōs\ *n* [F, fr. *Cellule* living cell. Fr. NL *cellula*] (1848) $(C_6H_{10}O_5)_n$. (1) A polysaccharide $(C_6H_{10}O_5)_x$ of glucose units that constitutes he chief part of the cell walls of plants, occurs naturally in such fibrous products as cotton and kapok, and is the raw material of many manufactured good as paper and rayon. (2) A natural carbohydrate polymer of high molecular weight, having the structure shown below. Cellulose is a constituent of most higher plants (*spermatophyta*). Cotton is the purest natural form, containing about 90%; it occurs to a small extent in the animal kingdom. Chemically, cellulose is 1–4 glucan of high degree of polymerization. It is desirable to apply "cellulose" to this material only and to designate the predominantly cellulosic residue obtained by subjecting woody tissues to various pumping processes as "cellulosic residues," "cellulosic pumps," or the like. Cotton linters and wood pulp are the major sources of cellulose for ▶ Cellulosic Plastics.

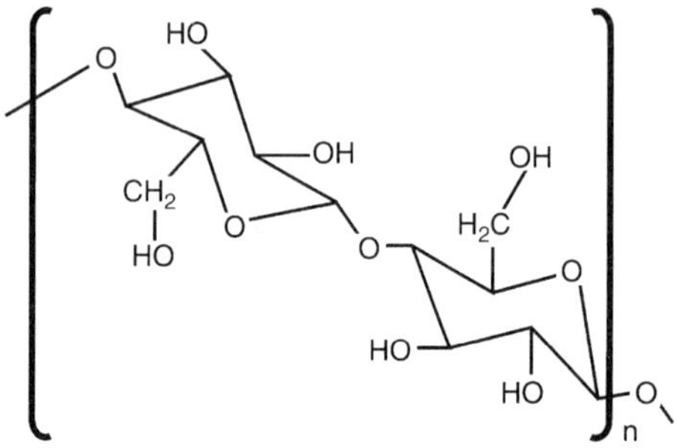

Cellulose

Cellulose Acetate *n* (1895) (CA) An acetic-acid ester of cellulose, forming a tough, transparent thermoplastic material when compounded with plasticizers. It is obtained by the action, under rigidly controlled conditions, of acetic acid and acetic anhydride on purified cellulose, usually obtained from cotton linters. All three available hydroxyl groups in each glucose unit of the cellulose can be acetylated, but in the material normally used for plastics it is usual to acetylate fully, then to lower the acetyl value by partial hydrolysis, leaving, on average, 2.4 acetate groups per C_6 unit. Cellulose acetate compounds are used when toughness, permanence, flame resistance, and transparency are required at moderate cost. However, they absorb up to 2.5% of atmospheric moisture, making them unsuitable for long-term outdoor exposure. (2) A clear thermoplastic material, usually in film form, made from cellulose and acetic acid. (3) Cellulose triacetate is derived from a reaction of acetic anhydride, acetic acid and little sulfuric acid; partial hydrolysis removes some of the acetate groups and degrades the chain to 200–300 repeat units and yields cellulose acetate (roughly a diacetate) which is the commercial product. (4) Thermoplastic esters of cellulose with acetic acid. Have good toughness, gloss, clarity, processability, stiffness, harness, and dielectric properties, but poor chemical, fire and water resistance and compressive strength. Processed by injection and blow molding and extrusion. Used for appliance cases, steering wheels, pens, handles, containers, eyeglass frames, brushes, and sheeting. Aslo called CA (Sepe MP (1998) Dynamic mechanical analysis. Plastics Design Library, Norwich).

Cellulose Acetate Butyrate *n* (CAB) A mixed ester produced by treating fibrous cellulose with butyric and acetic acids and anhydrides in the presence of sulfuric acid. CAB is generally supplied in the form of pellets prepared by mixing the molten ester with a plasticizer, then extruding and palletizing. It is one of the toughest of the cellulosic plastics, and has good transparency, colorability, weatherability, electrical properties and resistance to inorganic chemicals. It can be processed by extrusion, injection molding, blow molding, rotational molding, and thermoforming. Applications include pipe, tool handles, instrument housing, lighting, packaging film, and marine hardware.

Cellulose Acetate Propionate *n* (CAP, cellulose propionate) A thermoplastic formed by treating fibrous cellulose with propionic and acetic acids and anhydrides in the presence of sulfuric acid. CAP is easily extruded and injection molded, forming tough, flexible products with shock resistance close to that of ethyl cellulose. In properties and applications it resembles the acetate butyrate rather than the straight acetate.

Cellulose Acetobutyrate See ▶ Cellulose Acetate Butyrate.

Cellulose Esters *n* Any derivative of cellulose in which the free hydroxyl groups attached to the cellulose chain have been replaced wholly or in part by acidic groups, e.g., nitrate, acetate, propionate, butyrate, or stearate groups. Esterification is effected by the use of a mixture of an acid with its anhydride in the presence of a catalyst such as sulfuric acid. Mixed esters of cellulose, e.g., cellulose acetate butyrate, are prepared by using mixed acids and mixed anhydrides.

Cellulose Ethers *n* A cellulose derivative based on the etherification products of cellulose, such as ethyl cellulose, methyl cellulose, and sodium carboxymethyl cellulose. See ▶ Cellulose Thickeners.

Cellulose Ethers *n* Derivatives of cellulose in which one or more of the hydroxyl hydrogens have been replaced by alkyl groups.

Cellulose Fiber *n* The fibrous material remaining after non-fibrous components of wood have been removed by the pulping and bleaching operations. Used in making paper, etc.

Cellulose Glycolate See ▶ Carboxymethyl Cellulose.
Cellulose Gum See ▶ Carboxymethyl Cellulose.
Cellulose Lacquer See ▶ Lacquer.
Cellulose Nitrate *n* (1880) (CN, nitrocellulose, NC, pyroxylin) Cellulose nitrate, dating back to the work of French chemist Braconnet in 1833, is the oldest of the synthetic plastics. It is made by treating fibrous cellulose with a mixture of nitric and sulfuric acids, and was first used in the form of a lacquer (see ▶ Collodion). In 1870, John Wesley Hyatt and his brother patented the use of plasticized cellulose nitrate as a solid, moldable material, the first commercial thermoplastic (celluloid). Camphor was the first (and is still the best) plasticizer for CN, although many camphor substitutes have been developed. Alcohol is normally used as a volatile solvent to assist in plasticization, after which it is removed. Molded products of CN are extremely tough, but highly flammable and subject to discoloration in sunlight. CN is amendable to many decorative variations. Its principal uses today are in knife handles, table-tennis balls, and eyeglass frames. A mixture of nitric and sulfuric acids converts cellulose into cellulose nitrate; pyroxylin is a less nitrated material and it has been useful for photographic film, collodion, and celluloid plastics.
Cellulose Plastics *n* Plastics based on derivatives of cellulose, such as esters (cellulose acetate) and ethers (ethyl cellulose).
Cellulose Propionate *n* Ester of cellulose and propionic acid. See ▶ Cellulose Acetate Propionate.
Cellulose Thickeners *n* Aqueous thickeners based on cellulose compounds such as carboxymethyl cellulose, hydroxyethyl cellulose, and methyl cellulose.
Cellulose Triacetate *n* A member of the cellulosic plastics family made by reacting purified cellulose with acetic anhydride in the presence of a catalyst in such a manner that at least 92% of the hydroxyl groups are replaced by acetyl groups. Because of its high softening point this material cannot be molded or extruded. Its major use is for casting films or spinning fibers from solutions, such as in a mixture of methylene chloride and methanol.
Cellulosic Fiber *n* A fiber composed of, or derived from, cellulose. Examples are cotton (cellulose), rayon (regenerated cellulose), acetate (cellulose acetate), and triacetate (cellulose triacetate).
Cellulosic Plastic *n* (cellulosic resin) Any of a family of thermoplastics made by substituting various chemical groups fore the hydroxy groups in the cellulose molecules of cotton and purified wood pulp. See the following: ▶ Cellulose Acetate, ▶ Cellulose Acetate Butyrate, ▶ Cellulose Acetate Propionate, ▶ Cellulose Esters, ▶ Cellulose Nitrate, ▶ Cellulose Triacetate, ▶ Ethyl Cellulose, ▶ Hydroxyethyl Cellulose, and ▶ Regenerated Cellulose.
Cellulosic Plastics *n* Thermoplastic cellulose esters and ethers. Have good toughness, gloss, clarity, processability, and dielectric properties, but poor chemical, fire and water resistance and compressive strength. Processed by injection and blow molding and extrusion. Used for appliance cases, steering wheels, pens, handles, containers, eyeglass frames, brushes, and sheeting.
Cellulosics, Resins *n* Resins and polymers derived from cellulose $[C_6H_{10}O_5]_n$.
Celsius \ˈsel-sē-əs\ *adj* One hundredth of the thermometer scale, divided into 100°, in which 0°C is the freezing point of water, and 100°C is the boiling point at 1.013 bar. The preferred name according to the International System of Units (SI) is degree of Celsius (°C).
Cement *n* A material or a mixture of materials (without aggregate) which, when in a plastic state, possesses adhesive and cohesive properties and hardens in place. Frequently, the term is used incorrectly for concrete, e.g., a "cement" block for concrete block. *(n)* See ▶ Adhesive. *(v)* See ▶ Bond.
Cement-Asbestos Board *n* A dense, rigid, noncombustible board containing a high proportion of asbestos fibers which are bonded with Portland cement; highly resistant to weathering. *Also called Asbestos-Cement Board.*
Cement Coating See ▶ Cement Paint.
Cement Colorants *n* Term used to describe colors with sufficient tinctorial strength and alkali resistance to be suitable for coloring Portland cement or concrete. The natural earth colors, synthetic iron oxide colors, chromium oxide, ultramarine, and some of the organic pigments are used for this purpose.
Cementing *n* Joining plastics to themselves of dissimilar materials by means of solvents (see ▶ Solvent Cementing), dopes, or chemical cements. *Dope adhesives* comprise a solvent solution of a plastic similar to the plastic to be joined. *Chemical cements*, the only type suitable for thermosetting plastics, are based on monomers or semi-polymers or semi-polymers that polymerize in the joint to form a strong bond. See ▶ Adhesive.
Cementitious *n* Having cementing properties.
Cement Paint *n* Paint supplied in dry powder form, based essentially on Portland cement, to which pigments are sometimes added for decorative purposes. This dry powder paint is mixed with water immediately before use.
Cement Plaster *n* (1) Plaster with Portland cement as the binder; sand and lime are added on job. Used for

exterior work or in wet or high humidity areas. (2) In some regions, Gypsum Plaster.

Cement, Portland See ▶ Portland Cement.

Cement, Rubber *n* An adhesive that is a dispersion or solution of raw or compounded rubber, or both, in a suitable liquid.

Cenobrium See ▶ Mercuric Sulfide.

Cenospheres *n* Hollow microspheres in fly ash formed during combustion of coal in electric-power plants. They have had some use as a lower-cost substitute for glass microspheres in ▶ Syntactic Foams.

Centered Cell *n* A unit cell which has entities (atoms, molecules, ions) at locations in addition to the ell corners. A nonprimitive cell.

Center-Gated Mold *n* In injection molding, a mold in which each cavity is fed through an orifice at the center of the cavity. This type of gating is employed for items such as cups and bowls.

Centering Mark See ▶ Clip Mark.

Center Loop See ▶ Kink.

Center of Interest *n* In a room, the principal focal point, architecturally speaking, such as, for example, a fireplace. In a wallcovering design, the dominant motif, usually hung at eye level in the central area of a room to establish a starting point for handing the rest of the wallcovering.

Centi- (c) The SI-approved prefix signifying multiplication by 10^{-2}.

Centigrade See ▶ Celsius.

Centimeter \ˈsen-tə-ˌmē-tər\ *n* (Brit. *centimetre*) A measure of length equal to a hundredth part of a meter (Brit. *metre*), or 0.3937 in.; abbreviated cm; an inch equals 2.54 cm.

Centipoise \-ˌpȯiz\ *n* (cp) A deprecated, but still widely used viscosity unit, 0.01 POISE. Water at 20°C has a viscosity of 1.002 cp. The SI equivalent is: 1 cp = 0.001 Pa·s. Liquids of low viscosity are usually given in centipoise units.

Centistoke *n* (cs) (1) A deprecated, but still used unit of kinematic viscosity, 0.01 STOKE, the approximate kinematic viscosity of water at 20°C. The SI equivalent is: 1 cs = 10^{-6} m^2/s. (2) One one-hundredth of a stokes, which is the unit of kinematic viscosity:

$$v = \frac{\eta}{\rho}$$

where: η = viscosity ρ = density. Sin SI units 1 cSt = 1 mm^2/s.

Central Stop *n* The opaque stop usually placed in the objective back focal plane to give central stop dispersion staining.

Centrifugal Casting *n* A method of forming plastic in which the dry or liquid plastic is placed in a rotatable container. It is heated to a molten condition by the transfer of hear through the walls of the container, and rotated such that the centrifugal force induced will force the molten plastic to conform to the configuration of the interior surface of the container.

Centrifugal Clarifier *n* Apparatus somewhat similar to a cream separator, used for clarifying clear and colored solutions by throwing out solid particles by means of centrifugal force. Clarifiers of this type are operated at high speed. The bulky and apparently undissolved particles are thrown aside into a collector.

Centrifugal Coasting *n* The process of forming tubes or other hollow cylindrical objects by introducing a measured amount of fluid resin dispersion into a rotatable container or mold, rotating the mold about the cylinder's axis at a speed high enough to force the fluid against all parts of the mold by centrifugal force, maintaining such rotation while solidifying the plastic by applicable means such as heating, then cooling if necessary, and removing the formed part. The fluid resin may be a dispersion such as a plastisol, or an A-stage thermoset with or without reinforcing strands; This process should not be confused with ▶ Rotational Casting, which involves rotation at low speeds about one or more axes of rotation and gravity flow. See also ▶ Centrifugal Molding. Centrifugal casting is also used with metals, in particular, to manufacture ▶ Bimetallic Cylinders.

Centrifugal Impact Mixer *n* A device used for continuously mixing free-flowing dry blends, comprising a conical hopper in which are rotated at high speeds a rotor disk and a peripheral impactor. The material is fed to the center of the rotor which throws it against the impactor blades, which in turn throws the material against fixed impactors at the extremities of the cone. From there, the material flows downward to a discharge orifice. Compare: ▶ High-Intensity Mixer.

Centrifugal Molding *n* A process similar to Centrifugal Casting except that the materials employed are dry, sinterable powders such as polyethylene. The powders are fused by heating the mold then solidified by cooling it.

Centrifugal Pot See ▶ Pot Spinning.

Centrifugation *n* A method for determining the distribution of molecular weights by spinning a solution of the specimen at a speed such that the molecules are not removed from the solvent but are held at a point where the centrifugal force tending to remove them is balanced by the dispersive forces caused by the thermal agitation.

Centrifuge *n* Machine which exploits centrifugal force as a means of removing solid or semisolid particles from liquids.

Centripetal Force *n* The force required to keep a moving mass in a circular path. Centrifugal force is the name given to the reaction against centripetal force.

Cera Alba *n* Bleached or white beeswax.

Cera Flava *n* Yellow or unbleached beeswax.

Ceramic \sə-ˈra-mik, *esp Brit.* kə-\ *adj* [Gr *keramikos*, fr. *keramos* potter's clay, pottery] (1850) Technology of producing fired clay and porcelain articles, their glazes, pigments, and modifiers.

Ceramic Fiber *n* A term embracing all reinforcing fibers made of refractory oxides such as Al_2O_3, BeO, MgO, $MgO \cdot Al_2O_3$, ThO_2, and ZrO_2. Although glasses are also ceramic materials, glass fibers are not generally included. Ceramic fibers are produced by chemical vapor deposition, melt drawing, spinning, and extrusion. Their main advantages are high strength and modulus, and resistance to high temperatures.

Ceramic Fiber Reinforcements *n* Nonmetallic inorganic fibrous materials, available in a wide spectrum of forms, both continuous and discontinuous.

Ceramics *n* A general term applied to the art or technique of producing products by a ceramic process.

Ceraplast *n* Any reinforced thermoplastic, particularly polyethylene, containing ceramic or mineral particles that have been dispersed in the polymer melt to their ultimate size (no agglomerates) and completely enveloped in resin. Bonding of the envelope to the filler particles and the matrix polymer is aided by the addition of a small percentage of reactive monomer or resin precursor. It is believed that, in the extremely thin transition envelope, there is a smooth gradient of modulus from that of the particulate material to that of the polymer. The mechanical properties of ceraplasts are superior to those in which the same fillers have been conventionally incorporated.

Ceresine *n* Noncrystalline wax made by refining ozokerite. Mp, 60–71°C (140–160°F).

Cerium Naphthenate *n* Rare earth drier for air drying and baking finishes; sometimes used to replace lead naphthenate.

Cermet \ˈsər-met] *n* [*cer*amic + *met*al] (1948) (1) Composite materials consisting of two components, one being either an oxide, carbide, boride or similar inorganic compound and the other a metallic binder. (2) Any refractory composition made by bonding grains of ceramics, metal carbides, nitrides, etc., with a metal. Codeposition of cermets with nickel in the electroless-nickel process provides excellent wear resistance and

chemical resistance to molds, dies, extruder screws and other tooling components used in the plastics industry.

Cerotic Acid *n* A common name for either heptacosanoic acid or hexacosanoic acid. The former has the formula: $C_{26}H_{53}COOH$, and a mp of 82°C; the latter, the formula: $C_{25}H_{51}COOH$, and a mp of 88°C. Cerotic acid is a constituent of natural waxes, in which it occurs as the cerotate esters. Neocerotic acid, formula, $C_{24}H_{49}COOH$, mp of 77.8°C, is known as a constituent acid of beeswax.

Cerulean Blue \sə-ˈrü-lē-ən\ *n* Very complex pigment consisting essentially of a combination of cobalt and tin oxides. It is bluish-green in color and very stable.

Cerussite \sə-ˈrə-sīt\ *n* [Gr *Zerussit*, fr L *cerussa*] (1850) $PbCO_3$. Natural lead carbonate, found in the upper zone of lead deposits. Colorless, white, and gray. Sp gr of 6.55. Effervesces in nitric acid.

Ceryl Alcohol *n* $C_{26}H_{53}OH$. Alcohol obtained from Chinese wax.

Cetoleic Acid *n* $CH_3(CH_2)_9CH=CH(CH_2)_9COOH$. One of the constituent fatty acids of many fish oils, It has a single double bond.

Cetraria See ▶ Iceland Moss.

CF *n* Cresol/formaldehyde resin.

CFK Man-made fiber-reinforced plastics.

CFRP See ▶ Carbon-Fiber-Reinforced Plastics.

C Glass *n* A type of glass fiber not quite as strong and stiff as E GLASS but having better chemical resistance. Its major constituents are SiO_2 65%, CaO 13%, Na_2O 8%, B_2O_3 5%, Al_2O_3 3%, and MgO 2%. Fiber density is 2.49 g/cm³, tensile modulus is 71 GPa, and tensile strength is 3.2 GPa.

CGPM *n* Abbreviation for CONFÉRENCE GÉNÉRATE DES POIDS ET MESURES, the international group that developed the system of weights and measures intended for worldwide use. The name *Systéme International des Unités* and the abbreviation SI were adopted by the 11th CGPM in 1960. For information on SI units see SI (2) and the Appendix.

cgs System Centimeter-gram-second system. Now replaced by SI units.

cgs-EM System of Units System for measuring physical quantities in which the basic units are the centimeter, gram and second, and the numerical value of the magnetic constant, m, is unity. It is being replaced by the SI units.

Chafe Mark See ▶ Abrasion Mark and ▶ Shuttle Chafe Mark.

Chafed End *n* A warp end that has been abraded during processing. It generally appears as a dull yarn often containing broken filaments.

Chaffer Fabric *n* A fabric, coated with unvulcanized rubber, that is wrapped around the bead section of the tire before vulcanization of the complete tire. The purpose of the chafer fabric is to maintain an abrasion-resistant later of rubber in contact with the wheel on which the tire is mounted.

Chai Oil *n* Mexican oil obtained from *Salvia hispanica*. It has excellent drying properties. A prominent characteristic of the oil is its high surface tension, which causes it to "crawl," cooking at 260°C (500°F) for a short time destroys this property. Its main constituent acids may make up more than 90% of the total acids. Properties: Sp gr, 0.9338/15°C; refractive index, 1.4855; iodine value, 196.3; saponification value 192.2.

Chain Binders *n* Yarns running in the warp direction on the back of a woven carpet which hold construction yarns together.

Chain Branching *n* In a chain reaction, a step which produces more chain carriers than it consumes.

Chain Dyeing See ▶ Dyeing.

Chain Expansion Factor *n* The ratio of the length of a polymer chain in any given solvent to the length of the polymer chain in a theta solvent.

Chain Extension *n* Linear polymers (such as urethanes) of relatively low molecular weight may be extended by reaction of isocyanate endgroups with diamines to form interchain urea linkages, which, in turn add to other unreacted isocyanate endgroups to form biuret branching sites; reaction if isocyanate endgroups with diols to form new urethane linkages and extend the polymer chain, then, under more vigorous reaction conditions, by causing some still unreacted isocyanate endgroups to react with internal urethane linkages to form allophanate branching sites and crosslinks; and by reaction of isocyanate endgroups with water to form urea linkages between linear chains, again followed by biuret crosslinking reactions.

Chain Flexibility *n* The ability of polymer molecules to assume a variety of configurations, arising from freedom of segments to rotate around C–C bonds. Polar side groups generally hinder rotation, making chains stiffer, while alkyl side chains tend to increase flexibility. Higher melting point materials are often characterized by relatively low chain flexibility (higher chains stiffness).

Chain, Folded *n* Chain folded is the conformation of a flexible polymer molecule when present in a crystal. The molecule exists and reenters the same crystal, frequently generating folds.

Chain Isomerism *n* Isomerism involving differences in skeletal chains of atoms in molecules.

Chain Kinetics *n* The study of rates and mechanisms of chemical reactions and of the factors on which they depend.

Chain Length *n* The number of monomeric or structural units in a linear polymer, or the main chain in a branched polymer. This can also be written as the total length of chain molecule, measured from atom to atom along the chain. Kinetic chain length v gives the average number of monomer molecules that can be added to a free radical before the free radical is destroyed by a termination reaction. NOTE − This term should not be used for the direct distance between the ends of the molecule. See also ▶ Degree of Polymerization.

Chain Polymerization *n* Polymerization processes are of two basic types: stepwise (or step reaction) and chain reaction. The kinetics of the two types of reaction are entirely different; the properties of the polymers they produce differ with respect to molecular weight distribution and usually, although not inevitably, differ in kind. The vast majority of chain reaction polymerizations are those generally known as addition polymerizations involving the conversion of vinyl-type monomers.

Chain Reaction *n* In general, any self-sustaining process, whether molecular or nuclear, the products of which are instrumental in, and directly contribute to the propagation of the process. Specifically, a *fission chain* reaction, where the energy liberated for particles produced (fission products) by the fusion of an atom cause the fusion of other atomic nuclei, which in turn propagate the fission reaction in the same manner. In other words, a reaction type characterized by the formation of products of a later step (chain carriers) which are reactants for an earlier step.

Chain Scission *n* A degradation mechanism (300–500°C) involving rupturing of bonds in the backboned of the polymer chain, while nonchain scission is concerned with all other bond-breaking reactions caused by pyrolysis.

Chain Stiffness See ▶ Chain Flexibility.

Chain Transfer *n* Refers to the termination of a growing polymer chain and the start of a new one. The process is mediated by a chain transfer agent, which may be the monomer, initiator, solvent, polymer, or some species that has been added deliberately to affect chain transfer.

Chain-Transfer Agent *n* A substance, used in polymerization that has the ability to stop the growth of a molecular chain by yielding an atom to the active radical at the end of the growing chain, but in turn being converted to another radical that n initiate the growth of a new chain. Examples are thiols and carbon tetrachloride. Such agents are useful for preventing the occurrence of too long chains and too high ▶ Weight-Average Molecular Weights.

Chair Rail *n* The topmost molding of a dado, placed on the wall at the height of a chairback as a protection.

Chalcogen \kal-kə-jən\ *n* [ISV *chalk-* bronze, ore (fr. Gk *chalkos*) + *-gen*; from the occurrence of oxygen and sulfur in many ores] (ca. 1961) Any of the elements oxytgen, sulfur, selenium and tellurium. A member of group VIA in the periodic table.

Chalk *n* (1) A soft white mineral consisting essentially of Calcium Carbonate occurring as the remains of sea shells and minute marine organisms. (2) Soft naturally occurring form of calcium carbonate. The whiter varieties are powdered, washed, and dried. The product is known as whiting; the finer varieties as Paris white. (3) The product of chalking. See ▶ Chalking.

Chalkiness *n* (1) A dull, whitened appearance sometimes associated with certain extra-dull colors. (2) A fillingwise fabric defect observed as bands varying luster or sheen.

Chalking *n* (1) Formation of a friable powder on the surface of a paint film caused by the disintegration of the binding medium due to disruptive factors during weathering. The chalking of a paint film can be considerably affected by the choice and concentration of the pigment (BSI). It can also be affected by the choice of the binding medium. (2) A condition of a printing ink in which the pigment is not properly bound to the paper and can be easily rubbed off as a powder. (3) Formation of a powdery surface condition due to oxidation of the surface of rubber and release of pigments and fillers at the surface. Not to be confused with bloom, which looks similar.

Chalking Resistance *n* The ability of a coating to resist the formation of a friable powder on the surface of its film caused by the disintegration of the binding medium due to degradative weather factors. The chalking of a coating can be considerably affected by the choice and concentration of pigment and binding medium. See ▶ Chalking.

Chalk Kauri *n* Soft, white, powdery form of kauri copal.

Chalk Masking *n* The color change due to chalking and not due to a change in the colorant.

Chalk, Precipitated *n* Obtained as a by-product in water softening by sodium carbonate. It consists chiefly of calcium carbonate, but is often contaminated by appreciable quantities of calcium sulfate.

Challis \sha-lē\ *n* [prob. fr. the name *Challis*] (ca. 1837) A very soft, lightweight, plain-weave fabric, usually printed with a delicate floral pattern. The name is derived from the Anglo-Indian term "shalee" meaning soft.

Chamber White Lead *n* Particular type of white lead pigment, made by hanging strips or straps of lead in chambers, and subjecting them to the corrosive action of moist air, carbon dioxide, and acetic acid vapor. Chamber white lead differs from the stack form, usually being finer and of a better color. See ▶ Carter Lead.

Chambray \sham-brā, -brē\ *n* [irreg. fr. *Cambrai*, France] (1814) (1) A plain woven-spun fabric, almost square (i.e., 80 × 76), with a colored warp and a white filling. Lightweight chambrays are used for shirts, dresses, and children's clothes. (2) A similar but heavier fabric of carded yam, used for work clothing.

Chameleon \kə-mēl-yən\ *n* [ME *camelion*, fr. MF, fr. L *chamaeleon*, fr. Gk *chamaileōn*, fr. *chamai* on the ground + *leōn* lion] (14c) A variable multicolored effect achieved by using warp yarns of one color and two filling yarns of different colors in each shed. It is sometimes used in taffeta, faille, or poplin made from silk or manufactured filament yarns.

Change-Can Mixer *n* (pony mixer) A type of planetary mixer comprising several paddle blades mounted on a vertical shaft rotating in one direction while the can or contained counter-rotates. The paddle shaft is usually mounted on a hinged structure so that it can be swung out of the can, permitting the can to b removed, emptied, and replaced easily. This type of mixer is employed for relatively small batches (12–480 L) of fluid dispersions and dry materials.

Change in Filling See ▶ Filling, ▶ Mixed End or Filling.

Change in Length on Untwisting *n* The increase or decrease in length measured when a specimen is untwisted. The change is expressed as the percentage extension or contraction of the nominal gauge length of the specimen, i.e., specimen length prior to untwisting.

Channel Black *n* Form of carbon black made from natural gas by the channel combustion process. The gas is burned with insufficient air in jets, and the flames are allowed to impinge on a cool, channeled metallic surface. The deposited carbon is then scraped from the channel after a certain period of burning. Because of air pollution control requirements, this type of black has been almost completely replaced by Furnace Black in the U.S.

Channel Depth *n* (h or H) Of an extruder screw, at any point along its length, the radial distance between the flight-tip surface and the screw-root surface. In a screw section of constant depth, half the difference between the other (major) diameter of the screw and its root diameter.

Channel-Depth Ratio *n* In an extruder screw, the ratio of the depth in the first turn of the screw at the feed end to the depth in the last turn at the delivery end. If the Lead of the screw is constant, the channel-depth ratio is slightly larger than the channel-volume ratio (which follows).

Channel-Volume Ratio *n* In an extruder screw, the ratio of the volume of the first turn of the screw at the feed end to the volume of the last turn at the delivery end. The term ▶ Compression Ratio is commonly used as a synonym in the extrusion industry.

Char *n* (l, n) Animal or vegetable Carbon Black used as a decolorant in the process industries. (2, v) To partly burn and blacken, especially the outside surface of a carbonaceous material.

Characterization
Tension: Perhaps the best indication of the properties of a material is obtained from a tensile test in which a specimen with parallel sides is caused to extend.
Compression: The compression test is essentially similar to the tension test except that the force is in the opposite sense (i.e., push instead of pull). An additional complication arises because of a further failure mode: for slender specimens buckling may occur. Most standard test techniques employ some form of anti-buckling guide to suppress this failure.
Flexure: If a beam is supported at the ends and the middle is then forced downwards, there will be a gradual change from maximum compression stress at the upper surface to maximum tensile stress at the lower surface. The shear stresses will be minimum at the surface and maximum on the neutral axis (this axis is normally the centre-line between the two surfaces). Where a beam is tested in three-point loading, there will be significant shear forces throughout the beam except at very high span/depth rations. In four-point bending, the region between the central rollers will be in pure bending with shear outboard of the rollers. The optimum configuration for the determination of flexural modulus would be four-point bending with the deflection of the central section referenced to the loading rollers. An alternative is strain gauges in this position.
In-plane shear properties: In simple shear loading, two parallel faces move in opposite parallel directions. In pure shear, the plane is subjected to tensile forces on one axis and compressive forces of equal magnitude on the orthogonal axis. Many different techniques have been proposed for the determination of the in-plane and through-plane shear properties with variations appropriate to composite plates, rods and tubes [1].
Inter-laminar shear strength: The inter-laminar shear strength (ILSS) test is a three-point bend test at very small span/depth ratio (typically 4/1 or 5/1). This parameter should not be used for design purposes. The test is often used to monitor the quality of the laminate: it normally measures the strength of the fiber/resin interface or of the resin rich area between the laminae. The value for a good composite may be around 50 MPa.
Creep: Creep is the time-dependent deformation of a material under a sustained load. It is normal to design the composite to avoid such deformation by:

- limiting the working temperature to below the glass transition temperature (Tg),
- using a thermoset matrix with a high cross-link density,
- maximizing the area to minimize the stress,
- using the highest fiber modulus economically practicable,
- using the maximum fiber volume fraction economically practicable,
- orienting the fibers parallel to the stress direction,
- using the longest (preferably continuous) fiber available.

Fatigue: Fatigue is the time-dependent deformation of a material under a cyclic load. It is normal to undertake fatigue testing with a specific waveform (especially sine wave) or with pre-recorded deformations from a practical situation. In general, fatigue testing is undertaken from a small to a large percentage (e.g., 10–60%) of the ultimate (failure) load for the mode of testing. This usually avoids the complex gripping arrangements that are necessary for reversed cycle (tension and compression) loading.
Impact: For metals (which have a uniform microstructure) it is appropriate to use a pendulum impact machine (Charpy/Izod) to determine impact strengths and toughness. However, in composites the crack often turns and runs parallel to the fibers until is finds a specimen edge. In such a situation, the machine will under-record the strengths. For composites, the preferred methods are drop-weight impact or ballistics using a larger sample such that the failure does not extend to the specimen edge.
Non-Destructive Testing (NDT): There are many situations where the sampling of material for discrete tests is inappropriate and non-destructive examination is preferred. An example would be the determination of the extent of damage in an impacted panel. There are a number of techniques that can give clues in this context.
X-radiography (pictures from X-rays): does not yield good images of carbon-fiber composites because the density difference between the fiber and the resin matrix

is small. Glass fiber composites can be imaged more clearly. Cracks running parallel to the X-ray beam will be clearly imaged in either material but are rare in composites. Where there are surface-breaking cracks these can be imaged used penetrant-enhancement (e.g., silver iodide solution) but there may be concerns arising from composite-penetrant interactions.

Thermography: The differences in heat flows in a material can be imaged using thermal/infrared cameras (as used by emergency rescue teams). A delamination in a composite will act as a barrier to the passage of heat. If viewed from the heated surface during a transient heating, the delaminated area will heat up more quickly than the surrounding area because of the broken thermal path to the other surface. If heated from the opposite side, that area will heat up more slowly.

Ultrasound: Sound can travel as compression waves in fluids and as compression (longitudinal) or shear (transverse) waves in solids. The different terminologies come from engineering (physics). To interrogate defects in composite materials, a high-frequency pulse (ultrasound) is input to the material and the returned signal analysed. The trace of voltage against time is known as A-scan. Surfaces and defects reflect or scatter the pulse and hence influence the peak amplitude of the returned signal. By scanning the transducer n the x-y coordinates, a color map of the attenuated signal strength against position can be built. This display is known as C-scan. For example, in the C-scan at ▶ Figure 2 light areas indicate high attenuation of the signal due to absorption and scattering (the instrumentation used stretches the image along the x-axis).

Vibration: When an object is struck it will vibrate with a characteristic frequency and damping. For example, a rail vehicle wheel will ring clear and long if it is defect free and sound dull and die quickly if there is a significant defect. This is known as wheel-tap testing. An alternative is to use less energy and excite a more discrete area. This technique is known as coin tapping after the excitation source used.

Other NDT techniques exist, but they do not find such common usage in composites structures [2–5].

Thermal Tests

TGA: ThermoGravimetric Analysis measures weight changes in a material as a function of temperature (or time) under a controlled atmosphere to determine the thermal stability and composition.

DTA (Differential Thermal Analysis) and DSC (Differential Scanning Calorimetry) measure the thermal transitions in materials by monitoring temperatures and heat flows. DTA is qualitative while DSC is quantitative. Glass transition temperatures, phase changes, heat capacity, cure kinetics and thermal degradation can be monitored by these techniques. Modulated DSC can be both faster and more accurate.

DMTA: Dynamic Thermo-Mechanical Analysis measures the mechanical properties of materials under stress as a function of time, temperature, and frequency.

DETA: Dynamic Electrical Thermal Analysis measures the electrical properties of materials as a function of time, temperature, and frequency.

TMA (Thermo-Mechanical Analysis) measures the change in dimensions or in a mechanical property of the sample while it is subjected to a temperature regime.

STA: Simultaneous application of TGA and DSC to a single sample to yield more information than separate tests in different instruments.

The Role of the Different Thermal Anaylsis Techniques

TGA	DTA	DSC	DMTA	DETA	TMA
Weight loss moisture Chemical composition Catalyst activity Thermal stability Chemical stability	Melt temperature Glass transition State-of-cure Crystallization Chemical composition Catalyst activity Thermal stability Chemical stability Oxidative stability	Melt temperature Heat capacity Heat of Fusion Purity Glass transition State-of-cure Cure kinetics Crystallization Chemical composition Catalyst activity Thermal stability Chemical stability Oxidative stability	Modulus compliance Viscosity Rheology Glass transition State-of cure Kinetics tan δ Crystallization	Dielectric constant Rheology Glass transition State-of-cure Cure kinetics tan δ Crystallization	Thermal expansion Rheology Glass transition Crystallization
Decomposition kinetics	Decomposition Kinetics	Decomposition Kinetics Polymer compatibility	Polymer compatibility	Polymer compatibility	

Microscopy, *Optical*: For optimum resolution in optical microscopy it is necessary to produce an accurate flat surface which will sit normal to the optical axis of the microscope. Specimens are normally individually potted in an epoxy casting resin and (at the University of Plymouth) prepared using a Buehler 2000 Metpol grinder/polisher with Metlap fluid dispenser according to the procedure shown in the Table.

Charcoal Black *n* Black pigment obtained from wood charcoal.

Charge *(n)* The amount of material used to load a mold at one time or for one cycle. The amount may be expressed in either mass or volume units. In injection and transfer molding, the charge includes sprues and runners.

Charge-Transfer *n* Refers to the termination of a growing polymer chain and the start of a new one.

Char Length *n* In flammability testing, the distance from the edge of the sample exposed to the flame to the upper edge of the charred or void area.

Charles' Law *n* (Gay-Lussac's Law) The temperature part of the IDEAL-GAS LAW as follows: at constant pressure, the volume of any gas is directly proportional to its absolute temperature (K).

Charpy Impact Test *n* A destructive test (ASTM D 256B) of impact resistance using a centrally notched test specimen 126 mm long and typically 12.7 mm^2. The specimen is supported horizontally near its ends and struck on the side opposite the notch by a pendulum, having sufficient kinetic energy to break the specimen with one blow. The result is expressed as the quotient of the energy absorbed from the pendulum divided by the specimen width (J/cm or ft-lb/in).

Chase *n* (shoe) An enclosure of any shape used to (1) Shrink-fit parts of a mold cavity in place, (2) Prevent spreading or distortion in hobbing, (3) Enclose an assembly of two or more parts of a split-cavity block and (4) A rectangular metal frame in which type and plates are locked up for letterpress printing.

Chaser *n* Mill for dispensing pigments in binder to form very stiff pastes or putty. It consists of a circular pan in which one or two massive rollers run about central axes.

Chatki *n* Stone mill for grinding shellac.

Chatter *n* Transverse roll marks on roll coat-painted strip with varying film thicknesses which are usually due to roll coater vibration or a nonconcentric roller.

CHC *n* Chlorohydrin copolymers (from epichlorohydrin and ethylene oxide).

Checking *n* (1) That phenomenon manifested in paint films by slight breaks in the film that do not penetrate to the underlying surface. The break should be called a crack if the underlying surface is visible. Where precision is necessary in evaluating a paint film, checking may be described as visible (as seen with the naked eye) or as microscopic (as observed under a magnification of 10 diameters). (2) Equals CRAZING. See ▶ Cold-Checking and ▶ Checking Resistance.

Checking Resistance *n* The ability of a coating to resist slight breaks in its film that do not penetrate to the underlying surface. The breaks should be called cracks if penetration extends through the underlying surface. See ▶ Cracking Resistance. Where precision is necessary in evaluating a coating film, checking may be described as visible (as seen with the naked eye) or as microscopic (as observed under minimum magnification of 10 diameters). See ▶ Checking.

Checks *n* (1) Rough surface due to fine cracks from weathering. (2) Roughness formed on calendered sheet when temperature of calender rolls is too low or when the sheet is chilled too suddenly.

Cheese *n* (1) A supply of glass fiber wound into a cylindrical mass. (2) A cylindrical package of yarn wound on a flangeless tube.

Cheesecloth *n* A low-count, plain weave, soft cotton, or cotton blend cloth also known as *gauze*.

Cheesy *adj* Character of a paint or varnish film which, although dry, is mechanically weak and rather soft.

Chelate \ ▮kē-▮lāt *also* ▮chē-\ *adj* (1826) A compound comprised of metallic ions bound by a ▶ Chelating Agent.

Chelating Agent *n* (1) A term derived from the Greek word *chele*, meaning claw. Thus, a chelating agent is a substance whose molecules are capable of seizing and holding metallic ions in a claw like grip. (2) A sequestering or complexing agent that, in aqueous solution, renders a metallic ion inactive through the formation of an inner ring structure with the ion. (3) Organic compound that can remove many of the heavy metal cations from solution by forming soluble chelate compounds. Ethylenediaminetetraacetetic acid (EDTA) is typical. Syn: ▶ Sequestering Agent and ▶ Chelate.

Chelation *n* Reversible reaction of a metallic ion with a molecule or ion to form a complex molecule which does not have all or most of the characteristics of the original metallic ion. Syn: ▶ Sequestration.

Chemical Analysis *n* Chemical analysis refers to the determination of chemical structure and chemically active species. It involves both direct measurements and use of specific compounds to achieve selective reactions of a component of the substance being analyzed; to produce a readily measurable species; or to determine a reactive end product.

Chemical Change (Reaction) *n* A change in which one or more substances are transformed into one or more new substances.

Chemical Composition *n* Chemical composition is basically organic polymers that are very large molecules composed to chains of carbon atoms generally connected to hydrogen atoms (H), and often also to oxygen (O), nitrogen (N), chlorine (Cl), fluorine (F), and sulfur (S).

Chemical Conversion Coating *n* A treatment, either chemical or electrochemical, of the metal surface to convert it to another chemical form which provides an insulating barrier of exceedingly low solubility between the metal and its environment, but which is an integral part of the metallic substrate. It provides greater corrosion resistance to the metal and increased adhesion of coatings applied to the metal. Examples are phosphate coatings on steel or zinc and chromate coatings on aluminum.

Chemical Crimping *n* A crinkled or puckered effect in fabric obtained by printing sodium hydroxide onto the goods in a planned design. When the material is washed, the part to which the paste has been applied will shrink and cause untreated areas to pucker. The same effect is obtained with a caustic resist print and a sodium hydroxide bath.

Chemical Fiber See ▶ Manufactured Fiber.

Chemical Finishing *n* Processes in which additives are applied to change the aesthetic and functional properties of a material. Examples are the application of antioxidants, flame-retardant, wetting agents, and stain and water repellents.

Chemically Foamed Plastic *n* A cellular plastic in which the cells are formed by thermal decomposition of a blowing agent or by the reaction of gas-liberating constituents. See also ▶ Cellular Plastic.

Chemical Polymeric Blends *n* Mixtures of thermoplastic resins, etc.

Chemical Pretreatment See ▶ Chemical Conversion Coating.

Chemical Properties *n* Material characteristics that relate to the structure of a material and its formation from the elements.

Chemical Property *n* A property of a substance which can be described only by referring to a chemical reaction.

Chemical Reactions *n* Reactions between atoms and molecules to produce chemical compounds different from reactants. Chemical reactions may be elementary reactions or stepwise reactions.

Chemical Resistance *n* (Reagent resistance) The ability of a plastic to maintain structural and esthetic integrity when exposed to acids, alkalis, solvents, and other chemicals. ASTM tests for chemical resistance of plastics include: C 581, Chemical Resistance of Thermosetting Resins Used in Glass-Fiber-Reinforced Structures; D 543, Resistance of Plastics to Chemical Reagents; D 1239, Resistance of Plastic Films to Extraction by Chemicals; D 1712, Resistance of Plastics to Sulfide Staining; D 21451, Test Method for Staining of Polyvinyl Chloride Compositions by Rubber-Compounding Ingredients; D 2299, Determining Relative Stain Resistance of Plastics; D 3615, Chemical Resistance of Thermoset Molding Compounds Used in the Manufacture of Molded Fittings; D 3681, Chemical Resistance of Reinforced-Thermosetting-Resin Pipe in a Deflected Condition; D 3753, Specification for Glass-Fiber-Reinforced Polyester Manholes; and D 4398, Determining the Chemical Resistance of Fiberglass-Reinforced Thermosetting Resins by One-Side Panel Exposure.

Chemical-Resistant Paint See ▶ Chemical Resistance.

Chemicals for Electroplating *n* Most plastics plated today are finished with a copper/nickel/chrome electroplate, but many other finishes are possible, such as bright brass antique brass, satin nickel, silver, black chrome, and gold. The actual composition of the electroplate is designed for the particular application of the electroplate.

Chemical Shifts *n* The variation of the resonance frequency of a nucleus in nuclear magnetic resonance (NMR) spectroscopy in consequence of its magnetic environment. through absorption arising from shielding and deshielding by electrons, value is ppm. The chemical shift of a nucleus, δ, is expressed in ppm by its frequency, ν_{cpd}, relative to a standard, ν_{ref}, and defined as $\delta = 10^{6}(\nu_{cpd} - \nu_{ref})/\nu_{o}$ where: ν_{o} is the operating frequency of the spectrometer. For ^{1}H and ^{13}C NMR the reference signal is usually that of tetramethylsilane ($SiMe_4$).

Chemical Saturation *n* Absence of double or triple bonds in a chain organic molecule such as that of most polymers, usually between carbon atoms. Saturation makes the molecule less reactive and polymers less susceptible to degradation and crosslinking. Also called chemically saturated structure.

Chemical Stability *n* Degree of resistance of a material to chemicals, such as acids, bases, solvents, oils, and oxidizing agents, and to chemical reactions, including those catalyzed by light.

Chemical Unsaturation *n* presence of double or triple bonds in a chain organic molecule such as that of some polymers, usually between carbon atoms. Unsaturation makes the molecule more reactive, especially in free-radical addition reactions such as addition polymerization, and polymers more susceptible to degradation, crosslinking and chemical modification. Also called polymer chain unsaturation.

Chemically Saturated Structure See ▶ Chemical Saturation.

Chemiluminescence \ˌke-mē-ˌlü-mə-ˈne-sᵊn(t)s, ˌkē-\ n [ISV] (1889) Emission of light during a chemical reaction.

Chemisorption \ˈke-mi-ˌsȯrb, ˈke-, -ˌzȯrb\ vt [chem.- + -sorb (as in adsorb)] (1935) Adsorption, particularly when irreversible, by chemical action rather than physical action.

Chemist \ˈke-mist\ n [NL chemista, short for ML alchimista] (1562) (1) obs. Alchemist, one trained in chemistry (2) Brit: Pharmacist. (3) A professional who possesses an earned bachelor's or higher degree with a major in a science from an accredited institution and who develops, applies, or communicates the principles of chemistry and exercises independent judgment and discretion in conceiving, planning, coordinating, or executing chemical projects, or who has experience in so doing. The chemical sciences deal with the composition, structure, and properties of substances and of the transformations they undergo.

Chemistry \ˈke-mə-strē\ n The science of the compositions and structures of substances and of the changes which these undergo. The study of molecules and the reactions they undergo.

Chenille \shə-ˈnē(ə)l\ n [F, literally, caterpillar, fr. L canicula, dim. of canis dog; fr. its hairy appearance] (ca. 1739) (1) A yarn with a fuzzy pile protruding from all sides, cut from a woven chenille weft fabric. Chenille yarns are made from all fibers, and they are used as filling in fabrics and for embroidery, fringes, and tassels. (2) Fabric woven with chenille yarn. Also see ▶ Tufted Fabric.

Cherry n An even-textured, moderately high-density wood of the eastern U.S.A., rich red-brown in color; takes a high luster; used for cabinetwork and paneling.

Cheviot \ˈshe-vē-ət, esp British ˈche-\ n (1815) A rugged tweed made from uneven yarn, this fabric usually has a rather harsh hand.

Chevron n \ˈshev-rən\ n [ME, fr. MF, rafter, fr. (assumed) VL caprion-, caprio rafter, akin to L caper goat] (14c) A broad term applied to prints in zigzag stripes or to herringbone weaves.

Chicle n \ˈchi-kəl, ˈchi-klē\ n [AS, fr. Nahautl tzictli] (ca. 1889) Raw material for chewing gum (mixture of trans-1,4-poly(isoprene) + triterpenes. A natural product.

Chiffon n A plain weave, lightweight, sheer, transparent fabric made from fine, highly twisted yarns. It is usually a square fabric, i.e., having approximately the same number of ends and picks and the same count in both warp and filling.

Chill-Back n A material capable of reducing the temperature of a batch of varnish or resin, either because of a considerable temperature difference between itself and the hot material, or because it absorbs heat in melting also.

Chillers n A self-contained (individual or central) system comprised of a refrigeration unit and a coolant circulation mechanism consisting of a reservoir and a pump.

Chilling n (1) In varnish or resin manufacture, the deliberate and rapid reduction of the temperature of the charge in a varnish or resin pot. Also called Quenching. (2) On a painted or varnished surface, a clouding of the surface or a reduction of luster as a result of the movement and cold air over the drying surface.

Chill Roll n A metal roll – a shell within a shell with a relatively small clearance between the two – with water or other heat-transfer medium circulating through the so-formed annular space, used to cool an extruded or cast film or sheet prior to winding or cutting. The surface of the roll may be polished or textured to impart a finish to the plastic.

Chill-Roll Extrusion n Film extrusion in which the molten film is drawn over cooled, polished rolls, imparting high gloss to the film. Syn: ▶ Cast Film Extrusion.

Chimb See ▶ Chime.

Chime n The top ring of a coating container into which the friction top is pressed in order to effect an air tight seal. Syn: ▶ Chimb and ▶ Chine.

Chimney n An incombustible vertical structure containing one or more flues to provide draft for fireplaces, and to carry off gaseous products of combustion to the outside air from fireplaces, furnaces, or boilers.

China Clay n $Al_2O_3 \cdot 2SiO_2 \cdot 2H_2O$. Pigment White 19 (77005). A complex hydrated aluminum silicate produced by the breakdown of the mineral feldspar. The finest grades are used by the color and paint trades – often under proprietary names – and are utilized as a base for lakes and as an extender and anti-settling agent in paints. See ▶ Aluminum Silicate. Also known as Kaolin.

China Wood Oil See ▶ Tung Oil.

Chinchilla Cloth n A heavy, twill weave, filling-pile fabric with a napped surface that is rolled into little tufts or nubs. The material is frequently double faced with a knitted or woven, plain or fancy back. Chinchilla cloth is used primarily in coats. The term is also used to refer to a knitted woolen fabric having a napped surface.

Chine See ▶ Chime.

Chinese Blue Term often applied to better grades of Prussian blue. See ▶ Milori Blue and ▶ Iron Blue.

Chinese Ink See ▶ India Ink.

Chinese Lac *n* Resinous exudation of the wild Chinese tree, *Angia sinensis* or *Tssichau*, from which it is obtained by making deliberate incisions in the trunk. It is very susceptible to exposure, changing in color gradually, through red to black. The resinous exudation has very good solubility, blending satisfactorily with turpentine and alcohol.

Chinese Red See ▶ Chrome Orange, Light and Deep.

Chinese Vegetable Tallow Fatty substances from the kernels of the Chinese vegetable tallow tree, *Stillingia sebifera*. This is the same tree which yields stillingia oil.

Chinese Vermillion See ▶ Mercuric Sulfide.

Chinese Wax An insect secretion, collected from certain evergreens which are indigenous to China. Properties: mp, 81°C; sp gr, 0.970; saponification value, 92; and a negligible iodine value. *Known also as Insect Wax.*

Chinese White (1) A paint using zinc oxide as the principal pigment. (2) Syn: ▶ Zinc Oxide.

Chino A cotton or cotton blend twill used by armies throughout the world for summer-weight uniforms. Chino is frequently dyed khaki.

Chinoiserie \shēn-ˈwäz-rē, -ˈwä-zə-; ˌshēn-ˌwäz-ˈrē, -ˌwä-zə-\ *n* [F, fr. *chinois* Chinese, fr. *Chine* China] (1883) Originally, European designs "in the Chinese taste." Now loosely applied to almost any oriental form of decoration.

Chinon Graft copolymer of 70% acrylonitrile on 30% casein. Manufactured by Toyoba, Japan.

Chintz \ˈchin(t)s\ *n* [earlier *chints*, plural of *chint*] (1614) A glazed fabric produced by friction calendering. Unglazed chintz is called cretonne.

Chip (1) The form of polymer feedstock used in fiber production. Also see ▶ Flake. (2) The feedstock for a pulp digestor. (3) A defect in a nonwoven fabric

Chip Board *n* Paperboard generally made from reclaimed paper stock and frequently used as a facing for partitions. The board is of moderately low density and should not be confused with particle board.

Chipping *n* (1) Removal of paint, or rust and scale, by mechanical means. (2) Total or partial removal of a dried paint film in flakes by accidental damage or wear during service; in traffic paints, this failure is usually characterized by sharp edges and definite demarcation of the base area. (3) Spreading of white or colored chips or flakes during the application of seamless flooring. This procedure is used to impart a decorative effect to the flooring.

Chipping Rating Standards *n* A set of photographic standards depicting the size and number of chips in each of several categories, as described in ASTM Method D-3170.

Chipping Resistance *n* The ability of a coating or layers of coatings to resist total or partial removal, usually in small pieces, resulting from impact by hard objects or from wear during service. See ▶ Chipping.

Chips *n* (1) Term used to describe the size of resin particles. It represents a form intermediate between bold lumps and dust. (2) Dry combination of binder and pigment which in solvent forms paint, or which can be added to color plastics (e.g., vinyl chips and nitrocellulose chips).

Chirality \ˈkī-rəl\ *adj* [*chir-* + *¹-al*] (1894) The property of an organic molecule of not being identical with its mirror image; a compound whose molecules are chiral can exist as enantiomers, but non-chiral compounds cannot be enantiomers. All asymmetric molecules are chiral; however, not all chiral molecules are asymmetric since some having axes of rotational symmetry are chiral. Chiral and prochiral atoms are sites or potential sites, respectively, of Stereoisomerism.

Chlorazol Dyes *n* Range of soluble dyestuffs which are used in the manufacture of colored inks.

Chlorendic Anhydride *n* (1,4,5,6,7,7-hexachloro-5-norbornene-2,3-dicar-boxylic anhydride, HET anhydride, a difunctional acid anhydride. HET anhydride is a white, crystalline powder used as a hardening agent and flame retardant in epoxy, alkyd, and polyester resins.

Chlorendic Polyester *n* A chlorendic anhydride-based unsaturated polyester.

Chlorinated Biphenyl *n* (chlorinated diphenyl) Any of a group of plasticizers ranging from liquids to hard solids, used with polyvinylidene chloride and polystyrene. They are also used in conjunction with DOP as coplasticizers for PVC, and in conjunction with polyvinyl acetate, ethyl cellulose and other thermoplastics, as adhesives.

Chlorinated Diphenyls *n* (PCB) Range of chlorinated hydrocarbons varying in properties from very liquid plasticizers, through thick syrups to solid resins. They are characterized by excellent compatibility, stability, absence of free acidity, high toxicity, and nonflammability.

Chlorinated Hydrocarbon *n* Any of a wide variety of liquids and solids resulting from the substitution or addition of chlorine in hydrocarbons such as methane, ethylene, and benzene. They are employed as solvents, plasticizers, and monomers for plastics manufacture.

Chlorinated Hydrocarbons *n* Powerful solvents which include such members as chloroform, carbon tetrachloride, ethylene dichloride, methylene chloride, tetrachorethane, trichloroethylene, etc. Generally speaking, they are toxic. Their main applications include nonflammable paint removers, cleaning solutions, and special finishes where presence of residual solvent in the film is a disadvantage. Many are not prohibited in most countries.

Chlorinated Paraffin *n* (chlorocosane) Any of a family of yellow to light amber liquids produced by chlorinating a paraffin oil, with uses as secondary plasticizers for vinyls, polystyrene, polymethyl methacrylate, and coumarone-indene resins. Chlorinated paraffins also impart flame resistance to polyolefins, polystyrene, PVC, natural rubber, and unsaturated polyester resins.

Chlorinated Paraffin Wax *n* Ordinary paraffin wax can be chlorinated, under certain conditions, to yield products which no longer resemble the parent wax, but which have assumed definite resinous characteristics. The amount of chlorine present determines the physical properties of the resultant waxes. These with smaller amounts of chlorine are liquids, whereas those with about 70% chlorine are brittle resins. Their applications are in fire-retarding compositions and as plasticizers.

Chlorinated Para Nitraniline Red See ▶ Chlorinated Para Reds.

Chlorinated Para Reds *n* Pigment Red 4 (12085). There are two varieties of chlorinated para red: (1) Ortho-chlor-papa nitraniline and (2) Para-chlorotho nitraniline; both are diazotized and coupled to beta naphthol. Both are light yellow shade reds of the toluidine red orange.

Chlorinated Polyether *n* [poly-3,3-bis-(chloromethyl) oxacyclobutane, Penton®] A corrosion-resistant thermoplastic obtained by polymerization of chlorinated oxetane monomer, the oxetane being derived from ▶ Pentaerythritol, to a high molecular weight (250,000–350,000). The polymer is obtained from pentaerythritol by preparing a chlorinated oxetane and polymerizing it to a polyether by means of opening the ring structure. The polymer is linear, crystalline, and extremely resistant to degradation at processing temperatures. It may be injection molded, extruded, or applied as a coating by the fluidized-bed method. The resin is widely used in valves, pumps, flow meters, etc., for chemical plants.

Chlorinated Polyethylene *n* (CPE) Any polyethylene modified by simple chemical substitution of chlorine on the linear backbone chain, CPEs range from rubbery amorphous elastomers at 35–40% Cl to hard, semicrystalline materials at 68–75% Cl. They are sometimes included with chlorinated natural and butyl rubbers under the term *chlorinated rubbers*. Certain CPEs are used as modifiers in PVC compounds to obtain better flexibility and toughness, particularly low-temperature toughness, greater latitude in compounding, and ease of processing.

Chlorinated Polyvinyl Chloride *n* (CPVC) A PVC resin modified by post-chlorination. A series of such resins, known as "Hi-Temp Geon," is available from B F Goodrich Chemical Co. Compared to conventional rigid PVC, CPVC withstands service temperatures 20–30°C higher, is stronger, and has better chemical resistance. CPVC is mildly hygroscopic, so requires predrying before processing. With that proviso, it can be processed by all the methods used for rigid PVC with few modifications.

Chlorinated Polyvinyl Chloride Resins *n* CPVC is a plastic produced by the post-chlorination of PVC.

Chlorinated Rubber *n* (rubber chloride) (1) Natural rubber in which about two-thirds of the hydrogen atoms have been replaced by chlorine atoms. The resin is formed by the reaction of rubber with chlorine at about 100°C in an inert solvent or as a latex. Unlike rubber, the resulting product is readily soluble and yields solutions of low viscosity. It is sold as white powder, fibers, or as blocks. Commercial products generally contain about 65% chlorine. It has good chemical resistance properties, however, it tends to cobweb when sprayed. Now mostly chlorinated polymers are used, as 1-butene, polyethylene, etc. It has adhesive properties and, because of its good fire resistance, is used in paints. (2) Chlorinated natural rubber. Manufactured by Bayer, Germany.

Chlorine Retention *n* A characteristic of several resins and textile finishes whereby they retain some of the chlorine from bleach. On heating of the goods, the chlorine forms hydrochloric acid, causing tendering of the cloth. This is especially true of certain wrinkle resistant finishes for cotton and rayon.

Chlorobenzene \ˈklȯr-ō-ˈben-ˌzēn, ˈklȯr-, -ben-ˈ\ *n* [ISV] (ca. 1889) (chlorbenzene, chlorbenzol, chlorobenzol, phenylchloride) C_6H_5Cl. A solvent, and an intermediate in the production of phenol.

Chlorobutanol *n* (chlorbutanol, 1,1,1-trichloro-2-methyl-2-propanol, acetone chloroform, trichloro-*tert*-butyl alcohol) $Cl_3CC(CH_3)_2OH$. A plasticizer for esters and ethers of cellulose.

Chlorodiphenyl Resin *n* (chlorobiphenyl resin) Any resin made from chlorinated biphenyl, rosin or rosin ester, and the higher fatty acids. These resins are used as plasticizers and modifying resins in plastics, and in lacquers and varnishes.

Chloroethane *n* Syn: ▶ Ethyl Chloride.

Chloroethene (chlorethylene) Syn: ▶ Vinyl Chloride.

Chlorofluorocarbon Resin *n* Any resin made by the polymerization of monomer(s) containing only carbon, chlorine, and fluorine. The principal member is Polychlorotrifluoroethylene (PCTFE).

Chlorofluorohydrocarbon Resin *n* A resin made by polymerization of monomer(s) containing only carbon, chlorine, fluorine, and hydrogen.

Chloroform \ˈklȯr-ə-ˌfȯrm, ˈklȯr-\ *n* [F *chloroforme*, fr. *chlor-* + *formyle* formyl; fr. its having been regarded as a trichloride of this group] (1838) (trichloromethane) $CHCl_3$. A pungent, toxic, dense liquid, useful as a solvent for epoxy resins and others. (2) Clean, colorless, volatile liquid. Sp gr of 1.485 (20/20°C); bp of 61.2°C; fp of 63.5°C; wt/gal of 12.29 lb (25°C); refractive index of 1.4422.

Chlorohydrin \ˌklȯr-ə-ˈhī-drən, ˌklȯr-\ *n* [ISV *chlor-* + *hydr-* + ^{1}-*in*] (ca. 1890) (α-chlorohydrin, 3-chloropropane-1,2-diol, glyceryl α-chlorohydrin) $ClCH_2CHOHCH_2OH$. A solvent, mainly for cellulosics.

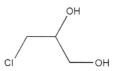

Chlorohydrin Rubber See ▶ Epichlorohydrin Rubber.

α-Chloro-*meta*-Nitroacetophenone *n* $O_2NC_6H_4CO$ OH_2Cl. A bacteriostat and fungistat for plastics.

Chloronaphthalene Oil *n* Any of several nearly colorless oils derived by chlorinating naphthalene, used as plasticizers and flame retardants.

Chloroprene \-ˌprēn\ *n* [*chlor-* + iso*prene*] (1931) $CH_2Cl–CHCl–CH=CH_2$. Polymerization of the monomer to poly(1-chloroprene) to form the synthetic chlorinated elastomer.

Chloroprene Polymer See ▶ Neoprene.

Chloropropylene Oxide *n* Syn: ▶ Epichlorohydrin.

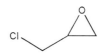

Chlorostyrenated Polyester *n* An unsaturated polyester resin made by reacting a fluid polyester with monochlorostyrene in place of styrene. (See ▶ Unsaturated Polyester.) Monochlorostyrene is less volatile and more reactive than styrene, providing faster cure rates and increased flexural strength and modulus in glass-fiber laminates.

Chlorosulfonated Polyethylene *n* (CSM, CSPR) Polyethylene which has been reacted with a mixture of chlorine and sulfur dioxide under ultraviolet light irradiation. Polymer may be vulcanized to form a product with good ozone, heat, oxygen and weathering resistance. It is produced by simultaneous treatment, with sulfur dioxide and chlorine, of dissolved, radicalized polyethylene. A commercial product (Hypalon®, DuPont) contains 22–26% Cl and 1.3–1.7% S.

Chlorosulfonated Polyethylene Rubber *n* Thermosetting elastomers containing 20–40% chlorine. Have good weatherability and heat and chemical resistance. Used for hoses, tubes, sheets, footwear soles, and inflatable boats.

Chlorotrifluoroethylene *n* $ClFC=CF_2$. A colorless gas. The monomer for ▶ Poly-Chlorotrifluoroethylene. It is obtained by either dehalogenation or dehydrohalogenation of saturated chlorofluorocarbons or chlorohydrocarbons, e.g., by reacting 1,1,2-trichlorotrifluoroethane with zinc. The monomer for the preparation of polychloro-trifluoroethylene by free radical polymerization in aqueous systems.

Choked Coiler *n* A condition in carding or drawing in which sliver is either puffy, badly condensed, or very uneven, leading to overloading of the coiler trumpets and causing work stoppage.

Choked Flyers *n* A situation in which roving will not pass through the flyer channels because of heavy or cockled conditions caused by such factors as uneven drafting, waste, overcut fibers, and improper finish.

Choker Bar *n* (restrictor bar) A bendable metal bar incorporated in a sheet-extrusion die for controlling flow distribution and lateral sheet thickness, and for reducing stagnation in the melt. The shape of a flow passage between the choker bar and the lower die body is altered by turning bolts connected to the bar.

Cholesteric \ kō-lə-▮ster-ik, kə-▮les-tə-rik\ *adj* [*cholesteric* relating to cholesterol, fr. F *cholesterique*] (1942) See ▶ Liquid-Crystal Polymer.

Chopped Strand *n* A type of ▶ Glass-Fiber Reinforcement consisting of strands of individual glass fibers that have been chopped into lengths from 1 to 12 mm. The individual fibers are bonded together within the strands so that they remain in bundles after being cut. See also ▶ Roving.

Chopped-Strand Mat *n* A mat formed from randomly oriented, chopped strands of glass and held together, just strongly enough for handling, by a binder.

Choppers *n* Chopper guns, long cutters, roving cutters cut glass fibers into strands and shorter fibers to be used as reinforcements in plastic.

CHR Chlorohydrin elastomer [poly(epichlorohydrin)].

Christiansen Effect *n* When finely powdered substances, such as glass or quarts, are immersed in a liquid of the same index of refraction complete transparency can only be obtained for monochromatic light. If white light is employed the transmitted color corresponds to the particular wavelength for which the two substances, solid and liquid have exactly the same index of refraction. Due to differences in dispersion the indices of refraction will match for only a narrow band of the spectrum.

Chroma \ ▮krō-mə\ *n* [Gk *chroma*] (ca. 1889) (1) Color intensity or purity of tone, being the degree of freedom from gray. (2) One of the three terms used in Munsell notations. It correlates approximately with the psychological dimension of saturation.

Chromatic Aberration *n* A lens receiving white light from an object will form the blue image closer to the lens than the red image, forming color fringes. This caused by refractive index variation with wavelength (dispersion), always blue greater than red.

Chromaticity *n* The quality of color expressed as a function of wavelength and purity

Chromaticity Coordinates *n*, **CIE** The ratios of each of the three tristimulus values X, Y and Z in relation to the sum of the three; designated as x, y, and z, respectively. They are sometimes referred to as the trichromatic coefficients. When written without subscripts they are assumed to have been calculated for Illuminant C and the 2° (1931) Standard Observer unless specified otherwise. If they have been obtained for other illuminants or observers, a subscript describing the observer of illuminant should be used. For example, x_{10} and y_{10} are chromaticity coordinates for the 10° observer and Illuminant C.

Chromaticity Coordinates *n*, **General** The two dimensions of any color order system which exclude the lightness dimension and describe the chromaticity. Thus, x and y, or a and b, or u and v, or YB and RG, may be considered as chromaticity coordinates in various color spaces. (Unless otherwise specified, the term chromaticity coordinates is assumed to refer to the CIE coordinates x, y, and z for Illuminant C and the 2° (1931) Standard Observer. *Also known as Color Coordinates.*

Chromaticity Diagram *n*, **CIE** A two-dimensional graph of the chromaticity coordinates, x as the abscissa and y as the ordinate, which shows the spectrum locus (chromaticity coordinates of monochromatic light, 380–770 nm). It has many useful properties for comparing colors of both luminous and nonluminous materials.

Chromaticity Diagram *n* General Plane diagram formed by plotting one of the chromaticity coordinates against the other.

Chromaticity Difference Diagrams *n* Plane diagram formed by plotting the differences of one of the chromaticity coordinates against the differences of the other chromaticity coordinate, the differences being taken from the neutral point or, for a specific color, from the chromaticity coordinates for the standard of that color.

Chromaticness *n* Combined hue and saturation of a color, with no reference to its lightness. Syn: ▶ Chromaticity.

Chromatography *n* (1937) In chemistry, analytical technique used for the chemical separation of mixtures and substances. A process in which a gas or liquid solution moves through a calibrated column containing a subdivided solid phase into which some components of the solution are absorbed, smaller molecules more quickly and thoroughly than larger ones. This is followed by pure carrier gas or solvent, the stream being monitored by a differential detector. Larger molecules emerge first, smaller ones later. The detector signal is proportional to the concentration of each species in the effluent. The process is mainly used for analysis of organic mixtures, but also for their separation. The name "chromatography" derives from the work of the Russian botanist N. Tswett, who first used the process to separate chloroplast pigments, obtaining colored bands on filter paper. Some variations of the process are *gas chromatography* (the gas mixture is passed through a porous bed, or through a capillary tube lined with an absorbent liquid or solid phase); *paper chromatography* (a drop of specimen is placed near one end of a porous paper); *ion-exchange chromatography; thin-layer chromatography* (the sample is placed on an absorbent cake spread on a smooth glass plate); and, important for plastics, ▶ Size-Exclusion Chromatography.

Chrome Green *n* (brunswick green) $PbCrO_4 \cdot xPbSO_4 \cdot yFeNH_4Fe(CN)_6$. Any of a family of pigments ranging from light yellow-green through dark green, based on physical mixtures of chrome yellow and iron blue (a complex ammonium iron hexacyanoferrate). The amount of iron blue determines the shade, about 2% being used for light yellow-green, and up to 64% being used for dark greens. Variation can be achieved by pigment blends with extenders. Noted for good hiding and tinting strength, poor alkali resistance. Density, 4.1–5.4 g/cm^3 (34.2–45.0 lb/gal); O.A., 13–25; particle size, 0.2–1.2 μm. Syn: ▶ Chromium Green, ▶ Lead Chrome Green, ▶ Green, Brunswick and ▶ Milori Green.

Chromel and Alumel *n* Special high-nickel alloys that, when their wires are joined, develop high thermoelectric power, and are therefore useful for temperature measurement. "Type K" thermocouples are usually chosen for high-temperature work in oxidizing atmospheres.

Chrome Ochre See ▶ Chromium Oxide Green.

Chrome Orange, Light and Deep *n* $PbCrO_4 \cdot PbO$. Pigment Orange 21 (77601). Chrome orange is chemically basic lead chromate. The hue ranges from light to deep orange, and the color is related to particle size. A pigment characterized by good opacity, fair lightfastness, and good gloss retention and durability. Disadvantages are poor alkali and acid resistance and poor high temperature stability. Density, 6.6–7.1 g/cm^3 (55.1–58.8 lb/gal); O.A., 9–12; particle size, 0.1–1.0 μm. Syn: ▶ Orange Chrome, ▶ Austrian Cinnabar, ▶ Chinese Red, ▶ Derby Red, ▶ American Vermillion, and ▶ Chrome Red.

Chrome-Orange Pigment *n* Any pigment based on basic lead chromate, $PbO \cdot PbCrO_4$, that is of a deep orange color.

Chrome Oxide Green *n* A stable pigment based on anhydrous chromium oxide, Cr_2O_3. The form based on hydrated chromium oxide is called *Guignet's Green*. See ▶ Chromium Oxide Green. Syn: ▶ Schnitzer's Green.

Chrome Red See ▶ Basic Lead Chromate, and ▶ Chrome Orange, Light and Deep.

Chrome Vermillion See ▶ Molybdate Orange.

Chrome Yellow *n* A light resistant opaque yellow pigment composed essentially of lead chromate.

Chrome Yellow, Light and Primrose *n* $xPbCrO_4 \cdot yPbSO_4$. Pigment Yellow 34 (77603). Yellow pigments based on combinations of lead chromate and lead sulfate. The hue and tint strength being controlled by the ratio of chromate to sulfate during precipitation. In the light and lemon hues, the approximate ratio of x/y is 2.5/1 (2.5 $PbCrO_4 \cdot PbSO_4$); in the primrose hues, the x/y ratio is approximately 3.2/1. The pigments are characterized by good opacity, fair lightfastness, good glass retention and durability, and poor resistance to alkali, acid, and high temperature. Density, 5.44–6.09 g/cm^3 (17–39 lb/gal); O.A., 17–39; particle size, 0.1–0.8 μm. Syn: ▶ Cologne Yellow, ▶ Golden Chrome, ▶ Leipsic Yellow, ▶ Lemon Yellow, and ▶ Middle Chrome.

Chrome Yellow, Medium *n* PbCrO4. Pigment Yellow 34 (77600). Approaches chemically pure lead chromate. Characterized by good opacity, fair to good lightfastness, good gloss retention and durability, poor alkali resistance. Density, 5.58–6.04 g/cm^3 (46.6–50.3 lb/gal); O.A., 16–35; particle size, 0.15–1 μm.

Chrome Yellow Pigment *n* (primrose chrome, permanent yellow) Any pigment based on normal lead chromate, $PbCrO_4$ that is characterized by a medium yellow color. Other shades ranging from light greenish-yellow to medium reddish-yellow are made by coprecipitating lead chromate with other insoluble compounds such as lead sulfate or lead phosphate.

Chromic Acid n (1800) H_2CrO_4 An acid analogous to sulfuric acid but known only in solution and in the form of its salts.

$$O=Cr(OH)_2(=O)$$

Chromic Chloride n (chromium chloride, chromium chloride hexahydrate) $CrCl_3$ or $[Cr(4H_2O)Cl_2]$ $Cl·2H_2O$. A catalyst for polymerizing olefins.

$$Cr^{+++}\ with\ 3\ Cl^-$$

Chromic Hydrate See ▶ Chromium Hydroxide.
Chromic Oxide See ▶ Chromium Oxide Green.

$$Cr^{+++}\ Cr^{+++}\ with\ 3\ O^{--}$$

Chromium Green See ▶ Chrome Green.
Chromium Hydrate See ▶ Chromium Hydroxide.
Chromium Hydroxide n $Cr(OH)_3$. Green pigment, manufactured by adding a solution of ammonium hydroxide to the solution of a chromium salt. Syn: ▶ Chromic Hydrate and ▶ Chromium Hydrate.

$$Cr^{+++}\ with\ 3\ OH^-$$

Chromium Oxide Green n Cr_2O_3. Pigment Green 17 (77288). Most permanent green pigment, almost pure chromium sesquioxide. May be manufactured by reducing a chromate (e.g., sodium dichromate) with sulfur or carbonaceous materials. Outstanding lightfastness and resistance to acid, alkali and high temperatures often used as a colorant in cementitious products. Density, 5.1–5.4 g/cm^3 (42.5–45.0 lb/gal); O.A., 12–24; particle size 2–6 μm. Syn: ▶ Arnaudon's Green, ▶ Chromium Sesquioxide, ▶ Chrome Oxide Green, ▶ Vert Emeraude, ▶ Anadonis Green, ▶ Chrome Ochre, ▶ Green Cinnabar, ▶ Green Rouge, ▶ Leaf Green, ▶ Oil Green, ▶ Dingler's Green, and ▶ Schnitzler's Green.
Chromium Oxide, Hydrated See ▶ Hydrated Chromium Oxide.
Chromium Plating n (chrome plating) An electrolytic process that deposits a hard, inert, smooth layer of chromium onto working surfaces of other metals for resistance to corrosion and wear. Extruder screws, chill rolls for sheet and film production, calendaring rolls, dies, and molds are commonly chromium plated.
Chromium Sesquioxide See ▶ Chromium Oxide Green.
Chromophore n (1) A group such as –NO, –NO$_2$, or –N=N– that, when present in a molecule, enables the molecule to be transformed into a dye upon the introduction of an acid group. (2) Certain groups of atoms such as, –C=Cl, –C=N–, –N=N–, –N=O, when present in an organic molecule, can give rise to colored compounds. Groups acting in this way are called chromophores. Although compounds containing chromophores are not necessarily colored. The hue is affected by the presence of other groups known as "auxochromes," such as –NH$_2$, –OH, –NO$_2$, –CH$_3$ and halogens –Cl, –Br, and –I.
Chrysotile \-ˌtīl\ n [Gr *Chrysotil*, fr. *chrys-* + *-til* fiber, fr. Gk *tillein* to pluck] (1850) (serpentine) $Mg_3Si_2O_5(OH)_4$ for $3MgO·2SiO_2·2H_2O$. A hydrated magnesium orthosilicate, the chief constituent of *serpentine* asbestos. Chrysotile-bearing asbestos has been the most used type, once accounting for over 90% of the world production. Its fine and silky fibers, and mats and felts made therefrom, where widely used as fillers and reinforcements for plastics, providing excellent resistance to chemicals and fire. Many of its former uses are now prohibited because of the carcinogenicity of some types of asbestos. See ▶ Fibrous Asbestos and ▶ Magnesium Silicate, Fibrous.
Chute-Feed System n Pneumatic fiber transport system used in linking textile processing equipment or operations, especially opening, blending, and carding.
CI n Abbreviation for Colour Index.
Cibanoid n Urea-formaldehyde resin. Manufactured by CIBA, Switzerland.
C.I.E n Abbreviation for COMMISSION INTERNATIONALE d'ÉCLAIR-AGE, the French name for the International Commission on Illumination. In older publications, the abbreviation ICI was used.
CIE Chromaticity Coordinates See ▶ Chromaticity Coordinates, ▶ CIE.
CIE Color Difference Equation n May refer to the color difference equation provisionally adopted by the CIE in 1964, one of four then recommended for study. It incorporates a linear transformation into a more nearly uniform color space. The following are the equations used:

$$U^* = 13W^*(u - u_o)$$
$$V^* = 13W^*(v - v_o)$$
$$W^* = 25Y^{1/3} - 17\ (1 \leq Y \leq 100)$$

$$u = \frac{4X}{X + 15Y + 3Z} \quad \text{or} \quad u = \frac{4x}{-2x + 12y + 3}$$

$$v = \frac{6X}{X + 15Y + 3Z} \quad \text{or} \quad v = \frac{6y}{-2x + 12y + 3}$$

u_o and v_o refer to a nominally achromatic color. For the 1931 Standard Observer and Illuminant C, $u_o = 0.2009$ and $v_o = 0.3073$. The total color difference is calculated as

$$\Delta E = [(\Delta U^*)^2 = (\Delta V^*)^2 + (\Delta W^*)^2]^{1/2}$$

$\Delta U^* = U^*$ sample $- U^*$ standard, indicating redder if positive and greener if negative.

$\Delta V^* = V^*$ sample $- V^*$ standard, indicating yellower if positive and bluer if negative.

This equation is properly titled the 1964 CIE U^*, V^*, and W^* equation (Sward GG (ed) (1972) Paint testing manual: physical and chemical examination of paints, varnishes and lacquers, and colors, 13th edn. ASTM Special Technical Publication No. 500, American Society for Testing and Materials, Philadelphia; Billmeyer FW, Saltzman M (1966) Principles of color technology. Wiley, New York).

CIE 1976 (L* a* b*) Color Difference Equation (from J Op Soc Am 64: 896 (1974) (Modified Adams-Nickerson)).

$$\Delta E_{CIE}(L^*, a^*, b^*) = [(\Delta L^*)^2 + (\Delta a^*)^2 + (\Delta b^*)^2]^{1/2}$$

where $L^* = 25 \left(\frac{100Y}{Y_o}\right)^{1/3} - 16 \quad (1 \leq Y \leq 100)$

$$a^* = 500 \left[\left(\frac{X}{X_o}\right)^{1/3} - \left(\frac{Y}{Y_o}\right)^{1/3}\right]$$

$$b^* = 200 \left[\left(\frac{Y}{Y_o}\right)^{1/3} - \left(\frac{Z}{Z_o}\right)^{1/3}\right]$$

X, Y, and Z are the tristimulus values of the sample, X_o, Y_o, and Z_o define the color of the nominally white object color stimulus (the illuminant). $\Delta L^* = L^*$ for the sample $- L^*$ for the standard; $\Delta a^* = a^*$ for the sample $- a^*$ for the standard; $\Delta b^* = b^*$ for the sample $- b^*$ for the standard.

CIE 1976 (L*, u*, v*) Color Difference Equation *n* (from J Op Soc Am 64 (1974)).

$$\Delta E_{CIE}(L^*, u^*, v^*) = [(\Delta L^*)^2 + (\Delta u^*)^2 + (\Delta v^*)^2]^{1/2}$$

where $L^* = 25 \left(\frac{100Y}{Y_o}\right)^{1/3} - 16 \quad (1 \leq Y \leq 100)$

$$u^* = 13L^*(u' - u'_o)$$

$$v^* = 13L^*(v' - v'_o)$$

$$u' = \frac{4X}{X + 15Y + 3Z}$$

$$v' = \frac{9Y}{X + 15Y + 3Z}$$

$$u'_o = \frac{4X_o}{X_o + 15Y_o + 3Z_o}$$

$$v'_o = \frac{9Y_o}{X_o + 15Y_o + 3Z_o}$$

X, Y, and Z are the tristimulus values of the sample. X_o, Y_o, and Z_o are the tristimulus values for the illuminant. $\Delta L^* = L^*$ for the sample $- L^*$ for the standard; $\Delta u^* = u^*$ for the sample $- u^*$ for the standard; $\Delta v^* - v^*$ for the sample $- v^*$ for the standard.

CIE Color Notation System *n* Colorimetric specification system based on stimulus-response characteristics adopted by the CIE in 1931. The current recommendations for the system may be obtained from the official publication, CIE Publication No. 15 (E-1.3.1) 1971, "Colorimetry Official Recommendation of the International Commission on Illumination," available from National Bureau of Standards and Technology, Washington, DC 20234.

CIE Luminosity Curve See ▶ Luminosity Curve.

CIE Standard Observer *n* The observer data adopted by the Commission Internationale d'Eclairage to represent the response of the average human eye, when light-adapted, to an equal-energy spectrum. Unless otherwise specified, the term applies to the data adopted in 1931 for a 2° field of vision. The data adopted in 1964, sometimes called the 1964 observer, were obtained for a 10°, annular field which excludes the 2° field of the 1931 observer functions.

CIE Tristimulus Values *n* See ▶ Tristimulus Values, ▶ CIE.

CIL Flow Test A capillary-rheometer test developed at Canadian Industries Ltd for characterizing the flow of thermoplastics. The reported flow unit is the amount of melt that is forced through a specified orifice per unit time when a suitably chosen force is applied. Similar to ▶ Melt-Flow Index.

Cill *n* Brit. term for sill.

CIM *n* Acronym for computer-integrated manufacturing.

Cimene See ▶ Dipentene.

Cinaper *n* An obsolete form of cinnabar.

Cinder Block *n* A lightweight masonry unit made of cinder concrete; widely used or interior partitions. See ▶ Clinker Block.

Cinnabar \ˈsi-nə-ˌbär\ *n* [ME *cynabare*, fr. MF & L; MF *cenobre*, fr. L *cinnabaris*, from Gk *kinnabari*, of non-Indo-European origin; akin to Arabic *zinjafr* cinnabar] (14c) See ▶ Mercuric Sulfide.

Cinoper *n* An obsolete form of cinnabar.

Cintz *n* Wallpapers resembling the printed cotton materials from India once known as "chints," featuring brightly colored flowers.

Circuit *n* In filament winding, one complete traverse of the fiber-feed mechanism of the winding machine; or a complete traverse of a winding band from one arbitrary point along the winding path to another point on a plane through the starting point and perpendicular to the axis.

Circular-Knit Fabric *n* A tubular weft-knit fabric made of a circular-knitting machine.

Circular Knitting See ▶ Knitting.

Ciré \sə-ˈrā, sē-\ *n* [F, fr. pp of *cirer* to wax, fr. *cire* wax, fr. L *cera*] (1921) A brilliant patent leather effect produced by application of wax, heat, and pressure.

cis- A chemical prefix (Latin: "on this side"), usually ignored in alphabetizing lists, denoting an isomer in which certain atoms or groups are on the same side of a plane. Opposite of *TRANS-*.

Cis Isomer *n* Any isomer in which two identical atoms or groups are adjacent to each other or on the same side of a structure.

Cissing *n* (1) A slight shrinkage of a glossy paint resulting in small cracks through which the undercoat may be seen; a mild form of crawling. *British Syn: Crawling*. (2) A process for preparing a wood surface for graining by wetting with a sponge. Also spelled "sissing."

Citrate Plasticizer *n* Any of a family of plasticizers derived from citric acid, $HO_2CCH_2C(OH)(CO_2H)CH_2CO_2H$, noted for their low order of toxicity. Included citrates are: triethyl, tri(2-ethylhexyl), tricyclohexyl, tri-n-butyl, acetyl triethyl, acetyl tri-n-butyl, acetyl tri-n-octyl, n-decyl, and acetyl tri(2-ethylhexyl).

Citrates *n* Esters derived from citric acid. The tri-ethyl, -butyl and –amyl citrates form a series of plasticizers.

Citric Acid *n* $C_3H_4(OH)(COOH)_3$. Tricarboxylic, monohydroxy acid. One molecule of water is normally included in the formula. It has an mp of 154°C, but decomposes on further heating.

Citronella Oil *n* Essential oil, known also as oil of lemongrass. It possesses a strong lemon-like smell, and is used to a small extent as a deodorant.

CL *n* Poly(vinyl chloride) fiber.

CLA *n* Abbreviation for Center Line Average (BS 1134); indication of surface roughness.

Clamping Capacity *n* The largest rated projected area of cavities and runners that an injection or transfer press can safely hold closed at full molding pressure.

Clamping Force *n* In injection and transfer molding, the force applied to the mold to keep it closed, and opposing the pressure exerted by the injected plastic acting upon the projected area of cavities and runners. Per square centimeter of cavities and runners, at least 3.5 kN of clamping force is required.

Clamping Plate *n* A plate, fitted to a mold that secures the mold to the frame of the molding machine.

Clamping Pressure *n* In injection and transfer molding with a hydraulically operated mold, the hydraulic-fluid pressure applied to the mold ram to keep the mold closed during the molding cycle. Compare ▶ Clamping Force.

Clamps *n* The parts of a testing machine that are used to hold a specimen while it is subjected to force. *Also called Jaws*.

Clamshell Molding *n* A term applied to the modern version of the oldest form of blow molding – preheating two sheets of plastic, placing them between halves of a split mold, closing the mold, drawing the sheets against their respective mold surfaces by means of vacuum, then completing the forming with air pressure between the sheets. The modern process, mechanized and conveyorized, is superior to blow molding from a parison for very large parts and for those in which uniformity of wall thickness is important.

Clapboard *n* A wood siding commonly used as an exterior covering on a building of frame construction; applied horizontally and overlapped, with the grain running lengthwise; usually thicker along the lower edge than along the upper. *Also known as Bevel Siding and Lap Siding*.

Claret Red *n* Azo pigment produced by coupling Tobias acid (2-naphthylamine-1-sulfonic acid) with 3-oxynaphthoic acid, followed by conversion to the calcium lake.

Clarifiers *n* Additives that increase the transparency of a plastic material.

Clarifoil *n* Cellulose acetate. Manufactured by British Celanese, Great Britain.

Clarity *n* (1) In general, the optical property of being clear. (2) In acetate manufacture, a measure of the appearance of dope solutions, indicating the quality of the acetylation mixture. (3) In printing, the sharpness or definition of a print pattern.

Clarity of Plastics *n* The ability of a transparent material to transmit a clear image of an object when viewed through it, without any aberration.

Clash-Berg Point *n* The rising temperature at which the apparent modulus of rigidity of a specimen falls to 931 MPa, the end point of "flexibility" as defined by Clash and Berg in their studies of low-temperature flexibility. In a similar test described in ASTM D 1043, the deciding shear modulus is one-third the C-B value.

Class I, II or III Liquids *n* Groupings of flammable and combustible liquids arbitrarily classified by boiling points and closed cup flash point methods, in order to specify particular procedures and types of electrical equipment to be used when handling them.

Classical Mechanics *n* A system of mechanics developed before the inception of quantum ideas; useful for describing the behavior of objects and particles much larger than atoms.

Clathrate \ ▪kla- ▪thrāt\ *adj* [L *clathratus*, furnished with a lattice, fr. *clathrin* (plural) lattice, fr. Gk *kl* ēithron bar, fr. *kleiein* to close] (1906) A "cage compound" in which atoms or molecules are trapped in a cage of covalently bonded atoms, but mot directly bonded to them.

Clay *n* Any naturally occurring sediment rich in hydrated silicates of aluminum, predominating in particles of colloidal or near-colloidal size. There are many types of clays and clay-like minerals. Those of particular interest to the plastics industry are varieties refined by nature and man to a state of good color and particle-size distribution, such as Kaolin (china clay). They are used as fillers in epoxy and polyester resins, PVC compounds, and urethane foams. *Calcined clays* are those that have been heated to a high temperature to drive off the chemically bound water, sometimes also surface-treated to improve their chemical inertness and moisture resistance. They are used primarily in vinyl insulation. See ▶ Aluminum Silicate,▶ Kaolinite, and ▶ China Clay.

Clean *n* A complete lack of any visible nonuniformity sometimes referred to as seeds, when viewed in thin films by any macroscopic or microscopic use of visible light.

Cleaner *n* (1) Detergent, alkali, acid or other cleaning material; usually water or steam borne. (2) Solvent for cleaning paint equipment.

Clear *n* A complete lack of any visible nonuniformity when viewed in mass, in bottles or test tubes, by strong transmitted light.

Clear Coating *n* Transparent protective and/or decorative film.

Clearcole, Clairecolle *n* (1) Glue size in appropriate dilution with the addition of small quantity of whiting. It is used to reduce porosity of ceiling and other surfaces prior to the application of size-bound distemper. (2) A primer consisting of glue, water, and white lead or whiting. (3) A clear coating used in application of gold leaf.

Clearing *n* The treatment of printed fabrics with a chemical solution to improve the appearance of the whites. In many cases the treatment also brightens the printed areas. Also see ▶ Reduction Clearing.

Clear Lacquer See ▶ Clear Coating.

Clear Point *n* With regard to vinyl plastisols, clear point is the rising temperature at which an unpigmented plastisol suddenly becomes transparent, signifying that the resin particles have completely dissolved in the warm plasticizer. This test is useful for determining the relative fusion temperatures of different plastisols.

Cleavage *n* (1) Breaking of a laminate due to the separation of the strata. (2) Portion of the material adhering to the side of a container after the major portion has been drained. (3) The property of a crystalline substance of splitting along definite crystal planes.

Cleveland Condensing Humidity Cabinet *n* (QCT) An accelerated weathering apparatus which operates on a condensation type of water exposure at elevated temperature.

Cleveland Open Cup *n* Device used in determining flash and fire points of petroleum products.

Cleve's Acid *n* Old name for α-naphthol-5-sulfonic acid. It was also known as *L acid* and is used in the manufacture of Helio Bordeaux BL.

Clicker Die *n* A cutting die for stamping out blanks from plastic sheet.

Clicker Press *n* A stamping press used with ▶ Clicker Dies to cut shapes from plastic sheet. Compare ▶ Die Cutting.

Clicking *n* See ▶ Die Cutting.

Climate Cabinet *n* Any enclosure used to emulate selected climatic conditions.

Clinker Block *n* British term for cinder block.

Clipmark *n* Visible deformation of selvage due to pressure from a tenter clip.

CLO *n* A unit of thermal resistance. The insulation needed to keep an individual producing heat at the rate of $58W/m^2$ comfortable of $21°C$ air temperature with air movement of 0.1 m/s. One clo is roughly equal to the insulation value of typical indoor clothing.

Clockspring *n* Slipping action in coil coating, usually at the center of a coil of the strip during the recoiling operation.

Cloisonné \ ▪klói-zə- ▪nā, ▪klwä-\ *adj* [F, fr. pp of *cloisonner* to partition] (1863) A surface decoration in which differently colored enamels or glazes are separated by fillets applied to the design outline. For porcelain enamel, the fillets are wire secured to the metal body; for tile and pottery, the fillets are made of ceramic paste, squeezed through a small-diameter orifice.

Cloqué Fabric \klō-ˈkā, ˈklō-ˌ\ *n* [F *cloqué*, fr. pp of *cloquer* to become blistered, fr. F dialect (Picard) *cloque* bell, bubble, fr. ML *clocca* bell] (1936) From the French term for blistered, it refers to any fabric whose surface exhibits an irregularly raised blister effect.

Closed Assembly Time See ▶ Assembly Time, ▶ Assembly.

Closed-Cell Foamed Plastic *n* (unicellular foam) A Cellular Plastic in which interconnecting cells are too few to permit the bulk flow of fluids through the mass.

Closed-Cell Foams *n* Individual cells are non-interconnecting. The ells are basically without access to the surrounding air of fluids; cells are not communicating.

Closed Coat *n* In coated abrasives, when the abrasive grains completely cover coatside surface of the backings. Closed coats are designed for severe service and are used for most applications. See also ▶ Open Coat.

Closed Loop Control See ▶ Feedback Control.

Close Drying *n* Material is said to be close drying when it does not show much fullness or body on the substrate. Often, the true fullness is not brought out until after rubbing. This is especially true with lacquer finishes.

Close Grain *n* Wood having narrow and inconspicuous annual growth rings. The term is sometimes used to designate wood having small and closely spaced pores, but in this sense the term "fine textured" is more often used.

Cloth *n* A generic term embracing all textile fabrics and felts. Cloth may be formed of any textile fiber, wire, or other material, and it includes any pliant fabric woven, knit, felted, needled, sewn, or otherwise formed.

Cloud Chamber *n* An apparatus containing moist air or other gas which on sudden expansion condenses moisture to droplets on dust particles or other nuclei. Thus charged particles or ions in the space become nuclei and their numbers and behavior, when properly illuminated, may be studied.

Cloudiness *n* The lack of clarity or transparency in a paint or varnish film.

Clouding *n* Development in a clear varnish or lacquer film or liquid of an opalescence or cloudiness caused b y the precipitation of insoluble matter or immiscibility of components.

Cloud Point *n* (1) In condensation polymerization, the temperature at which the first turbidity appears, caused by water separation when a reaction mixture is cooled. (2) In petroleum and other oils, the falling temperature at which the oil becomes cloudy, from precipitation, of wax or other solid. (3) Point at which a definite lack of clarity (cloudiness) appears when a liquid is subject to adulteration or when it is mixed with another substance, or the temperature at which a liquid becomes cloudy when it is cooled.

Cloudy Web *n* An uneven or irregular web from the doffer of a card.

Clumps In nonwoven fabrics, an irregularly shaped grouping of fibers caused by insufficient fiber separation.

Clupanodonic Acid *n* Acid found in many fish oils. The acid possesses an unusually high iodine value – well in excess of 200. *Also called Doscosapentanoic Acid.*

CMC *n* Abbreviation for ▶ Carboxymethyl Cellulose or for ceramic-metal composite.

CN *n* Abbreviation for ▶ Cellulose Nitrate.

CNR *n* Carboxynitroso rubber.

Co *n* Chemical symbol for the element cobalt.

Coacervate *n* \kō-ˈa-sər-ˌvāt\ *v* [L *coacervatus*, pp of *coacervare* to heap up, fr. *co-* + *acervus* heap] (1929) An aggregate of colloidal droplets held together by electrostatic attractive forces.

Coacervation *n* The separation of a polymer solution into two or more liquid phases, one of which is a polymer-rich liquid. The term was introduced to distinguish this phenomenon from the precipitation of a polymer solute in solid form. The process is used in ▶ Microencapsulation by emulsifying or dispersing the material to be encapsulated with a solution of the polymer. By changing the temperature or concentration of the mixture, or by adding another polymer or solvent, a phase separation may be induced and the polymeric portion forms a thin coating on the external surfaces of the particles. After further treatment to solidify the polymeric wall, the capsules can be isolated in powder form by filtration. It is an intermediate stage between sol and gel formation.

Coagulant *n* \kō-ˈa-gyə-lənt\ *n* (1770) A substance that (1) initiates the formation of relatively large particles in a finely divided suspension or (2) assists in the formation of a gel, thus accelerating settling of the particles or their deposition on a substrate.

Coagulation *n* (1) A physical or chemical action inducing transition from a fluid to a semi-solid or gel-like state, or the bringing together of small, individual particles into clumps. (2) Process whereby a fluid liquid is changed into a thickened, curdled or congealed mass. (3) Irreversible agglomeration of particles originally dispersed in a rubber latex.

Coagulation Bath *n* A liquid bath that serves to harden viscous polymer strands into solid fibers after extrusion through a spinneret. Used in wet spinning processes such as in rayon or acrylic fiber manufacture.

Coagulum *n* An agglomerate of particles.

Coalesced Filaments *n* Filaments stuck together by design or accident during the extrusion process.

Coalescence *n* The formation of a film of resinous or polymeric material when water evaporates from an emulsion or latex system, permitting contact and fusion of adjacent latex particles. Action of the joining of particles into a film as the volatile evaporates.

Coalescent (Coalescing Agent) *n* Solvent with a high bp which, when added to a coating, aids in film formation via temporary plasticization (softening) of the vehicle.

Coal Tar *n* Coal tar which is also known as *aniline* is an oily, colorless, toxic liquid, which darkens upon exposure to air. It is soluble in water, alcohol, and ether, and forms a number of salts. It can also be described as a dark brown to black cementitious material produced by the destructive distillation of bituminous coal.

Coal Tar Colors See ▶ Aniline Pigments.

Coal Tar Epoxy *n* Modified coal tar floor coating designed to bond to asphalt and concrete surfaces. High strength and resistant to most corrosive reagents.

Coal Tar Epoxy Coating *n* Coating in which binder or vehicle is a combination of coal tar with epoxy resin.

Coal Tar Hydrocarbons *n* Aromatic hydrocarbons derived from coal tar, including benzene, toluene, xylene, naphtha, etc.

Coal Tar Pitch *n* Distillation residue from coal tar. It varies considerably from a very soft to a very hard product. Fusion points vary from as low as 27°C (80°F) to as high as 232°C (450°F).

Coal Tar Pitch Coatings *n* Most plastics plated today are finished with a copper/nickel/chrome electroplate, but many other finishes are possible, such as bright brass, antique brass, satin nickel, silver, black chrome, and gold. The actual composition of the electroplate is designed for a particular application of the electroplate.

Coal-Tar Resin See ▶ Coumarone-Indene Resin.

Coal Tar-Urethane Coating *n* Coating in which binder or vehicle is a combination of coal tar with a polyurethane resin.

Coarse End See ▶ Coarse Thread.

Coarse Filling See ▶ Coarse Thread.

Coarse Grain *n* Wood with wide and conspicuous annual rings having considerable difference between springwood and summerwood. The term is sometimes used to designate wood and large pores, such as oak, ash, chestnut and walnut, but in this sense the term "coarse texture" is more often used.

Coarse Pick See ▶ Coarse Thread.

Coarse Thread *n* A yarn larger in diameter than other yarns being used in the fabric.

Coat *n* Paint, varnish or lacquer applied to a surface in a single application (one layer) to form a properly distributed film when dry. A coating system sexually consists of a number of coats separately applied in a predetermined order at suitable intervals to allow for drying or curing. It is possible with certain types of material to build up coating systems of adequate thickness and opacity by a more or less continuous process of application, e.g., wet-on-wet spraying. In this case, no part o the system can be defined as a separate coat in the above sense.

Coated Abrasive *n* A flexible-type backing upon which a film of adhesive holds and supports a coating of abrasive grains. The backing may be paper, cloth, vulcanized fiber or a combination of these materials. Various types of resin and hide glues are used as adhesives. The abrasives used are flint, emery, crocus, garnet, aluminum oxide and silicon carbide.

Coated Fabric *n* A cloth that has been impregnated and/or coated with a plastic material in the form of a solution, dispersion, hot melt, or powder. The term is sometimes used when a preformed film is applied to the fabric by calendaring, although such products are more properly termed *laminates*.

Coated Paper *n* A paper coated with clay, other white pigments and a suitable binder.

Coater Apparatus which applies paint.

Coathanger Die *n* A sheet-, or film-extrusion die whose melt-distribution manifold has the obtuse-isoceles outline of a coathanger. This popular die design is said to yield uniform distribution of material across the full width of the extruded web, thus producing sheet of laterally more uniform thickness. Side-fed blow-molding dies and spiral-type dies for blown film may also be spoken as a coathanger dies.

Coating *n* (1) Generic term for paints, lacquers, enamels, printing inks, etc. (2) A liquid, liquefiable or mastic composition which is converted to a solid protective, decorative, or functional adherent film after application as a thin layer. (3) A composition, which when applied in thin layers, forms a non-tacky, adherent film that hides, protects, and/or decorates the substrate.

Singleton-albright salt spray cabinet

Coating, Dip n The process in which a substrate is immersed in a solution (or dispersion) containing the coating material and withdrawn.

Coating Methods See:

▶ Air-Knife Coating	▶ Gravure-Coating
▶ Calender Coating	▶ Intumescent Coating
▶ Cascade Coating	▶ Painting of Plastics
▶ Curtain Coating	▶ Plasma-Spray Coating
▶ Decorating	▶ Printing on Plastics
▶ Dip Coating	▶ Reverse-Roll Coating
▶ Electrophoretic Deposition	▶ Roller Coating
▶ Electroplating on Plastics	▶ Silver-Spray Process
▶ Electrostatic Fluidized-Bed Coating	▶ Sinter Coating
▶ Electrostaticspray Coating	▶ Solution Coating
▶ Extrusion Coating	▶ Spray Coating
▶ Flame-Spray Coating	▶ Spread Coating
▶ Flocking	▶ Strippable Coating
▶ Flow Coating	▶ Transfer Coating
▶ Fluidized-Bed Coating	▶ Urethane Coatings
▶ Friction Calendering	▶ Vacuum Metalizing

Coating Powders n Finely divided, solid plastic materials which are heat fusible and form relatively smooth, tough, electrical insulating coatings upon application to metal surfaces. See ▶ Powder Coating.

Coating, Spray n The process in which a substrate is sprayed with the coating material.

Coating System See ▶ Coat.

Cobalt \ˈkō-ˌbólt\ n [Gr *Kobalt*, alter. of *Kobold*, literally, bobbin, fr. MHGr. *kobolt*; fr. its occurrence in silver ore, believed to be due to globins] (1683) A tough lustrous silver-white magnetic metallic element that is related to and occurs with iron and nickel and is used in alloys.

Cobalt Aluminate See ▶ Cobalt Blue.

$$\begin{array}{c} ^{-}O \diagdown \!\!\!\!\! Al \diagup\!\!\!\!\!\diagdown O \\ Co^{++} \\ ^{-}O \diagup\!\!\!\!\!\diagdown \\ Al \diagup\!\!\!\!\!\diagdown O \end{array}$$

Cobalt Blue n (1835) $C_oO \cdot Al_2O_3$. Pigment Blue 28 (77346). Bright blue pigment, which is a complex product derived from oxides of cobalt and aluminum. It has excellent lightfastness and good chemical and heat resistance and with low opacity. Density, 4.2–4.3 g/cm^3 (35.0–35.8 lb/gal); O.A., 27–36; particle size, 0.5–1.0 μm. Syn: ▶ Azure Blue, ▶ Cobalt Aluminate, ▶ Cobalt Ultramarine, ▶ Dumont's Blue, ▶ Enamel Blue, ▶ Gahn's Ultramarine, ▶ Hungary Blue, ▶ Leyden Blue, and ▶ Zaffre.

Cobalt Chloride n (1885) A chloride of cobalt. The dichloride $CoCl_2$ that is blue when dehydrated, turns red in the presence of moisture, and is used to indicate humidity.

$$Cl^- \; Co^{++} \; Cl^-$$

Cobalt Drier n One of the many organic cobalt salts (cobalt, naphthenate, cobalt octoate, etc.) which are soluble in paints and varnishes and used to speed the drying and hardening of the oil vehicle. They are also used to accelerate oxidation and polymerization of an ink film. Also used as an initiator in polyesters.

Cobalt Green n Similar to cobalt blue, except that zinc oxide replaces wholly or partly the aluminum oxide in the latter. It does not contribute to good hiding power. *Also known as Rinmann's Green or Zinc Green.*

Cobalt Naphthenate See ▶ Naphthenic Acid.

Cobalt 60 n (1946) A heavy radioactive isotope of cobalt of the mass number 60 produced in nuclear reactors and used as a source of gamma rays (as for radiotherapy).

Cobalt Ultramarine n Red shade cobalt blue. See ▶ Cobalt Blue.

Cobalt Yellow See ▶ Aureolin.

Cobwebbing n Production of fine filaments instead of the normal atomized particles when some coatings are sprayed. Although generally considered a defect in ordinary lacquers, use is made of this property to provide textured coating or a protective covering for equipment during storage. A cocoon is formed around the article by the pronounced cobwebbing action.

Cobwebbing *n* (in gravure) A filmy, web-like build-up of dried ink or clear material on the doctor blade, ends of impression roll, or engraving.

Cobweb Coating See ▶ Cobwebbing.

Cocatalyst *n* Chemicals which themselves are rather weak catalysts, but which greatly increase the activity of a given catalyst; also called promoters.

Cochineal \ˈkä-chə-ˌnēl, ˈkō-\ *n* [MF & Sp; MF *cochenille*, fr. OSp *cochinilla* cochineal insect] (1582) Natural organic dyestuff made from the bodies of the female insect. *Coccus Cacti*, which lives on plants in Central and South America. The coloring principle is known as carminic acid and is generally laked. It is not fast to light.

Cochin Oil *n* Refined grade of coconut oil.

Cockled Yarn *n* Spun yarn in which some fibers do not lie parallel to the other fibers but instead are curled and kinked, forming a rough and uneven surface on the yarn. The general cause is fiber overcut to the extent that the drafting rolls catch and hold both ends of the fiber at the same time while attempting to draft, resulting in slippage or breakage. (Also see ▶ Overcut.)

Cockling *n* A crimpiness or pucker in yarn or fabric usually caused by lack of uniform quality in the raw material used, improper tension on yarn in weaving, or weaving together yarns of different numbers.

Coconut Oil *n* Expressed from the nut kernels of the coconut palms, the oil is composed of glycerides of lauric acid, capric acid, myristic, palmitic and oleic acids. Its main use is in nonyellow alkyds. Properties: sp. gr 0.9190/25°C; refractive index, 1.4545/40°C; iodine value, 9; and saponification value, 255.

Co-Cure *n* To cure two or more different materials in one step.

Coefficient of Cubical Expansion See ▶ Coefficient of Expansion.

Coefficient of Elasticity *n* Reciprocal of Young's modulus in a tension test. A rarely seen Syn: ▶ Modulus of Elasticity.

Coefficient of Expansion *n* Ratio of increase in length, area, or volume of a substance for a 1°C rise in temperature. Also known as *Coefficient of Cubical Expansion*.

Coefficient of Friction *n* Measure of the resistance to sliding of one surface in contact with another surface. See ▶ Friction.

Coefficient of Friction, Kinetic See ▶ Kinetic Coefficient of Friction.

Coefficient of Friction, Static See ▶ Static Coefficient of Friction.

Coefficient of Plastic Viscosity See ▶ Viscosity, ▶ Plastic.

Coefficient of Scatter See ▶ Scatter, ▶ Scatter, Coefficient of.

Coefficient of Thermal Conductivity See ▶ Thermal Conductivity.

Coefficient of Thermal Expansion *n* The fractional change in length (or sometimes in volume, when specified) of a material for a unit change in temperature, as given by the equation:

$$CE = (1/L) \cdot dL/dT = dlnL/dT \approx \Delta L/\Delta T$$

See ▶ also Volume Coefficient of Thermal Expansion.

Coefficient of Twist Contraction *n* The shortening of a yarn, due to twist, per unit length of untwisted yarn, usually expressed in percent.

Coefficient of Viscosity *n* (1866) See ▶ Viscosity (2).

Coextrusion The process by which the outputs of two or more extruders are brought smoothly together in a ▶ Feed Block to form a single multiplayer stream that is fed to a die to produce a layered extrudate. The extruder streams may be split within the feed block to form dual layers, usually in a symmetrical arrangement about the center plane of the final sheet. Sheeting containing up to nine layers is commercially produced. Coextrusion is employed in film blowing, sheet and flat-film extrusion, blow molding, and extrusion coating. The advantage of coextrusion is that each ply imparts a desired characteristic property, such as stiffness, heat-sealability, impermeability, or resistance to some environment, all of which properties would be impossible to obtain with any single material. Layers of poorly compatible plastics can be coextruded by including a thin adhesive layer between them.

Coextrusion Blow Molding *n* A variant of extrusion ▶ Blow Molding in which the parison contains two or more layers of at least two materials. See ▶ Coextrusion.

Cogswell Rheometer See ▶ Extensiometer.

Cohesion *n* Propensity of a single substance to adhere to itself; the internal attraction of molecular particles toward each other; the ability to resist partition from the mass; internal adhesion; the force holding a single substance together.

Cohesive Energy *n* (Density) For liquids the heat of vaporization per unit mass, divided by the specific volume, or the same quantity based on molar properties. Cohesive energy density is also equal to the square of the ▶ Solubility Parameter.

$$CED = \Delta E_v / V_1$$

For liquids, the heat of vaporization per unit mass, divided by the specific volume, or the same quantity based on molar properties. Cohesive energy density is also equal to the square of the ▶ Solubility Parameter.

Coil Coating n High speed process which applies paint and other coatings to a continuous flat coil of metal. A continuous coil of metal is unwound, cleaned, surface-treated, coated, heat-cured, cooled and rewound in one operation. The coated coil is subsequently unwound and formed into any number of products, such as house siding, venetian blinds, and automotive and appliance parts. *Also called Strip Coating.*

Coiling n The depositing of sliver into cylindrical cans in helical loops. This arrangement permits easy removal for further processing.

Coiling Soup n In coating or spreading operations, a defect caused by the curling or turning motion of a bank of lacquer or other "soup" compound in front of a doctor knife, or similar application device, wherein streaks are formed on the surface of the coated fabric due to uneven application of the soup.

Coil Yarn See ▶ Textured Yarns.

Coining n A term borrowed from the metal-stamping industry for a process of forming integral hinges from plastics. In the case of a polypropylene article, the hinge is produced by molding a thin section between the two parts of the article to be hinged. Such a thin section cannot be molded easily in articles of nylon or acetal resin because of the difficulty of filling the half of the mold cavity opposite the gated half through the thin section. In the coining process, the area to be formed into a hinge is molded in a thickness suitable fore the molding process. Subsequently, the article is placed between bars in a press that quickly squeezes the plastic to the desired thickness. The material must be deformed beyond its compressive yield point but short of failure, so that is remains essentially stable with little recovery from the deformation. This rapid cold pressing produces a high degree of orientation that imparts high strength and flexibility to the integral hinge area.

Coinjection n A process similar in its results to ▶ Coextrusion but accomplished by modifications of the injection-molding process. By means of various nozzle and valving arrangement, two or more materials, can be injected either simultaneously or sequentially to form an article with an outer shell of one material with certain desired properties, the shell filled with another material to attain other desired properties such as reduced cost. Coinjection, like coextrusion, basically means that two or more different plastics are formed into a composite or laminated structure.

Coin Marking See ▶ Metal Marking.

Coking n In the running of copals, or sweating of asphaltums, and other materials, which naturally contain a proportion of infusible dirt or mineral matter, this extraneous matter is likely to escape the action of the stirrer. As a consequence, any organic matter associated with it is liable to be overheated, and carbonization occurs. This ultimately results in the formation of carbonaceous lumps, and it is then said to have coked.

Cold n Dull, flat surface where a bright, lustrous one is desired.

Cold-Bend Test n A test for measuring the flexibility of a plastic material at low temperatures. A specimen in a series is chilled to one of several specified low temperatures, then bent to a predetermined radius until the temperature at which half the specimens tested do not survive the bend has been identified.

Cold-Checking n Surface defect on lacquer, especially furniture finishes. It appears as a pattern of fine cracks developing on film surface, when the finished article is subjected to alternate hot and cold conditions. See ▶ Checking and ▶ Temperature Checking.

Cold Cracking n Crazing and cracking of a coating subjected to low temperature or cold/ambient cycling.

Cold-Curing n Process of curing at normal atmospheric temperature. (i.e., air dry).

Cold Cut n Dissolving a resin or other material in a suitable solvent by mechanical agitation without the application of heat.

Cold Cycle n Low temperature cycle during many accelerated tests.

Cold Drawing n (cold stretching) A stretching process performed at a temperature below a thermoplastic's melting range to orient the material and improve the tensile modulus and strength.

Cold Flex n The lowest temperature at which the test strip can be twisted through a 200° arc without breaking.

Cold Flow n Continuing distortion, deformation, or dimensional change which take place in materials under continuous load (constant stress) at ambient temperatures. See ▶ Creep. Syn: ▶ Compression Set.

Cold Forming n A group of processes by which sheets or billets of thermoplastics are formed into three-dimensional shapes at room temperature by processes used in the metal-working industry such as forging, brake-press bending, deep drawing, rolling, stamping, heading, and coining. The materials used, generally in relatively thick sections, include ABS, polycarbonate, polyolefins, and rigid PVC. When either the material or the forming dies are preheated, the preferred term is ▶ Solid-Phase Forming.

Cold Heading *n* A process for forming short plastic rods into rivets by uniformly loading the projecting shaft end in compression while holding and containing the shaft trunk. All thermoplastics can be cold-heated but acetal and nylon are particularly suitable.

Cold Molding *n* A process similar to Compression Molding except that no heat is applied during the molding cycle. The formed part is subsequently cured by heating and cooling. A-stage phenolic resins and bituminous plastics are sometimes molded by his process.

Cold-Parison Blow Molding See ▶ Blow Molding.

Cold Pressing *n* (1) Assembly method in which the bonded structures are held in place by pressure without the application of heat or drying air until the adhesive interface has solidified and reached proper shear proportions. (2) An extraction method for oils such as fish oil.

Cold-Process Roofing *n* A bituminous roofing membrane which consists of layers of coated felts that have been bonded with cold-applied asphalt roof cement and surfaced with an emulsified or cutback asphalt roof coating.

Cold-Rolled Steel *n* Low-carbon, cold-reduced sheet steel.

Cold Rubber *n* A synthetic rubber made at a relatively low temperature (about 40°F or 4°C) and having greater strength and durability than that made at the usual temperature (about 120°F or 49°C).

Cold-Runner Injection Molding *n* (runnerless injection molding) Whereas in injection molding thermoplastics the runners are sometimes kept hot to reduce scrap (see ▶ Hot-Runner Mold), with thermosets the runners are kept *cooler* than the cavities to prevent material from curing within the runner system. In cold-runner injection molding of thermosets the mold is divided into two sections: a heated section containing the cavities, and an insulated manifold section containing the injection sprue and runners. Material is fed from runners to cavities through very short gates or sub-sprues. The insulated manifold is maintained at a temperature high enough to soften the uncured material, generally in the vicinity of 90°C, but well below the curing temperature prevailing in the cavity section.

Cold-Setting *n* Resin or lacquer products which, under the influence of suitable catalysts, are able to form suitable film properties without the application of heat.

Cold-Setting Adhesive *n* Synthetic resin adhesive capable of hardening at normal room temperature in the presence of a hardener. See ▶ Adhesive, ▶ Cold-Setting.

Cold Setting Inks *n* Solid inks which must be melted and applied on a hot press. They solidify again on contact with unheated paper.

Cold Slug *n* The first material to enter an injection mold, so called because in passing through the sprue orifice it is cooled below the effective molding temperature.; In some molds, a small well in the mold opposite the sprue catches the cold slug and thereby prevents it from entering the runner system.

Cold-Slug Well *n* A small circular cavity, directly opposite the sprue opening in an injection mold that traps the ▶ Cold Slug.

Cold Stretch *n* Pulling operation, usually on extruded filaments, to improve tensile properties.

Cold Test *n* Refers to pour point. A low cold test means that the product has a low temperature pour point.

Cold Water Paint *n* Paint in which the binder or vehicle portion is composed of soybean or other vegetable protein, glue, resin emulsion, or other similar material dispersed in water.

Collacral K *n* Industrial poly(vinyl pyrrolidone). Manufactured by BASF, Germany.

Collage *n* Literally, pasting. Employed to describe the practice of some twentieth-century artists in introducing paper or other nonpigmented materials in painting. Basically, a pasting technique whereby pictorial images or patterns, and pieces of colored, textured material are superimposed on each other.

Collapse *n* (1) Inadvertent densification of cellular material during manufacture resulting from breakdown of cell structure (ASTM D 883). (2) Inward contraction of the walls of a molded container, e.g., while cooling, resulting in permanent indentation.

Collet \\ˈkä-lət\\ *n* [MF, dim. of *col* collar, fr. L *collum* neck] (1528) Rigid lateral contained for the mold-forming manual; a dam, a restriction box; the drive wheel that pulls glass fibers from the bushing. [A forming tube is placed on the collet and a package of strand (forming cake) is wound up on the tube.]

Colligative Properties *n* (**Polymers**) Vapor pressure lowering, freezing point depression, boiling point elevation and osmotic pressure, measurements or properties dependent on the number of molecules (i.e., pressure) and not the chemistry of the molecule.

Colligative Property *n* A property of a material or solution that depends mainly on the number, rather than the nature, of the molecules, atoms, or ions present. Examples are gas pressure and osmotic pressure of solutions.

Collimated Roving *n* Roving with strands that are more nearly parallel than those in standard roving, usually made by parallel winding.

Collision Frequency Factors *n* The collision frequency factor p (number of collisions ($\sim 10^{11}$/s) is included in the "collision theory" to describe the temperature dependence of the rate constants of elementary reactions:

$$ki = pZ \exp(-E/RT) = A \exp(-E/RT)$$

where,
Z = steric factor,
E = Arrhenius activation energy,
$(-E/RT)$ = Boltzmann factor,
and often p and Z are incorporated into one constant A; the second theory is the "transition state theory" (Houston PL (2001) Chemical kinetics and reaction dynamics, McGraw-Hill).

Collodion \kə-ˈlō-dē-ən\ *n* [mod. of NL *collodium*, fr. Gk *kollōdēs* glutinous, fr. *kola* glue] (1851) A solution of cellulose nitrate in alcohol and ether used as a coating lacquer or to cast very thin films of cellulose nitrate on water ("microfilm").

Collodion Cotton See ▶ Cellulose Nitrate.

Colloid \ˈkä-ˌlóid\ *n* [ISV *coll-* + *-oid*] (ca. 1852) A dispersion of one phase in another in which the particles or units of the dispersed phase have at least one dimension which is larger than usual molecular dimensions, but two small to be observed visually. In other words, a phase dispersed to such a degree that the surface forces become an important factor in determining its properties. In general particles of colloidal dimensions are approximately 10 Å to 1 μm in size. Colloidal particles are often best distinguished from ordinary molecules due to the fact that colloidal particles cannot diffuse through membranes which do allow ordinary molecules and ions to pass freely. Pertaining to colloid.

Colloidal Clay See ▶ Bentonite.

Colloidal Dispersion See ▶ Colloid.

Colloidal Particle *n* Large molecules or aggregates of smaller molecules in a size range of 1–500 nm (10^{-9} m).

Colloidal Solution *n* Any solution containing colloidal particles. Colloidal solutions are sometimes called sols.

Colloidal State *n* Particular state in which any substance may exist under the proper conditions, determined by fineness of particle subdivision. The colloidal state is defined by a more or less well-marked ultramicroscopic zone in the scale of subdivision, the lower extreme of the zone approaching molecular dimensions, and the upper end gradually passing over into molecular aggregates (suspensions) visible under the ordinary microscope (Becher P (1989) Dictionary of colloid and surface science. Marcel Dekker, New York).

Colloid, Colloidal State *n* A state of a substance in the form of small particles dispersed in a medium; and colloid size particles that do not precipitate from a media; smoke and fog are colloidal dispersions.

Colloid, Linear *n* Any colloid with fibrous particles or macromolecules. The long, asymmetrical particles may be coiled up or, in some instances, they may be branched.

Colloid, Micellar *n* Colloid composed of micelles. A micellar colloid is an aggregation of molecules, usually arranged in a definite order, whose dimensions are less than 0.05 μm. Examples: a colloidal solution of soap in water.

Colloid Mill *n* A device for preparing emulsions and reducing particle size, consisting of a high-speed rotor and a fixed or counter-rotating element in close proximity to the rotor. The liquid is conveyed continuously from a hopper to the space between the shearing elements, and then discharged into a receiver. See also ▶ Homogenizer.

Colloid, Molecular *n* Any colloid in which the colloidal particles are macromolecules.

Colloid, Protective See ▶ Protective Colloid.

Colloid, Reversible *n* Colloid substance that can be converted from the gel to the solid form without the expenditure of chemical energy, that is, by simply warming the gel form or redissolving the dried substance in water. It is a colloid which forms a hydrophilic solid, such as gelatin, agar, gum arabi, starch; a protective colloid.

Colloid, Sphere *n* Colloid possessing more or less symmetrically-shaped, corpuscular, and relatively compact particles.

Collotype \ˈkä-lə-ˌtīp\ *n* [ISV] (1883) (1) A printing process utilizing a glass plate with a gelatin surface carrying the image to be reproduced. (2) A print made by this process. *Also called Photogelatin Process.*

Colmonoy See ▶ Hard-Facing Alloy.

Cologne Earth, Cologne Brown *n* A pigment, sometimes referred to erroneously as Vandyke brown, made from roasted American clays which contain ochre and bituminous matter. See ▶ Vandyke Brown.

Cologne Spirits See ▶ Ethyl Alcohol.

Cologne Yellow See ▶ Chrome Yellow and ▶ Chrome Yellow, Light and Primrose.

Colonial Spirit *n* Another name for methyl alcohol.

Colophony, Colophonium *n* Old name for rosin; used in Europe. See ▶ Rosin.

Colorimeter, Tristimulus See ▶ Tristimulus Colorimeter.

Color *n* Color is a basic specification for plastic parts; one of its major properties is its ability to be integrally colored, almost without restriction; in viewing a colored object three factors observed: (1) The quality of light illuminating the object; (2) The ability of the object to absorb certain portions of the light and to reflect others; and (3) The sensitivity of the eye to the reflected illumination (NcDonald R (ed) (1997) Colour physics for industry; colour physics for industry, 2nd edn. Society of Dyers and Colourists, West Yorkshire; Color, universal language and dicionary of names (NBS Special Publication 440, Stock No. 003-003-01705-1; Computer colorant formulation, Kuehni, Lexington Books, Farnsborough; Hardy CH (1936) Handbook of colorimetry. MIT Press, Cambridge).

Color Abrasion *n* Color changes in localized areas of a garment resulting from differential wear.

Color Amplitude Test *n* Test designed to examine any of the factors involved in color aptitude. Most widely used is the ISCC Color-matching Aptitude Text, or CAT, based on differences in chroma.

Colorant *n* Any dye or pigment that can impart color to a plastic. The dyes are natural or synthetic compounds of submicroscopic or molecular size, soluble in most common solvents, yielding transparent colors. Their generally poor heat resistance and tendency to migrate limit their use as additives to a few families that are superior in heat resistance. However, dyes are sometimes used to post-color finished parts such as buttons and fibers. The pigments are organic and inorganic substances with larger particle sizes, rarely below 1 μm, and usually insoluble in the common solvents (Herbst W, Hunger K (2004) Industrial organic pigments. Wiley; Kirk-Othmer encyclopedia of chemical technology: pigments-powders (1996). Wiley, New York) Organic pigments produce translucent and nearly transparent colors, resist migration better than dyes, and are somewhat more, heat-resistant. Inorganic pigments are, with few exceptions, opaque and superior to organics in light-fastness, heat resistance and resistance to migration. Colorants are added to plastics by dry-coloring (simply tumbling the colorant with the base or compounded resin power or pellets); by extrusion coloring (extruding a dry-colored mixture and chopping it into pellets to be reprocessed); by masterbatching (see ▶ Color Concentrate); or by stirring colorants or dispersions thereof into liquid plastisols or resin systems. See also:

▶ Bon pigment	▶ Pearlescent Pigment
▶ Fluorescent Pigment	▶ Perylene Pigment
▶ Flushed Pigment	▶ Phosphorescent Pigment
▶ Glitter	▶ Phthalocyanine Pigment
▶ Inorganic Pigment	▶ Quinacridone Pigment
▶ Luminescent Pigment	▶ Rhodamine
▶ Metallic-Flake Pigment	▶ Ultramarine-Blue Pigment
▶ Organic Pigment	

(Kirk-Othmer encyclopedia of chemical technology: pigments-powders (1996). Wiley, New York; Paint/coatings dictionary (1978) Compiled by Definitions Committee of the Federation of Societies for Coatings Technology.)

Colorant Dispenser *n* Mechanical device used to disperse precisely measured volume amounts (usually) of colorants for the purpose of tinting or shading coatings bases.

Colorant Match *n* Color match made by using the same colorants in the match as were used in the standard.

Colorant Mixture *n* Mixture of colorants containing pigments, or dyestuffs, or both. The color of a mixture may be predicted by subtractive colorant mixture theories. See ▶ Complex and ▶ Subtractive Colorant Mixture.

Colorants *n* Colored pigments and dyes.

Color Aptitude *n* Ability to work with color, involving inborn factors as well as acquired abilities.

Color Bleeding See ▶ Bleeding.

Color Blindness *n* (Deprecated) An incorrect term applied to defective color vision. An extremely small number of persons having complete lack of color response (are achromatopes). Most so-called color-blind persons are anomalous trichromats, seeing all three primaries but having responses which are weaker than normal to one of the primaries. Persons who lack response to one primary are called dichromats, seeing only two primaries.

Color Burn-Out *n* An objectionable change in the color of a printing ink which may occur either in bulk or on the printed sheet. In the former case it is associated primarily with tints, and is caused by a chemical reaction between certain components in the ink formulation. In the latter case it is generally caused by heat generated in a pile of printed material during the drying of an oxidizing type of ink.

Color Center *n* A kind of point defect in a crystal.

Color Code *n* A system of coloring piping and parts of equipment with various precoded colors for identification purposes.

Color Compounding Formulation and preparation of colored materials for incorporation into molding of plastics.

Color Concentrate *n* A plastic compound that contains a high percentage of pigment, to be blended in precise amounts with the base resin or compound so that the correct final color will be achieved. The concentrate provides a clean and convenient method of obtaining accurate color shades in extruded and molded products.

The term *masterbatch* is sometime used for color concentrate, as well as for concentrates of other additives. See also ▶ Multifunctional Concentrate.

Color Concentrates *n* Concentrated color dyes and pigments.

Color Constancy *n* Relative independence of perceived object color to changes in color of the light source.

Color Coordinates See ▶ Chromaticity Coordinates.

Color Difference *n* Magnitude and character of the difference between two colors under specified conditions.

Color Difference Equations *n* Equations that transform CIE coordinates into a more uniform matrix such that a specified distance between two colors is more nearly proportional to the magnitude of an observed difference between them regardless of their hue. The total color difference is generally designated as ΔE, the lightness difference as ΔL, the total chromaticity difference as ΔC, the redness-greenness difference as Δa, $\Delta \alpha$, or ΔRG, and the yellowness-blueness difference as Δb, $\Delta \beta$, and ΔYB. The directional differences, such as ΔL, ΔRG, and ΔYB are frequently designated as the components of ΔE. Color difference equations may also be based on Munsell notations, although such equations are seldom used today (McDonald, Roderick (1997) Colour physics for industry, 2nd edn. Society of Dyers and Colourists, West Yorkshire; Billmeyer FW, Saltzman M (1996) Principles of color technology. Wiley, New York).

Colored Cement *n* A cement to which color pigment has been added.

Colorfast *n* Fade-resistant.

Colorfastness *n* Resistance to fading; i.e., the property of a dye to retain its color when the dyed (or printed) textile material is exposed to conditions or agents such as light, perspiration, atmospheric gases, or washing that can remove or destroy the color. A dye may be reasonably fast to one agent and only moderately fast to another. Degree of fastness of color is tested by standard procedures. Textile materials often must meet certain fastness specifications for a particular use. See ▶ Light Resistance and ▶ Lightfastness.

Color Floating See ▶ Floating.

Color Gamut See ▶ Color, ▶ Gamut, Color.

Color Harmony Manual *n* Orderly array of colors spaced (approximately) according to the Ostwald Color System, and made by the Container Corporation of America. Several editions were issued, each slightly different. It is no longer available. See ▶ Ostwald Color System.

Colorimeter \ˌkə-lə-ˈri-mə-tər\ *n* [ISV] (ca. 1872) Instrument used to measure light reflected or transmitted by a specimen. Two general types of colorimeters are commonly used: (1) for measuring concentrations of colored materials for analytical purposes, and (2) for measuring quantities which can be correlated with a psychophysical description of color. Generally, the second type should properly be referred to as a tristimulus colorimeter.

Colorimetric *n* Adjective used to refer to measurements converted to psychophysical terms describing color or color relationships.

Colorimetric Purity See ▶ Colorimetric, ▶ Purity, Colorimetric.

Colorimetry *n* (color-identification testing) Light measurements converted to a psychophysical description or notation which can be correlated with visual evaluations of color and color differences.

Color Index Name See ▶ Colour Index Name.

Color Index Number See ▶ Colour Index Number.

Coloring Aids *n* Bisbenzoxazoles, triazinephenyl-coumarins, and bis(styrl)-bisphenyls comprise the most widely used structural types of coloring additives. Used in conjunction with dyes and pigments, coloring aids absorb ultraviolet radiation and emit a blue-violet fluorescence, making polymer surfaces appear bright or cleaner.

Color in Oil See ▶ Colour Index Name.

Color in Varnish See ▶ Colorant.

Colorist *n* (1686) A person skilled in the art of color matching (colorant formulation) and knowledgeable concerning the behavior of colorants in a particular material; a tinter (in the American usage) or a shader. The word "colorist" is of European origin.

Coloristically *n* A general term which describes the nature of the comparison of two colorants in terms as used by a colorist. It may imply hue, saturation, or strength of any combination of these, as well as transparency, undertone, etc.

Coloristic Properties *n* A general term originating in Europe which is derived from the characteristics which a colorist sees and describes.

Color Match Pair of colors exhibiting no perceptible difference when observed under specified conditions. The quality of an attempted match is described by the closeness of this ideal match.

Color Matching See ▶ Color, ▶ Matching, Color.

Color Matching Functions *n* Relative amounts of three additive primaries required to match each wavelength of light. The term is generally used to refer to the CIE Standard Observer color matching functions designated $\bar{x} + \bar{y} + \bar{z}$. See ▶ Observer, ▶ Standard.

Color Measurement *n* Color is a manufactured object is normally obtained by applying a colorant (dye or pigment) to al polymer substrate, such as textile, paper or paint medium. The appearance of such surface colors depends on (1) the nature of the prevailing illumination, (2) the interaction of the illuminating radiation with the colored species in the surface layers, and (3) the ability of the radiation that is transmitted, reflected and scattered fromj the colored surface to induce the sensation of color in the human eye/brain system (NcDonald R (ed) (1997) Colour physics for industry, 2nd edn. Society of Dyers and Colourists, West Yorkshire). Color measurement consists of the physical measurement of light radiated, transmitted, or reflected by a specimen under specified conditions, and mathematically transformed into standardized colorimetric terms which can be correlated with visual evaluations of colors relative to one another. (Physical and chemical examination of paints, varnishes and lacquers and colors (1996). Gardner Laboratory Inc., Bethesda).

Color Migration n The movement of dyes or pigments through or out of a material.

Color Mixture, Additive See ▶ Additive Color Mixture.

Color Notion n Orderly system of numbers, letters, or a combination of both, which serves to describe the relationship of colors in three-dimensional space. Thus, three dimensions must be included; for example, hue, value, and chroma of the Munsell System. Single dimensional notations, such as a yellowness sale, can be used only if the other two dimensions are fixed or described; two-dimensional notations can be used only if the third is fixed or described. See ▶ Color Order Systems.

Color Order Systems n Systems used to describe an orderly three-dimensional arrangement of colors. Three bases can be used for ordering colors: (1) an appearance basis, i.e., a psychological basis; in terms of hue, saturation, and lightness – an example is the Munsell System; (2) an orderly additive color mixture basis, i.e., a psychophysical basis – examples are the CIE System and the Ostwald System; and (3) an orderly subtractive color mixture basis – an example is the Plochere Color System, based on an orderly mixture of inks.

Color Pigments n Color materials for imparting color to concentrates and plastics parts.

Color Proofs n Prints, in color, either from the engraving or from the mounted plates.

Color Rendering Index n Method for describing the effect that a particular light source has on the color appearance of objects, in comparison with their color appearance under a reference light source. The Illuminating Engineering Society of the U.S. and CIE Committee 1.3.2 have developed a method which is currently under study. A color rendering index of 100 indicates perfect agreement between the test source and the reference source.

Color Retention n The property that a material has when it is exposed to the elements and shows no signs of changing color.

Color Space n Three-dimensional solid enclosing all possible colors. The dimensions may be described in various geometries giving rise to various spacings with the solid. See ▶ Color Order, Systems and ▶ Uniform Color Space.

Color Specification n Term used loosely to describe either (1) the notation for a color standard using one of the color order systems, or (2) the allowable tolerance of samples from a standard in terms of color difference units. For (1), the best specification is a sample of the color desired.

Color Stability n The constancy of the characteristics of color in a plastic compound – hue, intensity, and saturation – in its products over their service lives in their design environments. See ▶ Light Resistance.

Color Standard n An ink, wet sample, or printed proof to which another similar material is compared.

Color Strength n In printing ink, the effective concentration of coloring material per volume. *Also known as Tinctorial Strength.*

Color Stripper n A chemical used to remove some or all of the dyestuffs from a fiber, yarn, or fabric so that a dyeing defect can be corrected, a shade lightened, or another color applied.

Color Temperature n Term used to describe the color of a Planckian (black-body) radiator. The color is expressed in chromaticity coordinates, x and y, determined from theoretical spectral power distribution curves, and the temperature is expressed in the absolute (Kelvin) scale. This term is sometimes used incorrectly to describe the color of "white" light sources.

Color Testing Instruments n Colorimeters or color spectrophotometers for measuring wavelength and tristimulus values for characterizing color.

Color Tolerance n Limit of color difference from a standard which is acceptable. It is generally expressed in terms of a particular color difference equation, which must be described, and may consist of a total color difference, ΔE, or directional color differences, or both. See ▶ Color Difference Equations.

Color Vision, Anomalous n Term "anomalous" is used to imply the defective vision of an anomalous trichromat, an observer who sees three primaries but has weaker than normal response to one. The degree of the anomaly varies from slight to severe. See ▶ Color Blindness.

Color Vision, Defective n General term used to describe abnormal vision. The term is preferred to the common term "color blindness." See ▶ Color Blindness and ▶ Color Vision, Anomalous.

Color Vision, Normal *n* Vision of observers requiring mixtures of three independent primary colors to color-match all colors, using quantities of the primaries sufficiently close to those required for the average observer. The observer with normal vision is referred to as a normal trichromat.

Color Wash *n* Each pigments, with or without whiting, lightly bound in glue size so as to facilitate ready removal since frequent removal is necessary, e.g., tinted lime wash.

Colour *n* British spelling of "color"; used in English-speaking countries of the old British Empire, including Canada, Australia, New Zealand, etc.

Colour Index *n* (**C.I.**) A publication of the Society of Dyers and Colourists (Great Britain) and the American Association of Textile Chemists and Colorists which includes periodic additions and amendments. It provides C.I. generic names for classifying commercial dyes and pigments with respect to usage and C.I. numbers for classifying them with respect to chemical composition. It gives basic chemical and usage information, performance characteristics, and manufacturers of them.

Colour Index Name *n* Consists of the category, hue, and an identifying number. For example, one phthalocyanine green pigment is C.I. Pigment Green 7. See Preface for Details.

Colour Index Number *n* A five digit number which describes the chemical constitution. For example, phthalocyanine green, C.I. Pigment Green 7, is C.I. 74260. (All phthalocyanines are, or will be, numbered 74000–74999.)

Colza Oil \ˈkäl-zə, ˈkōl-\ *n* [F, fr. D *koolzaad*, fr. MDu *coolsaet*, fr. *coole* cabbage + *saet* seed] (1712) Another name for rapeseed oil, usually reserved for refined grades.

Coma *n* An aberration of spherical lenses, occurring in the case of oblique incidence, when the bundle of rays forming the image is unsymmetrical. The image of a point is comet shaped, hence the name.

Comb *n* Thin spring-steel tool used in graining or combing.

Combed Sliver *n* A continuous band of untwisted fiber, relatively free of short fibers and trash, produced by combing card sliver.

Combed Yarn *n* A yarn produced from combed sliver. Also see ▶ Combing.

Combination *n* The termination of a polymer chain by combining chains which doubles the molecular weight.

Monomers of different chemical structure which are initiated to polymerize and form copolymers. See ▶ Comonomers.

Combination Fabric *n* A fabric containing: (1) different fibers in the warp and filling (e.g., a cotton warp and a rayon filling), (2) ends of two or more fibers in the warp and/or filling, (3) combination yarns, (4) both filament yarn and spun yarn of the same or different fibers, or (5) filament yarns of two or more generic fiber types. Combination fabrics may be either knit or woven. They should not be confused with blend fabrics. Although blend fabrics also contain more that one fiber, the same intimately blended spun yarn is present in both warp and filling.

Combination Frequencies *n* Two vibrations of arbitrary frequencies f_1 and f_2 when applied simultaneously to a nonlinear (distorting) device will excite it to a motion containing not only the original frequencies, but also members of a set of "combination" frequencies given by $f_c = mf_1 + nf_2$ where m and n are integers. A resonator sharply tuned to any one of these frequencies which may be produced in the nonlinear device will resound to it with an amplitude depending on the type of nonlinearity. The superheterodyne radio receiver depends on this phenomenon.

Combination Mold Syn: ▶ Family Mold.

Combination Yarn *n* A piled yarn containing two or more yarns that vary in fiber composition, content, and/or twist level; or plied yarn composed of both filament yarn and spun yarn.

Combined-Oxide Formula *n* $3Na_2O \cdot 2CaO \cdot 3Al_2O_3 \cdot 2CO_2 \cdot 6SiO_2$, $Na_2O \cdot Al_2O_3 \cdot 4SiO_2 \cdot 2H_2O$. A formula which represents the constituents as "oxides," with the metallic oxides preceding the acid anhydrides, each arranged in the order of increasing valence. Water of composition appears last. Examples: cancrinite and analcite.

Combined Water *n* The water chemically held, as water of crystallization, by the calcium sulfate dehydrate or hemihydrate crystal.

Combined Yarn See ▶ Combination Yarn.

Combing *n* (1) Act of partially removing a coat of wet paint with combs to imitate the grain of wood or other pattern. (2) A step subsequent to carding in cotton and worsted system processing which straightens the fibers and extracts neps, foreign matter, and short fibers. Combing produces a stronger, more even, more compact, finer, smoother yarn.

Combining Volumes *n* Under comparable conditions of pressure and temperature the volume ratios of gases involved in chemical reactions are simple whole numbers.

Combining Weight *n* The combining weight of an element or radical is its atomic weight divided by its valence. Syn: ▶ Equivalent Weight.

Combining Weights, Law of *n* If the weights of elements which combine with each other be called their "combining weights," then elements always combine either in the ratio of their combining weights or of simple multiples of these weights.

Combustible Liquid *n* Any liquid having a flp at or about 37.8°C (100°F). Combustible liquids are divided into two classes: *Class II* and *Class III*. *Class II* liquids – Those with a flp at or above 37.8°C (100°F) and below 60°C (140°F), except a mixture having components with flp of 93.3°C (200°F) or higher, the volume of which make up 99% or more of the total volume of the mixture. *Class III* liquids – Those with flp at or above 60°C (140°F). *Class III* is subdivided into two classes: *Class IIIA* – Those with flp at or above 60°C (140°F) and below 93.3°C (200°F), except a mixture having components with flp of 93.3° (200°F) or higher, the volume of which make up 99% or more of the total volume of the mixture; *Class IIIB* – Those with flp at or above 93.3°C (200°F). When a combustible liquid is heated for use to within 16.7°C (30°F) of its flp, it is handled in accordance with the requirements for the next lower class of liquids. See ▶ Flammable Liquid (Wypych G (ed) (2001) Handbook of solvents. Chemtec Publishing, New York).

Combustion Analysis *n* Any of several methods for quantitatively determining, by burning, the elemental composition of organic compounds, including plastics. First introduced in the 1830s by J. von Liebig, it was refined to permit accurate analysis of small samples (10–50 mg) by F. Pregel, who led the development of the microbalance. Modern combustion analysis is highly automated, but still relies on the microbalance.

Comfort *n* Performance parameter of apparel referring to wearability. Encompasses such properties as wicking, stretch, hand, etc.

Comic Inks See NEWS INKS.

Commercial Allowance *n* The commercial moisture regain plus a specific allowance for finish used in calculating the commercial or legal weight of a fiber shipment.

Commercial Blast See ▶ Nace No. 3.

Commercial Moisture Regain *n* An arbitrary value adopted as the moisture regain to be used in calculating the commercial or legal weight of a fiber shipment.

Commercial Weight *n* (1) In natural fibers, the dry weight of fibers or yarns plus the commercial moisture regain. (2) In manufactured fibers, the dry weight of staple spun yarns or filament yarns after scouring by prescribed methods, plush the commercial moisture regain.

Commercial Xylene See ▶ Xylene.

Commingled Yarn \kə-ˈmiŋ-gəl, kä-\ *v* (ca. 1626) In aerospace textiles, two or more continuous multifilament yarns, the filaments of which have been intermixed with each other without adding twist or otherwise disturbing parallel relationship of the combined filaments. Usually consists of a reinforcing yarn, such as graphite or glass, and a thermoplastic matrix yarn.

Comminute \ˈkä-mə-ˌnüt, -ˌnyüt\ *v* [L *comminutus*, pp of *comminuere*, fr. com- + minuere to lessen] (1626) To pulverize or reduce to very small sizes, as by grinding.

Comminution *n* Process by which aggregates are reduced to small size.

Common-ion Effect *n* The shifting of an ionic equilibrium due to the addition of an ion involved in the equilibrium; usually refers to the repression of dissociation of a weak electrolyte or the decrease in the solubility of an electrolyte brought about by the addition of an ion which is a dissociation product.

Comoforming *n* A fabrication process that combines vacuum-formed thermoplastic shapes with cold-molded fiberglass-reinforced resin to produce parts having excellent surface appearance and weatherability.

Comonomer \(ˌ)kō-ˈmä-nə-mər, -ˈmō-\ *n* [co- + monomer] (1945) A monomer that is mixed with one or more other monomers for a polymerization reaction, to make a ▶ Copolymer.

Compact See ▶ Powder Compact.

Compacted Yarns *n* Air-jet interlaced yarns. Since the entanglement serves only as a substitute for twist, the degree of interlace or tangle is not as great as in air-jet bulked yarns.

Compacting *v* Compacting of a plastic material and the forcing of it through an orifice in more or less continuous fashion. See ▶ Extrusion.

Compaction See ▶ Intermingling.

Compactor *n* A machine developed by Fabric Research Laboratories which is used to compact fabrics or to produce warp-stretch fabrics by means of forced crimp and/or shrinkage of the warp yarn.

Compact Spinning Process *n* A term generally referring to a spinning process carried out using any one of the several small spinning machines of compact design offered by equipment vendors as "packaged" units in which spinning and subsequent processing (drawing, crimping, cutting, etc.) are linked.

Compatibility *n* The ability of two or more substances to mix together or be joined without objectionable separation. In plastics technology the term is most often used in connection with plasticizers, but is also applied to resin pairs or to a resin and prospective compounding ingredients. ASTM tests for compatibility of plasticizers and PVC resins are D 2383 and D 3291. See also ▶ Loop Test.

Compatibility of Plasticizers *n* Miscibility and stability of one or more plasticizers to plastize a polymer or resin.

Compatibilizer *n* A material that, added to a blend of ordinarily incompatible polymers, suppresses phase separation.

Compatible Shrinkage *n* A term used for bonded fabrics to indicate that the face fabric and lining have similar shrinkage. This is necessary to avoid puckering.

Compensating Eyepiece *n* An overcorrected eyepiece designed to compensate for certain undercorrections in fluorite and apochromatic objectives.

Compensator *n* An anisotropic substance of known retardation superimposed on the field of view with its vibration directions 45° from the vibration directions of the polarizer and analyzer. When an anisotropic particle is positioned so its slow component is parallel to the slow component of the compensator, the retardations are added. When the particle and compensator slow components are perpendicular, the retardations are subtracted. Compensator retardations may be fixed or variable.

Compensator, Full-Wave Plate (First Order Red) *n* A layer of quartz, selenite, calcite, or an oriented polymer film of the proper thickness to produce a retardation equivalent to about 530 nm, the first-order red.

Compensator, Quarter-Wave Plate A thin mica (or other crystal) plate or an oriented polymer film of uniform thickness having a retardation of about 130 nm, first-order gray.

Compensator, Quartz Wedge *n* A wedge, cut from quartz, having continuously variable retardation extending over several orders (usually 3–7) of interference colors. The retardation which exactly compensates that of a crystal can be found by pushing in the wedge while counting orders until it reaches a position at which the interference color of the crystal appears black. Retardation can be compensated only when the flow component of the crystal and the wedge are perpendicular.

Complementary Colors *n* Two colors which, when mixed together in the proper proportions, result in a neutral color. Colored lights which are complementary, when mixed additively, form white light and follow the laws of additive color mixture. Colorants which are complementary, when mixed together, form black or gray and follow the laws of subtractive colorant mixture. With the exception of spectrophotometric complementaries, which exist only in theory, colors which are complementary depend on the illuminant chromaticity considered as the neutral point (McDonald, Roderick (1997) Colour physics for industry, 2nd edn. Society of Dyers and Colourists, West Yorkshire).

Complete Hiding See ▶ Hiding, Complete.

Complex *n* An ion or, sometimes, molecule which consists of a central atom or ion surrounded by some peripheral atoms, ions, or molecules (ligands) bonded to it.

Complexation *n* The formation of a complex.

Complex Dielectric Constant *n* Vectorial sum of the dielectric constant and the loss factor; analogous to complex shear modulus and to complex Young's modulus.

Complexes, Catalyst- Cocatalyst *n* Stereospecific chemical complexes, usually derived from a transition metal halide and a metal hydride or a metal alkyl. An example is in stereospecific polymerization of propylene to crystalline polypropylene.

Complexing Agent See ▶ Chelating Agent Sequestering Agent.

Complex Modulus *n* (complex dynamic modulus) A property of viscoelastic materials subjected to periodic variation or reversal of stress (stress cyling). The stress may be any of the three principal types; the material may be in the solid or liquid state (molten or concentrated solution). In such materials, the strain lags the applied stress in time and when the stress is periodic, the time lag is characterized by a phase angle, θ. The modulus – the ratio of stress to strain – is resolved into two parts, a "real" or in-phase part and an "imaginary" part lagging the real part by $\pi/2$ radians (90°). The resultant "envelope" that develops between the stress and lagging strain is call the hystereis of the stress-strain relationship. For example, in shear, the complex modulus is stated as $G = G' + iG''$, where $i = (-1)^{0.5}$. The vector sum of the two components is called the absolute (dynamic) modulus. The real part is often referred to as the "modulus" while the imaginary part is called the "dynamic viscosity." Both parts vary with frequency, both diminish with rising temperature, and, like static moduli, they are different for different modes of stress (Sepe MP (1998)

Complex Shear Modulus *n* Vectorial sum of the shear modulus and the loss modulus; analogous to complex dielectric constant.

Complex Subtractive Colorant Mixture See ▶ Subtractive Colorant Mixture.

Complex Young's Modulus Vectorial sum of Young's modulus and the loss modulus; analogous to the complex dielectric constant.

Compliance *n* The degree to which a material deforms under stress; the reciprocal of the modulus. Thus, in each mode of stress, the material is characterized by three moduli and their reciprocals, three compliances. However, when the stress is varying, the "real" and "imaginary" parts of the complex compliances are *not* equal to the reciprocals of their counterparts in the ▶ Complex Modulus. Tensile compliance; the reciprocal of Young's modulus; shear compliance: the reciprocal of shear modulus.

Compliance, Elastic *n* Symbol S. An elastic constant which is the ratio of a strain or strain component to a stress or stress component. For a perfectly elastic material it is the reciprocal of the elastic modulus. For a viscoelastic material the modulus and compliance are not reciprocally related due to their different time dependencies (Sepe MP (1998) Dynamic mechanical analysis. Plastics Design Library, Norwich).

Component Substances, Law of *n* Every material consists of one substance, or is a mixture of two or more substances, each of which exhibits a specific set of properties, independent of the other substances.

Composite \käm-ˈpä-zət, kəm-ˈ, *esp British* ˈkäm-pə-zit\ *adj* [L *compositus*, pp of *componere*] (1563) (1) An article or substance made up of two or more distinct phases of different substances. In the plastics industry the term applies broadly to structures of reinforcing members (*dispersed phase*) incorporated in compatible resinous binders (*continuous phase*). Such composites are subdivided into classes on the basis of the reinforcing constituents: Laminate, *particulate* (the dispersed phase consists of unlayered fibers); *flake* (flat flakes forming the dispersed phase); and *skeletal* (composed of a continuous skeletal matrix filled by a resin). (2) Hard or soft constructions in which the fibers themselves are consolidated to form structures rather than being formed into yarns. Rigidity of these constructions is controlled by the density, the modulus of the load-bearing fibers, and the fraction of fusible fibers. Strength is controlled by adhesion and shear-yield strength of the matrix unless fibers are bonded in a load-transferring matrix. (3) A structure made by laminating a nonwoven fabric with another nonwoven, with other materials, or by impregnating a nonwoven fabric with resins (Pittance JC (ed) (1990) Engineering plastics and composites. SAM International, Materials Park).

Composite Fibers *n* Fibers composed of two or more polymer types in a sheath-core or side-by-side (bilateral) relation.

Composite Laminate *n* A term sometimes applied to a laminated plastic bonded to a nonplastic material such as copper, vulcanized fiber, rubber, asbestos, lead, aluminum, etc. An example is the copper-clad laminated plastic used for printed-circuit boards.

Composite Mold *n* A mold in which different shapes are produced in one cycle from the several cavities. See also ▶ Family Mold.

Composite Molding *n* The process of molding two or more materials in the same cavity in the same shot, but a combination of transfer and compression molding. For example, in making a ring gear, a loose nylon-fiber-filled material is loaded into an open mold around the tooth circle, the mold is closed, and then molten nylon is injected by transfer molding.

Composite Pigment *n* Pigment usually made by the mechanical operation of intermixing or blending two or more pigments.

Composites Manufacturing Association *n* (CMA/SME) Formerly a subgroup of the Society of Manufacturing Engineers, CMA is now a separate entity with 3,000 members to promote composites, publish books and tapes, and hold conferences. Its office is at 1 SME Dr, P.O. Box 390, Dearborn, MI 48121.

Composite Structures *n* Any material made of more than one component. Composite structures are made from polymers, or from polymers along with other kinds of materials.

Composition *n* (1) A synthetic material containing resins or/and elastomers, and perhaps other components, in specified percentages. (2) The list of constituents and their percentages in such a material.

Composition Rollers *n* Printing press rollers made primarily of glue and glycerin.

Composition Roofing See ▶ Built-Up Roofing.

Composition Siding See ▶ Hardboard.

Compound \kəm-ˈ, ˈkäm-ˌ\ *n, v* [ME *compounen*, MF *compondre*, fr. L *componere*, fr. *com-* + *ponere* to put] (14c) (1) A mixture of resin and the ingredients necessary to modify the resin to a form suitable for processing into finished articles having the desired performance properties. (2) In chemistry, a combination of atoms ionically or covalently bonded in fixed ratios to form a molecule. (3) To produce a plastic compound by blending ingredients with resins in intensive mixers, extruders, or dry-blenders. (4) A pure substance composed of atoms of different elements, whose components cannot be separated by physical means (Wickson EJ (ed) (1993) Handbook of polyvinyl chloride formulating. Wiley, New York).

Compound Curve *n* A surface having curvature in two principal directions. Simply curved surfaces, such as cylinders and cones, having only one direction of curvature, may be cut along an element and laid flat. Compound curves, such as spheres and hyperbolic paraboloids, cannot be laid flat without distortion no matter how they are cut. Structures having compound curvature (reinforced-plastic roofs, for example) have high stiffness for their mass.

Compounders *n* Chemical combinations of materials which include all the materials necessary for the finished product. They include BMC (Bulk Molding Compounds), SMC (Sheet Molding Compounds) and TMC (Thick Molding Compounds). Approximately 1/3 of all U.S. polymer production undergoes subsequent compounding.

Compounder's Modulus *n* Stiffness measurement extensively used by rubber technologists, expressed as "modulus at 300%" or "300% modulus" (any other percent elongation may be indicated, but 300% is commonly used). By this is meant the tensile stress at the indicated elongation. See ▶ Modulus of Elasticity.

Compounding *n* Mixing basic resins with additives such as plasticizers, stabilizers, fillers, and pigments in a form suitable for processing into finished articles. In some areas of the industry the term includes fusion of the polymer, for example in the production of molding powders by extrusion and palletizing. In the plastisol industry, the compounding step ends with the preparation of the dispersion, fusion being part of the molding step.

Compounding Extruders Extruders that make use of compounded polymer melt. The melt is produced through either distributive or dispersive mixing techniques.

Compounding, Plastics *n* Upgrading of polymers or polymer systems through melt mixing. A compounded plastic has hybrid properties, such as high gloss and good impact strength, or precision moldability and good stiffness.

Compounding Screws *n* A typical screw-extruder device which make use of mixing enhancements such as barriers, flutes, waves, pins, or cavities to overcome the inherent shortcomings in distributive and dispersive mixing. Typically, the screw rotates and axially oscillates. Each turn in the screw's spiral is interrupted by three gaps. The resulting kneading flights are continuously wiped by stationary teeth in the machine's barrel. Uniform shear is introduced at low pressures; making the process particularly well-suited for heat- and shear-sensitive polymers.

Compounds *n* Substances containing more than one constituent element and having properties, on the whole, different from those which their constituents had as elementary substances. The composition of a given pure compound is perfectly definite, and is always the same no matter how that compound may have been formed.

Compreg *n* A contraction of "**com**pressed im**preg**nated wood," usually referring to an assembly of veneer layers impregnated with a liquid resin and bonded under high pressure.

Compregnate *n* To impregnate and simultaneously or subsequently compress, as in the production of compregs.

Compregnated Wood *n* Consolidation of the term "compressed-impregnated wood," referring usually to an assembly of layers of veneer impregnated with a liquid resin and bonded under very high pressures. *Also known as Compreg.*

Compressed-Air Ejection *n* The removal of a molding from its mold by means of a jet of compressed air.

Compressibility *n* The relative change in volume per unit change in pressure; the reciprocal of ▶ Bulk Modulus.

Compression & Transfer Molding *n* In TM the mold halves are brought together under pressure as in Compression molding. The charge of molding compound is then put into a pot and is driven from the pot through runners and gates into the mold cavities by means of a plunger.

Compression Modeling *n* A technique principally for thermoset plastic molding in which the molding compound is placed in the heated open mold cavity, mold is closed under pressure, causes the material to flow and compress.

Compression Modulus *n* The ratio of compressive stress to compressive strain below the proportional limit. While it is theoretically equal to Young's modulus determined from tensile testing, compressive modulus is usually somewhat greater in plastics.

Compression Mold *n* A mold used in the process of ▶ Compression Molding.

Compression Molding *n* A method of molding in which a thermosetting molding material, generally preheated, is placed in an open, heated mold cavity, the mold is closed with a top force or plug member, pressure is applied to force the material into intimate contact with all mold surfaces, and heat and pressure are maintained until the material has cured and solidified. The process most often employs thermosetting resins in a partly cured stage, either in the form of granules or putty-resins like masses, or sometimes in preformed shapes roughly conforming to the shape of the mold. Compression molding has also been used with thermoplastics, must notably phonograph records. In this process, the mold is cooled following the compression-flow stage. Compression molding is the oldest form of processing phenolic composite material. In its simplest form, it consists of a force and a cavity that make up a two-piece mold. The mold contains one or more cavities in the shape of the part to be molded. Its heated to 320–380°F, depending on the part's geometry and the closing speed of the press. Then the correct amount of composite material is added, and the two halves of the mold are brought together under 2,000–6,000 lb of pressure per square inch of molding area. The heat of the mold softens the phenolic material and, under pressure, it flows to form the shape of the mold (Strong AB (2000) Plastics materials and processing. Prentice Hall, Columbus; Harper CA (2000) Modern plastics encyclopedia. McGraw Hill, New York).

Compression-Molding Pressure *n* (1) The force per unit of projected area applied to the molding material in a compression mold. The area is projected from all parts of the material under pressure during the complete closing of the mold. (2) The *hydraulic* pressure applied to the compression ram during molding.

Compression Ratio *n* In an extruder, the ratio of the volume of the first turn of the feed section to that of the last turn of the metering section. This ratio is a rough indication of the total compaction performed on the feedstock. More precisely called the *channel-volume ratio* or, for a screw of constant pitch, *channel-depth ratio*.

Compression Set *n* A permanent deformation remaining after release of a compressive stress. It is a property of interest in elastomers and cushioning materials, such as plastic foams. See, for example, ASTM tests D 395 and D 1565. Sometimes used to mean CREEP. See ▶ Cold Flow.

Compression Testing *n* A method of determining behavior of a material subjected to a uniaxial compressive load.

Compression Zone *n* (compression section) The part of an extruder screw, connecting the feed and metering sections, in which the volume per turn is decreasing because of decreasing channel depth, or lead, or both. In some older designs, rarely made today, there were no distinct feed and metering sections and the volume per turn decreased over the entire screw length, so the compression zone was the entire screw. In two-stage screws used for vented operation, the rear metering zone is normally followed by a deep zone of *decompression*, then a second, short compression zone and second metering zone. Where decreasing channel depth is the means of volume reduction, it may be done with a conical screw-root profile or, more usual today, a helical root profile.

Compressive Strength *n* The load at which a test specimen fails in compression, divided by the original cross-sectional area perpendicular to the load. For rigid plastics, ASTM test D 695 is used; for rigid foams, D 1621. These tests also prescribe procedures for estimating compressive moduli. The actual mode of failure of a stiff material in a test of compressive strength is usually by diagonal shear.

Compressive Stress *n* The compressive load per unit area of perpendicular cross section carried by the specimen during a compression test.

Compton Effect *n* (Compton Recoil Effect) Elastic scattering of photons by electrons results in decrease infrequency and increase of wavelength of x-rays and gamma-rays when scattered by free electrons.

Computer-Aided Design *n* (CAD) Using a computer, and appropriate programs (software), in the engineering design – even to the production of finished working drawings – of parts, tools, molds, and assemblies.

Computer-Aided Manufacturing *n* (CAM) Using a computer – usually dedicated and typically a minicomputer – with appropriate programs (software), to control parts or all of a manufacturing operation.

Computer Numerical Control *n* (CNC, numerical control) The use of a dedicated small computer to implement the control of a process or, typically, a machining task.

Concavity Factor *n* The entire stress-strain curves of rubbers and elastomers that have no elastic limit are typically concave toward the stress axis (and convex to the strain axis). The concavity factor is the ratio (less than 1) between the energy beneath the extension curve to that beneath the straight line to the same final point.

Concentration *n* (1) Amount of a substance expressed in relationship to the whole. (2) Act or process of increasing the amount of a given substance in relationship to the whole.

Concentration Cell *n* Electrolytic cell consisting of an electrolyte and two electrodes of the same metal or alloy that develop a difference in potential as a result of a difference in concentration of ions (most often metal ions) or oxygen at different points in a solution.

Concentration Units *n* The units for measuring the content of a distinct material or substance in a medium other than this material or substance, such as solvent. Note: The concentration units are usually expressed in the units of mass or volume of substance per one unit of mass or volume of medium. When the units of substance and medium are the same, the percentage is often used.

Concentricity *n* The characteristic of circles or circular cylindrical surfaces of different radii having a common center. More loosely, the property, in *any* annular shape, of constant radial wall thickness. Concentricity is important in blown film, pipe and tubing, wire-coating, and many noncircular extrusions.

Conchoidal Fracture *n* Type of fracture seen when a mineral or other substance, such as glass, breaks to give irregularly curved, usually striated surface, and no cleavage along planes.

Concrete *n* A composite material which consists essentially of a binding medium within which are embedded particles or fragments of aggregate; in Portland cement concrete, the binder is a mixture of Portland cement and water. See ▶ Cement.

Concrete Block *n* A hollow or solid concrete masonry unit consisting of Portland cement and suitable aggregates combined with water. Lime, fly ash, air-entraining agents, or other admixtures may be included. Sometimes incorrectly called cement block.

Concrete Bond Plaster See Bond Plaster.

Concrete Gun *n* A spray gun used in applying freshly mixed concrete; compressed air forces the concrete along a flexible hose and through a nozzle.

Condensate *n* Product obtained by cooling a vapor, such that it is converted either to a liquid or a solid.

Condensation *n* (1) The process of reducing a gas or vapor to liquid or solid form. (2) A chemical reaction in which two or more molecules combine with the separation of water or some other simple molecule. If a polymer is formed the process is called *polycondensation*.

Condensation Agent *n* A chemical compound that acts as a catalyst and also furnishes a complement of material necessary for a polycondensation reaction to proceed.

Condensation Polymer *n* A polymer made by condensation polymerization.

Condensation Polymerization *n* A polymerization reaction in which water or some other simple molecule is eliminated from two or more monomer molecules as they combine to form the polymer or crosslinks between polymer chains. Examples of resins so made are alkyds, phenol-formaldehyde and urea-formaldehyde, polyesters, polyamides, acetals, and polyphenylene oxide.

Condensation Polymers *n*, by Chain Mechanisms Step-growth polymers produced by a polymerization reaction in which the elimination of a small molecule, often water, has occurred, e.g., produced by a condensation polymerization. Important examples include the polyesters, polyamides and phenol-, urea- and melamine-formaldehyde polymers (Odian GC (2004) Principles of polymerization. Wiley, New York).

Condensation Resin *n* A resin formed by Condensation Polymerization.

Condenser *n* The lens system mounted under the stage of the microscope to furnish a cone of light to the specimen. There are two basic types of condensers: brightfield and darkfield.

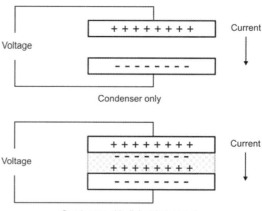

Condensers in Parallel and Series *n* If c_1, c_2, c_2, etc. represent the capacitances of a series of condensers and C their combined capacitance,
when in parallel, $C = c_1 + c_2 + c_3 \ldots$
when in series, $\frac{1}{C} = \frac{1}{c_1} + \frac{1}{c_2} + \frac{1}{c_3} \ldots$
(Serway RA, Faugh JS, Bennett CV (2005) College physics. Thomas, New York).

Conditional Match *n* A pair of colors which appear to match only under limited conditions, such as under a particular light source and a particular observer; a metameric match.

Conditioners *n* A process of allowing textile materials (staple, tow, yarns, and fabrics) to reach hygroscopic equilibrium with the surrounding atmosphere. Materials may be conditioned in a standard atmosphere (65% RH, 70°F) for testing purposes or in arbitrary conditions existing in manufacturing or processing areas.

Conditioning *n* Subjecting a test specimen to standard environmental and/or stress history prior to testing. Several "ambient" conditioning atmospheres, approved by ISO, are listed in ASTM specification E 171, and ASTM D 618 provides detailed information on conditioning. Plastics test specimens are usually conditioned at 23°C and 50% RH for several days or more. Test D 638 for tensile properties of plastics specifies "... not less than 40 hours...." at these conditions.

Conditioning *n* Process of bringing the material or apparatus to a certain condition, e.g., moisture content or temperature, prior to further processing, treatment, etc. Also called conditioning cycle.

Conditioning Cycle See ▶ Conditioning.

Conditioning Time See ▶ Time, Joint Conditioning.

Conductance *n* (1885) (1) The electrical term for the reciprocal of resistance, measured by the ratio of current flowing through a conductor to the difference of potential between its ends. The SI unit is the *siemens* (S), replacing the non-deprecated mho, to which it is exactly equal. ASTM D 257 prescribes tests for the resistance and conductance of electrical-insulating materials. (2) In heat transfer, the ratio of heat *flux* through a solid wall or a stagnant fluid film divided by the difference in temperature through the wall or film. The SI unit is $J/(m^2 \cdot K)$. The reciprocal of resistance, is measured by the ratio of the current flowing through a conductor to the difference of potential between its ends. The practical unit of conductance, the mho, the conductance of a body through which one ampere of current flows when the potential difference is one volt. The conductance of a body in mho is the reciprocal of the value of its resistance in ohms. Dimensions, $[\varepsilon\, l\, t^{-1}]$, $[\mu^{-1} l^{-1} t]$. (Weast RC (ed) (1971) Handbook of chemistry and physics, 52nd edn. The Chemical Rubber Co., Boca Raton).

Conductimetric Analysis *n* A method of analysis for certain ions in solution based on measurement of the solution's electrical conductivity.

Conducting Polymer *n* One of a class of polymers that, unlike most organic polymers, have high electrical conductivity. Polyacetylene is the best known member and at least one type has been prepared by doping that has conductivity almost equal to that of copper. While these polymers are expected to be useful in a wide range of devices, only a few commercial applications exist. Also, see ▶ Polyelectrolyte.

Conductive Composite *n* A composite material having a ▶ Volume Resistivity less than 500 ohm-cm. Such composites may be created by adding metal or graphite powders to ordinary resins. Others are made from inherently ▶ Conducting Polymers. They are useful for static elimination, RF shielding, and in storage-battery components.

Conductive Heat Transfer See ▶ Heat Transfer.

Conductivity *n* (1) The current transferred across per unit potential gradient, K = amperes/cm^2 divided by volts/cm. Reciprocal of resistivity. *Also known as Specific Conductance*. (2) The conductivity of any material is related to its atomic or molecular structure. The electrons of an atom exist in different energy states corresponding to different orbits around the nucleus. Quantum mechanical considerations permit only a certain number of electrons in a given radial orbit. The electrons in a given radial orbit comprise a shell. If an orbit is filled, additional electrons cannot enter the shell or have an energy associated with that level. The outermost shell contains the valence electrons, whose orbits are usually not filled and are responsible for chemical bonding. As two atoms bond to each other, their electrons interpenetrate each other, forming molecular orbitals. The energy levels of these interpenetrating electrons may split into valence bands and conduction bands. These bands are separated by an energy difference or gap. The magnitude of the energy gap determines the conductivity of the material. In highly conductive compounds (i.e., metals) the valence band is not filled and overlaps the conduction band. Therefore, there are numerous energy levels into which an electron can be excited. Under these conditions, the addition of a small external energy raises some of the valence electrons in to the conduction band, where they are free to flow. This type of conduction is referred to as band conduction. For less conductive materials (i.e., graphite) the valence band is filled, but the energy gap is so small that it can be easily jumped. Materials conducting by these mechanisms are referred to as semiconductors. When the valence band is filled and the energy gap is too large to be easily jumped, the materials are insulators. Source: Bhattachacharva SK (1986) Metal filled polymers. Marcel-Dekker, New York.

Conductivity (Electrical) *n* The reciprocal of volume resistivity; the CONDUCTANCE of a unit cube of material. The SI unit is siemens per meter (S/m). It is measured by the quantity of electricity transferred across unit area, per unit potential gradient per unit time. Reciprocal of resistivity. *Volume conductivity* or *specific conductance*, $k = 1/\rho$ where ρ is the volume resistivity. *Mass conductivity* = k/d where d is density. *Equivalent conductivity* $\Lambda = k/c$ where c is the number of equivalents per unit volume of solution. *Molecular conductivity* $\mu = k/m$ where m is the number of moles per unit volume of solution. Dimensions: volume conductivity,

$[\varepsilon\ t^{-1}]$; $[\mu^{-1}l^{-2}t]$, mass conductivity, $[\varepsilon\ m^{-1}l^3t^{-1}]$; $[\mu^{-1}m^{-1}lt]$ (Ku CC, Liepins R (1987) Electrical properties of polymers. Hanser Publishers, New York).

Conductivity (Thermal) *n* Time rate of transfer of heat by conduction, through unit thickness, across unit area for unit difference of temperature. It is measured as calories per second per square centimeter for a thickness of one centimeter and a difference of temperature of 1°C. Dimensions, $[m\,l\,t^{-3}\theta^{-1}]$. If the two opposite faces of a rectangular solid are maintained at temperatures, t_1 and t_2 the heat conducted across the solid of section a and thickness d in a time T will be,

$$Q = \frac{K(t_2 - t_1)at}{d}$$

K is a constant depending on the nature of the substance, designated as the specific heat conductivity. K is usually given for Q in calories t_1 and t_2 in °C, a in square centimeters, T in seconds, and d in centimeters. See ▶ Thermal Conductivity (Ready RG (1996) Thermodynamics. Pleum Publishing Company, New York; Pethrick RA, Pethrick RA (eds) (1999) Modern techniques for polymer characterization. Wiley, New York).

Conductors *n* A class of bodies which are incapable of supporting electric strain. A charge given to a conductor spread to all parts of the body.

Cone *n* A conical package of yarn, usually wound on a disposable paper core.

Cone Mill *n* Old type of mill used for dispersing pigments in media. They usually have conical hopes, and the grinding surfaces are also inclined like the surface of a cone.

Configuration *n* The arrangement of atoms that characterizes a particular stereoisomer is called its configuration. Related chemical structures produced by the cleavage and reforming of covalent bonds; arrangement of polymers along a plastic molecule chain and the structural makeup of a chemical compound, especially with reference to the spatial relationship of the constituent atoms; *trans-* and *cis-*configurations of the carbon-carbon double bond are geometric isomers. See ▶ Conformation.

Configurational Base Unit *n* A molecular repeating unit or mer whose configuration is defined at least at one site of stereoisomerism in the main chain of a polymer molecule. NOTE – In a regular polymer, a configurational base unit corresponds to the mer. For example, in regular polypropylene, the mer is –CH(CH$_3$)CH$_2$– and the configurational base units are shown in Configurational Base Unit.tiff, and these two configurational base units are enantiomeric to each other (mirror images).

Configurational Unit A molecular unit having one or more sites of defined stereoisomerism (IUPAC).

Configuration Repeating Unit *n* The smallest set of one, two, or more successive configurational base units that prescribes configurational repetition at one of more sites of stereoisomerism in the main chain of a polymer molecule (IUPAC).

Conformal Coating See ▶ Encapsulation.

Conformation *n* Different arrangements of atoms that can be converted into one another by rotation about single bonds are called conformations (e.g., *anti-*, *gauche-* conformations).

Congeal \kən-ˈjē(ə)l\ *v* [ME *congelen*, fr. MF *congeler*, fr. L *congelare*, fr. *com-* + *gelare* to freeze] (14c) To change from a liquid or soft state to a solid or rigid state.

Congo Copal *n* Most important of the copals used for oil varnishes. It originates in the Belgian Congo (now Zaire). It becomes soluble in vegetable oils after running.

Congocopalic Acid See ▶ Bengucopalic Acid.

Congo Gum *n* Fossilized gum resin obtained from the Congo region of Africa (now Zaire). Used mostly in manufacturing varnishes.

Congolene *n* Bicyclic hydrocarbon, obtained from the oily layer resulting from the running of Congo copal.

Conical Dry-Blender *n* A device consisting of two hollow cones joined at their bases by a short cylindrical section, mounted on a central shaft perpendicular to the conicylindrical axis. Material is charged and discharged at openings in the apexes of the cones. Mixing is accomplished by cascading, rolling, and tumbling of the charge as the chamber rotates bout the shaft.

Conical Transition *n* In a metering-type extruder screw, the root surface of the screw between the feed section and metering section having the shape of a cone whose diameter increases from that of the deeper feed section to that of the shallower metering section.

Coning The transfer of yarn from skeins or bobbins or other types of packages to cones.

Conjugate Acid-Base Pair *n* An acid and the base formed by removal of a proton from the acid, or a base and the acid formed by the addition of a proton to the base (Brønsted-Lowry).

Conjugated *n* In organic chemistry, referring to the regular alternation of single and double bonds between carbon atoms. For example, in the conventional representation of the benzene molecule shown in Conjugated Molecule.tiff where each single bond represents one pair of shared electrons, each double bond, two pairs.

Conjugated Diene Polymerization *n* Conjugated dienes often polymerize as bi-functional monomers with 1,4 addition or 1,2 addition. In this process, one double bond remains in the main chain for each monomer.

Conjugated Double Bonds *n* A chemical term denoting double bonds separated from each other by a single bond. An example is the bonding in 1,3-butadiene, $CH_2=CH–CH=CH_2$.

Conjugated Yarn *n* A yarn made from conjugate filaments.

Conjugate Fibers *n* A two-component fiber with specific ability to crimp on hot or hot/wet treatment because of differential shrinkage. Also see ▶ Bilateral Fibers.

Conjugate Foci *n* Under proper conditions light divergent from a point on or near the axis of a lens or spherical mirror is focused at another point. The point of convergence and the position of the source are interchangeable and are called conjugate foci.

Conjugate Focil *n* In an image-forming system, two fields are said to be conjugate with each other when one or more object fields are simultaneously in focus in a single plane, e.g., in Köhler illumination the field diaphram, specimen and ocular front focal plane.

Conjugation *n* The location of the π orbital in such a way that it can overlap other orbitals within the molecule (Morrison RT, Boyd RN (1992) Organic chemistry, 6th edn. Prentice Hall, Englewood Cliffs).

Conoscopic Observation *n* The study of the back focal plane of the objective by removing the eyepiece, by inserting a Bertrand lens, by examining the image at the eyepoint above the eyepiece with a magnifier or by using a phase telescope is called conoscopic because the observations are associated with the cone of light furnished by the condenser and viewed by the objective (cf., orthoscopic).

Consensus Standard *n* A standard developed according to a consensus agreement or general opinion among representatives of various interested or affected organizations and individuals.

Conservation of Energy *n* (Chem) In a chemical change there is no loss or gain but merely a transformation of energy from one form to another.

Consistency *n* That property of a liquid adhesive by virtue of which it tends to resist deformation. NOTE – Consistency is not a fundamental property but is comprised of viscosity, plasticity, and other phenomena. (See also ▶ Viscosity and ▶ Viscosity Coefficient).

Consistency *n* The density, firmness, viscosity, or resistance to flow of a substance, slurry, or aggregate. See ▶ Viscosity.

Consistency *n* The property of a material or composition which is evidenced by its resistance to flow, represented by an undefined composite of properties, each measurable from the complete, force-rate flow curve as plastic viscosity, yield value and thixotropy. The term is applied to a variety of materials. For Newtonian liquids, consistency is simply viscosity. While "consistency" is an accepted rheological term, it has qualitative meaning only, and is used with qualifying adjectives as "buttery," "thin," "high," etc., in describing plastic flow. Usually measured in an empirical manner and in arbitrary units. Often used as a synonym for ▶ Viscosity.

Consistometer *n* An instrument for measuring the flow characteristics of a viscous or plastic material. See ▶ Viscometer and ▶ Rheometer.

Consolidation *n* Application of heat and pressure to form composite structures.

Constantan \ˈkän(t)-stən-ˌtan\ *n* [fr. the fact that its resistance remains constant under change of temperature] (1903) An alloy containing about 55% copper and 45% nickel and having a low thermal coefficient of resistivity. Its main use in the plastics industry is in thermocouple wire with either iron or copper as the mating element. Iron-constantan, Type J, and chromel-alumel. Type K, are widely used to sense temperatures in plastics-processing equipment.

Constant White *n* Alternative name for blanc fixe.

Constitutional Formula *n* Device used to illustrate the composition of a chemical compound by displaying the individual atoms and radicals, joined together by valency linkages.

Constitutional Repeating Unit *n* The smallest molecular unit whose repetition describes a regular polymer (IUPAC).

Constitutive Equation *n* In material science, an equation that relates stress in a material to strain or strain rate. Simple examples are (1) Hooke's law, which states that, in elastic solids, strain is directly proportional to stress, and (2) Newton's law of flow, which states that, in laminar shear flow, the shear rate is equal to the shear stress divided by the viscosity. Few plastic solids and liquids obey either of these laws.

Constitutive Property *n* A property which depends on the constitution or structure of the molecule.

Consumer's Risk *n* In quality control and acceptance sampling, the risk of making a Type II error, i.e., of accepting, under a given sampling plan, a lot that is of definitely *un*acceptable quality.

Contact Adhesive *n* A liquid adhesive that dries to a film that is not sticky to other materials but very sticky to itself. A typical contact adhesive is a neoprene elastomer mixed with either an organic-solvent vehicle or an aqueous dispersion medium. The adhesive is applied to both surfaces to be joined and dried at least partly. When pressed together with light to moderate pressure a bond of high initial strength results. Some definitions of "contact adhesive" stipulate that, for satisfactory bonding, the surfaces to be joined shall be no further apart than about 0.1 mm (Skeist I (ed) (1990) Handbook of adhesives. Van Nostrand Reinhold, New York).

Contact Angle *n* The angle between the edge of a liquid meniscus or drop and the solid surface with which it is in contact. A droplet placed on a horizontal solid surface may remain spherical or spread to a degree that is related to the surface energies of the two materials. The angle between the solid surface and the tangent to the droplet at the curve of contact with the surface is the contact angle, and an example of this measurement is shown.

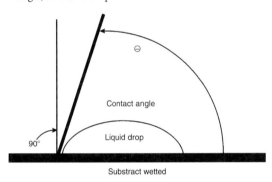

Measurement of contact angle of a soid matrial using a goniometer

By virtue of measuring such droplet-contact angles for droplets of several liquids of different surface energies, the surface energy of the solid may be calculated as shown in the plot (Mittal KL (January 2003) Contact angle, Wettability and adhesion, vol 1–3, VPS International Science Publishers; Oss CJ (1994) Interfacial forces in aqueous media., Marcel Dekker, New York; ASTM Test D 724–99, 2003).

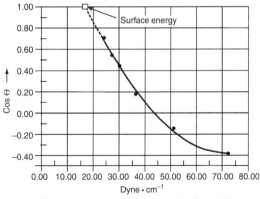

Surface energy determination of polytetrafluoroethylene (Teflon)

Contact Laminating See ▶ Contact-Pressure Molding.

Contact Molding *n* A process for molding reinforced plastics in which the reinforcement and plastic are placed on a mold.

Contact-Pressure Molding *n* (contact molding) This term encompasses processes for forming shapes of reinforced plastics in which little or no pressure is applied during the forming and curing steps. It is usually employed in connection with the processes of ▶ Sprayup and ▶ Hand-Layup when such processes do not include the application of pressure during curing.

Contact-Pressure Resin *n* (contact resin, impression resin, low-pressure resin) A liquid resin that thickens or crosslinks on heating and, components are an unsaturated monomer such as an allyl ester, or a mixture of styrene or other vinyl monomer with an unsaturated polyester or alkyd.

Contact Resin *n* A liquid resin that thickens or crosslinks on heating and, when used for bonding laminates, requires little or no pressure. Typical components are an unsaturated monomer such as an allyl ester, or a mixture of styrene or other vinyl monomer with an unsaturated polyester or alkyd. See ▶ Contact-Pressure Resin.

Contact Resins *n* Liquid resins which thicken on heating and, when used for bonding laminates, require little or no pressure.

Container Lining *n* Protective coating applied to the inner walls of a container to prevent interaction between the contents and the material of construction of the container or to make the walls of the container impervious to the contents.

Continuous Filament *n* A single, flexible, small-diameter fiber of indefinite length. See ▶ Filament.

Continuous-Filament Yarn *n* A yarn formed by twisting together two or more – typically scores – of continuous filaments.

Continuous Maximum Service Temperature *n* Maximum temperature at which a material can perform reliably in a long-term application.

Continuous Phase *n* (1) In a ▶ Suspension or ▶ Emulsion the continuous phase refers to the liquid medium in which the solid or second-liquid particles are dispersed. The solid particles or droplets are called the *disperse phase*. (2) In a plastic filled or reinforced with solid particles, flakes or fibers, the binding resin is the continuous phase.

Continuous Phase *n* The medium or continuum in which the dispersed phase is contained. *Also called External Phase.*

Continuous Polymerization *n* In polymer manufacture, linkage of the various stages of polymerization so that materials flow without interruption from the addition of raw materials to delivery of the finished polymer from the system. Extrusion as film, chip or fiber may be linked to a continuous polymerization line. Because there is no break in the process while the transition from low molecular weight to high occurs, multiple stage reaction vessels may be required and accurate process control is critical.

Continuous Roving See ▶ Roving.

Continuous Tone *n* Tonal gradation without use of halftone dots.

Contour Length *n* Also known as *Displacement Length*. The maximum value of the end-to-end distance of a polymer chain. If the chain consists of n links of each length l with valence angle θ, the chain length is $nl \sin(\theta/2)$.

Contraction See Take-Up (Twist) or Take-Up (Yarn In Fabric).

Contraction Allowance See ▶ Shrinkage Allowance.

Contrast Ratio *n* (1) Ratio of the reflectance of a dry paint film over a black substrate of 5% or less reflectance, to the reflectance of the same paint, equivalently applied and dried, over a substrate of 80% reflectance. (2) The ratio of the luminous reflectance, Y measured on a film over a black substrate, to the Y measure on the same film over a white substrate. The Y of the white and black must be specified. The contrast ratio will vary depending on the thickness of the film and on the concentrations of colorants (Sward GG (ed) (1972) Paint testing manual: physical and chemical examination of paints, varnishes and lacquers, and colors, 13th edn. ASTM Special Technical Publication No. 500, American Society for Testing and Materials, Philadelphia).

Controlled-Atmosphere Packaging *n* The packaging of a product in a gas other than air, typically an inert gas such as nitrogen.

Convalent Bond *n* A bond consisting of a pair of electrons shared between the bonded atoms.

Convection \kən-ˈvek-shən\ *n* [LL *convection-*, *convectio*, fr. L *convehere* to bring together, fr. *com-* + *vehere* to carry] (ca. 1623) Any process by which heat energy or material exchange is effected by flow of the medium. Convection may be *natural*, driven by gravity and density or concentration differences at different points the medium; or *forced*, driven by pumps, blowers, or vibrating devices. Forced convection is almost always faster and more efficient because of the higher velocities and greater turbulence produced.

Conventional Base Unit of a Polymer *n* Base unit, defined without regard to steric isomerism (IUPAC).

Convergent Die *n* An extrusion die in which the internal channels of the die leading to the die orifice are decreasing in cross section in the direction of flow.

Conversation of Energy *n*, **Law of** Energy can neither be created nor destroyed and therefore the total amount of energy in the universe remains constant.

Conversation of Mass *n* In all ordinary chemical changes, the total of the reactants is always equal to the total mass of the products.

Conversation of Momentum *n* Law of For any collision, the vector sum of the moment of the colliding bodies after collision equals the vector sum of their moment before collision. If two bodies of masses m_1 and m_2 have, before impact velocities v_1 and v_2 and after impact velocities u_1 and u_2

$$m_1 u_1 + m_2 u_2 = m_1 v_1 + m_2 v_2$$

Conversion *n* (1) In a chemical process, the molar percentage of any reactant, often the primary or most costly reactant, that is changed into product. (2) In the packaging industry, the intermediate processing and fabrication of plastic film or sheeting into useful forms by slitting, die cutting, heat-sealing into bags, etc., for resale to packagers.

Conversion Coating See ▶ Chemical Conversion Coating.

Converted Fabric *n* A finished fabric as distinguished from greige fabric.

Converter An individual or organization which buys greige fabrics and sells them as a finished product to cutters, wholesalers, retailers, and others. The converter arranges for the finishing of the fabric, namely bleaching, mercerizing, dyeing, printing, etc., to the buyers' specifications.

Conversion Coating See ▶ Chemical Conversion Coating.

Convertible Coating *n* Irreversible transformation of a coating after its film formation to a film insoluble in the solvent from which it was deposited. This can be effected by oxidation, thermal crosslinking or catalytic curing. See ▶ Nonconvertible Coating. This should not be confused with

Convolution \ˈkän-və-ˈlü-shən\ *n* (1545) (1) An irregular spiral or twisted condition characteristic of mature cotton fiber. It is visible under a microscopic. The finer fibers are generally more twisted than the coarser fibers. (2) Coil and curl in certain types of textured yarns which provide bulkiness to the yarn (Tortora PG (ed)

(1997) Fairchild's dictionary of textiles. Fairchild Books, New York).

Cooling *n* Processing highly radioactive materials to attain less radioactivity for subsequent use or handling.

Cooling Channel *n* A passageway provided in a mold, platen or die for circulating water or other cooling medium, in order to control the temperature of the metal surfaces in contact with the plastic being molded or extruded. Proper sizing and placement of cooling channels can do much to speed processing and optimize properties.

Cooling Fixture *n* (shrink fixture) A structure of wood or metal shaped to receive and restrain a part after its removal from a mold, so as to prevent distortion of the part while it is cooling.

Coordinate Bond *n* A covalent bond consisting of a pair of electrons donated by only one of the two atoms it joins.

Coordinate Covalent Bond *n* A covalent bond in which both shared electrons appear to have been contributed by one atom. (2) coordinate link, coordinate covalent bond, dative bond, semipolar bond, or dipolar bond is a description of covalent bonding between two atoms in which both electrons shared in the bond come from the same atom. The distinction from ordinary covalent bonding is artificial, but the terminology is popular in textbooks, especially those describing coordination compounds. Once such a bond has been formed, its strength and description is no different from that of other polar covalent bonds. An example of a dipolar bond in the ammonium ion.

$$NH_4^+$$

Ammonium ion

Dipolar bonds occur when a Lewis base (an electron pair donor or giver) donates a pair of electrons to a Lewis acid (an electron pair acceptor) to give a so-called adduct. The process of forming a dipolar bond is called coordination. The electron donor acquires a positive formal charge, while the electron acceptor acquires a negative formal charge. Classically, any compound that contains a lone pair of electrons is capable of forming a dipolar bond. The bonding in diverse chemical compounds can be described as coordinate covalent bonding. Carbon monoxide (CO) can be viewed as containing one coordinate bond and two "normal" covalent bonds between the carbon atom and the oxygen atom. This highly unusual description illustrates the flexibility of this bonding description. Thus in CO, carbon is the electron acceptor and oxygen is the electron donor. Beryllium dichloride ($BeCl_2$) is described as electron deficient in the sense that the triatomic species (which does exist in the gas phase) features Be centers with four valence electrons. When treated with excess chloride, the Be^{2+} ion binds four chloride ions to form tetrachloroberyllate anion, $BeCl_4^{2-}$, wherein all ions achieve the octet configuration of electrons. Dipolar bonding is popularly used to describe coordination complexes, especially involving metal ions. In such complexes, several Lewis bases "donate" their "free" pairs of electrons to an otherwise naked metal cation, which acts as a Lewis acid and "accepts" the electrons. Dipolar bonds form and the resulting compound is called a coordination complex, and the electron donors are called ligands. A more useful description of bonding in coordination compounds is provided by Ligand Field Theory, which embraces molecular orbitals as a description of bonding in such polyatomic compounds. Many chemical compounds can serve as ligands. Often these contain oxygen, sulfur, nitrogen, and halide ions. The most common ligand is water (H_2O), which forms coordination complexes with metal ions (like the hexaaquacopper(II) ion, $[Cu(H_2O)_6]^{2+}$). Ammonia (NH_3) is a common ligand. So are some anions, especially fluoride (F^-), chloride (Cl^-), and cyanide (CN^-). Terminology of these bonds follow: The atom which donates the pair of electrons is termed as Donor; The atom which accepts the pair of electrons is termed as Acceptor; The pair of electrons shared in this type of bond is the lone pair of the Donor; The Donor develops a positive charge and the Acceptor develops a negative charge; Although coordinate compounds do not ionize in water their physical properties are intermediate between those of ionic and covalent compounds; The coordinate bond is rigid and directional, like covalent bond; and Coordinate compounds exhibit stereoisomerism. References: International Union of Pure and Applied Chemistry. "dipolar bond." Compendium of Chemical Terminology Internet edition; International Union of Pure and Applied Chemistry. "coordinate link." Compendium of Chemical Terminology Internet edition; International Union of Pure and Applied Chemistry. "coordinate covalent bond." Compendium of Chemical Terminology Internet edition; International Union of Pure and Applied Chemistry. "dative bond." Compendium of Chemical Terminology Internet edition.

Coordination Catalyst *n* A catalyst comprising a mixture of an organo-metallic compound, e.g., triethylaluminum, and a transition-metal compound, e.g., titanium tetrachloride. Often called Ziegler or Ziegler-Natta catalysts, they are used in polymerizing olefins and dienes.

Coordination Catalysts *n* Catalysts comprising a mixture of (a) an organo-metallic compound such as triethylaluminum or a transition-metal compound, such as titanium tetra-chloride. Known as Ziegler or Ziegler-Natta catalysts, they are used for the polymerization of olefins and dienes.

Coordination Compound *n* (Werner complex) A complex compound whose molecular structure contains a central atom bonded to other atoms by coordinate covalent bonds based on a shared pair of electrons, both of which are from a single atom or ion. A ▶ Chelate is a special type of coordination compound.

Coordination Number *n* The number of atoms, ions, or molecules surrounding a central atom or ion in a complex, or sometimes, in a solid.

Coordination Polymerization *n* Polymerization of vinyl monomers using a catalyst comprising a transition metal salt and a metal alkyl; Ziegler-Natta catalysts for polymerization of ethylene to produce polyethylene.

Cop \kӓp\ *n* [ME, fr. OE *copp*] (before 12c) (1) A headless tube upon which yarn or thread is wound. (2) Thread or yarn wound into the shape of a hollow cylinder with tapered ends. (3) Filling yarn wound upon a tapered tube (generally paper).

Copal \kō-pəl\ *n* [sp, fr. Nahuatl *copalli* resin] (1577) A fossil resin used in printing ink vehicles.

Copal Esters *n* These are normally regarded as the glyceride esters of run Congo copal, although the term can also be applied to esters of other copals.

Copal Oil *n* Another name for the condensed fumes obtained during the running of natural copals.

Copals *n* Gum resins exuded from living plants and fossilized in the ground. Some of the opals are: Zanzibar, amber, kauri, Manila and Congo.

Copal Varnish *n* A high-gloss varnish made with a drying oil, such as linseed oil, and copal.

Coping *n* A protective cap, top, or cover of a wall, parapet, plaster, or chimney; often of stone, terra-cotta, concrete or wood.

Copolyamides *Also known as Polyether Block Amide Elastomers* – based on a block copolymer and a polyether. A wide range of grades and performance characteristics can be achieved by varying the polyamide and polyether blocks. These high-performance thermoplastic elastomers can withstand high heat and offer good heat-aging characteristics, long flex life, and excellent chemical resistance. Copolyamides are used in demanding medical applications such as catheters and in wire-and-cable jackets, automotive parts, and sporting goods.

Copolycondensation *n* The copolymerization of two or more monomers by ▶ Condensation Polymerization.

Copolymer \( )kō-¹pä-lə-mər\ *n* (1936) This term usually, but not always, denotes a polymer two chemically distinct monomers. It is sometimes used for terpolymers, etc., containing more than two types of mer units. Three common types of copolymers are ▶ Block Copolymers, ▶ Graft Copolymers, and ▶ Random Copolymer. The IUPAC term for a polymer derived from two species of monomer, *bipolymer* eschews the foregoing ambiguity but is nevertheless rarely seen or heard. See also ▶ Bipolymer and ▶ Terpolymer.

Copolymer Equation $nF_A/F_B = (r_A f_A/f_B + 1)/(r_B f_B/f_A + 1)$

f_A = mole fraction of monomer A in feed
f_B = mole fraction of monomer B in feed
F_A = mole fraction of A in copolymer
F_B = mole fraction of B in copolymer
r_A = reactivity ratio of A
r_B = reactivity ratio of B

Copolymerization *n* The building up of linear or nonlinear macromolecules (copolymers) in which many monomers, possessing molecules have one or more double bonds, have been located in every macromolecule of different size which constitutes the copolymerizate, following alternations which may be regular or not. See ▶ Polymerization.

Copolymers *n* Polymers constructed from two different materials

Copper Acetoarsenite *Also known as Paris Green, Emerald Green and Schweinfurt Green.*

Copperas \¹kä-p(ə-)rəs\ *n* [ME *copperas*, fr. OF *couperose*, fr. ML *cuprosa*, prob. fr. *aqua cuprosa*, literally, copper water, fr. LL *cuprum*] (14c) See ▶ Ferrous Sulfate.

Copper Blue *n* Another name for *cupri sulfide* (copper (II) sulfide). Syn: ▶ Covellite.

Copper Bronze *n* Name given to a range of metallic powders ranging in color from "pale gold" to the deep "coin bronze." They consist generally of alloys of copper, but recently dyed aluminum powders of similar shades have been produced. See also ▶ Copper Powder.

Copper Carbonate $CuCO_3$. Material frequently used as a poison in antifouling paints.

$$Cu^{++}$$
$$\text{-O-C(=O)-O}^-$$

Copper-Clad Laminate *n* A laminated plastic surfaced with copper foil or plating, used for preparing printed circuits.

Copper Driers (Deprecated) Copper salts of acids generally used for driers, such as naphthenic and 2-ethyl hexoic, which are used as fungicides and preservative additives in coatings, but do not function as driers. They are deep green and cannot be used in whites or light colors.

Copper Powder *n* Metallic copper and copper bronze powders have some application in anti-fouling paints, and in special primers. Copper bronzes of various colors are produced according to the amount of alloy metal present. In producing finishes from copper and copper bronze powders, great care in formulation is necessary, for these pigments are apt to cause inhibition of drying, discoloration of media, and gelation.

Copper Resinate *n* Green compound formed by dissolving copper acetate, verdigris or other copper salt in Venice turpentine, balsam, or similar resinous solutions.

Copper Soap *n* Combination of copper with a fatty acid.

Copper Staining *n* Usually caused by corrosion of copper screens, gutters or downspouts washing down on painted surfaces.

Copper Sub-Oxide Cu_2O. Another name for cuprous oxide [copper(I) oxide].

$$Cu^+$$
$$Cu^+ \quad O^{--}$$

Copper Sulfate *n* $CuSO_4 \cdot 5H_2O$. Used occasionally in pigment manufacture, e.g., Para Brown. *Also known as Blue Vitriol and Blue Stone.*

$$Cu^{++} \quad \underset{\underset{O}{\|}}{\overset{\overset{O}{\|}}{{}^-O-S-O^-}}$$

Copy *n* Material, including art and text, submitted for reproduction. The term is also used to refer to the final printed result.

Coral Rubber *Cis*-1,4-poly(isoprene). Manufactured by Goodrich, U.S.

Cord *n* (1) The product formed by twisting together two or more plied yarns. (2) A rib on the surface of a fabric (e.g., corduroy and whipcord).

Corded Selvage See ▶ Loop Selvage.

Corduroy \ ˈkór-də-ˌrói\ *n* (ca. 1791) (1) In wallpaper, a very narrow strip imitating the fabric. The name is derived from the French "corde du roi" or king's cord. (2) A filling-pile fabric with ridges of pile (cords) running lengthwise parallel to the selvage.

Core *n* (1) The central member of a laminate to which the faces of the sandwich are bonded. (2) A channel in a mold, extruder screw or cast-in heating element for circulation of heat-transfer media. (3) Part of a complex mold that forms undercut sections of a part, usually withdrawn to one side before the main members of the mod are opened. (4) The central member of a die for extruding pipe, tubing, wire-coating, or a parison to be blow-molded. (5) The central conductor in a coaxial cable. (6) All of an atom except for its valence shell of electrons; also called *the Kernel.*

Core and Separator See ▶ Core (4) above.

Core-Bulked Yarn See ▶ Textured Yarns.

Cored Screw *n* An extruder screw bored centrally from the rear to permit circulation of temperature-controlled liquid within all or part of the screw's length.

Core Spinning *n* The process of making a corespun yarn. It consists of feeding the core yarn (an elastomeric filament yarn, a regular filament yarn, a textured yarn, or a previously spun yarn) into the front delivery roll of the spinning frame and of covering the core yarn with a sheath of fibers during the spinning operation.

Core-Spun Yarn *n* A yarn made by twisting fibers around a filament or a previously spun yarn, thus concealing the core. Core yarns are used in sewing thread, blankets, and socks and also to obtain novelty effects in fabrics.

Corfam® *n* Permeable artificial leather from polyurethane/polyester/polyester fleece, manufactured by DuPont, U.S. See ▶ Poromeric.

Cork *n* The outer bark of *Quercus suber*, a species of oak native to Mediterranean countries, having density in the range 0.22–0.26 g/cm³. Cork is used as a core material (see ▶ Core (1)) in sandwich structures and, in ground form, as a density-lowering filler in thermoplastic and thermosetting compounds for special applications such as flooring, ablative plastics, insulating compositions, and shoe inner soles.

Cork Composite *n* A compound consisting of ground CORK, resins, and other additives and reinforcements, formed into rods, sheets, etc. Cork composites have relatively low density and are used in sporting goods, for thermal insulation, and ion ablative material.

Cork Dust *n* Very finely divided cork which is used in anti-condensation paints.

Corkscrew Twist *n* A place in yarn or cord where uneven twist gives a corkscrew-like appearance.

Cornice *n* (1) A horizontal molding or combination of moldings to finish the top of a wall. (2) Paper simulating the same.

Corn Oil *n* Oil obtained from the kernels of Indian corn, maize, *Zea mays.* It is semidrying, lying between cottonseed and soybean oils.

Corona \kə-ˈrō-nə\ *n* [L, garland, crown, cornice] (1563) Cracks on the surface caused by ozone, usually spearing at right angles to a stress, checking, cutting or cracking.

Corona Discharge *n* (1) The flow of electrical energy from a conductor to or through the surrounding air or gas. The phenomenon occurs when the voltage difference is sufficient (> 5,000 V) to cause partial ionization of the gas. The discharge is characterized by a pale violet glow, a hissing noise, and the odor of ozone formed when the surrounding gas contains oxygen. Corona discharge occurs around high-voltage cables, thus making ozone resistance an important factor in compounding plastics for insulation of electrical wire and cables. (2) Curing method involving the bombardment of organic vapors with high-energy electrons at very low pressures in contact with the substrate to be coated. See also ▶ Casing, ▶ Corona-Discharge Treatment and ▶ Glow Discharge.

Corona-Discharge Treatment *n* A method of rendering inert plastics, primarily polyolefins, more receptive to inks, adhesives, and decorative coatings by subjecting their surfaces to a corona discharge. A typical method of treating film is to pas the film over a grounded metal cylinder above which is located a sharp-edged, high-voltage electrode spaced so as to leave a small air gap between the film and the electrode. The corona discharge oxidizes the film, forming polar groups on vulnerable sites, increasing the surface energy and making the film receptive to inks, etc. See also ▶ Flame Treating.

Corona, Internal *n* In electrical cable, fault due to ionization of air between conductor and insulation.

Corona Resistance *n* Ability of a material to withstand the effects of corona discharge.

Correlated *n* Different types of merchandise systematically related in color and design, as wallpaper with a fabric, or a group of papers designed to be used together.

Correlated Color Temperature Term used to describe the color of "white" light sources. Specifically, it is the temperature of the Planckian (black body) radiator which produces the chromaticity most similar to that produced by the light source in question. The temperature is expressed in degrees on the absolute or Kelvin scale, or in mireds (micro-reciprocal degrees), $10^6/T$. See ▶ Color Temperature.

Correlation *n* Relationship, degree of association, or index of prediction between two scores or sets of data. Measures the tendency of one score or set of data to vary concomitantly with the other (e.g., the tendency of students with high IQ to be above average in reading ability). The existence of a strong relationship – i.e., a high correlation – between two variables does not necessarily indicate that one has any causal influence on the other. Usually expressed as a decimal coefficient between −1.00 and +1.00 (Pearson v), where −1.00 indicates a perfect negative relationship, 0 indicates no relationship, and +1.00 indicates a perfect positive relationship. The Pearson v coefficient of correlation can assume any value on a continuum between −1.00 and +1.00.

Corrosion \kə-ˈrō-zhən\ *n* [ME, fr. LL *corrosion-*, *corrosio* act of gnawing, fr. L *corrodere*] (14c) The deterioration of metal or of concrete by chemical or electrochemical reaction resulting from exposure to weathering, moisture, chemicals, or other agents in the environment in which it is placed (Corrosion, National Association of Corrosion Engineers, Houston; Uhlig HH (1971) Corrosion and corrosion control. Wiley, New York; Uhlig HH (1948) Corrosion handbook. Wiley, New York).

Corrosion Barriers *n* A broad term applying to the ability of plastics to resist many environments, but in particular, attack by acids, bases, and oxidants.

Corrosion Coating See ▶ Chemical Conversion Coating.

Corrosion Fatigue *n* Reduction by corrosion of the ability of a metal to withstand cyclic or repeated stresses.

Corrosion-Inhibiting Paint See ▶ Anticorrosion Paint or Composition and ▶ Composition.

Corrosion-Inhibitive Pigment *n* A pigment which when made into a paint has the property of minimizing corrosion of the substrate to which it is applied.

Corrosion Inhibitor *n* Any of a number of materials used to prevent the oxidation of metals; may be a coating applied to the surface, a paint undercoat, an additive or an element alloyed with the metal.

Corrosion Potential *n* Potential that a freely corroding metal or alloy exhibits in a particular solution.

Corrosion Rate *n* Speed at which a metal or alloy is wasted away because of corrosion.

Corrosion Resistance *n* A broad term applying to the ability of plastics to resist many environments, but in particular, attack by acids, bases, and oxidants. See ▶ Acid Resistance, ▶ Alkali Resistance, ▶ Artifi-Cial Weathering, ▶ Chemical Resistance, ▶ Deterioration, ▶ Permanence, ▶ Solvent Resistance, ▶ Stain Resistance, ▶ Sulfide Straining, ▶ Light Resistance, ▶ Volatile Loss, and ▶ Weathering.

Corrugation Mark *n* A fabric defect consisting of a crimped, rippled, wavy, pebbled, or cockled area in the fabric spoiling the uniformity of the texture.

Corundum \kə-ˈrən-dəm\ *n* [Tamil *kuruntam*; akin to Sanskrit *kuruvinda* ruby] (1804) Natural aluminum oxide (including many gemstones), extremely hard, used as a filler in plastics to impart hardness, heat resistance, and abrasion resistance.

Corvic Poly(vinyl chloride). Manufactured by ICI, Great Britain.

Cosanic Acids *n* A series of fatty acids beginning with eicosanic acid ($C_{20}H_{40}O_2$) and ending with nanacosanic acid ($C_{29}H_{58}O_2$), each acid differing from its predecessor by one CH_2 group.

Cosmic Rays *n* Highly penetrating radiations which strike the earth, assumed to originate in interstellar space. They are classed as: primary, coming from the assumed source, and secondary, those induced in upper atmospheric nuclei by collision with primary cosmic rays.

Cosmotron A particle accelerator capable of given them energies to billions of electron volts.

Cosolvent See ▶ Coupling Agent.

COT *n* The covering material used on various fiber-processing rolls, especially drawing rolls.
Leather, cork, rubber, and synthetic materials are frequently employed.

Cottage Steamer *n* A chamber used for batch steaming of printed or dyed textiles. Cloth is looped on "poles" on a special cart which fits into the steamer for processing.

Cotton *n* Staple fibers, surrounding the seeds of various species of Gossypium.

Cotton Count *n* The yarn numbering system based on length and weight originally used for cotton yarns and now employed for most staple yarns spun on the cotton, or short-staple, system. It is based on a unit length of 840 yards, and the count of the yarn is equal to the number of 840-yard skeins required to weigh 1 lb. Under this system, the higher the number, the finer the yarn. Also see ▶ Yarn Number.

Cotton Fiber *n* A unicellular, natural fiber composed of almost pure cellulose. As taken from plants, the fiber is found in lengths of 3/8–2 in. For marketing, the fibers are graded and classed for length, strength, and color.

Cotton-free See ▶ Dust-Free.

Cotton-free Dry See ▶ Dust-Free.

Cotton Linters See ▶ Linters.

Cottonseed Oil *n* A semidrying oil obtained from the seeds of many types of plants of the genus, *Gossypium*. Its main constituent acids are linoleic (46%), palmitic (29%) and oleic (24%). As oil, it is rarely used in paint, but its fatty acids are used in the manufacture of alkyd resins. See ▶ Becchi-Millian Test.

Cottonseed Pitch See ▶ Stearin (Stearine) Pitch.

Cotton System *n* A process originally used for manufacturing cotton fiber into yarn, and now also used extensively for producing spun yarns of manufactured fibers, including blends. Processing on the cotton system includes the general operations of opening, picking, carding, drawing, roving, and ring or mule spinning in the production of carded yarns. For combed yarns, three steps, culminating in combing, are included after the carding operation. There have been many modifications of this process, especially in recent years for the so-called long draft, or Casablancas, system. The cotton system is also proving to be the basis of many hybrid systems for handling wool yarns and for manufacturing other long-staple yarns.

Couette Flow *n* Shear flow in the annulus between two concentric cylinders, one of which is usually stationary while the other turns. By measuring the relative rotational velocity and the torque required to maintain steady flow, one can infer the viscosity of the liquid. See ▶ Rotational Viscometer. Flow in the metering section of a single-screw extruder resembles Couette flow, modified by the presence of the flight and, normally, by the pressure rise along the screw.

Coulomb (C) \ˈkü-ˌläm\ *n* Charles A. de Coulomb] (1881) (1) A quantity of electricity defined in the SI system as equal to a current of one ampere flowing for one second, i.e., 1 C = 1A·s. (2) Before SI, the quantity of electricity that must pass through a circuit to deposit 0.0011180 gram of silver from a solution of silver nitrate. (3) The quantity of electricity on the positive plate of a one-farad capacitor when the potential difference between the plates is one volt.

Coulomb's Law *n* (1854) The force of attraction or repulsion acting along a straight line between two electric charges is directly proportional to the product of the changes and inversely to the square of the distance between them.

Coumarin \ˈkü-mə-rən\ *n* [F *coumarine*, fr. *coumarou* tonka bean tree, fr. S or P; S *cumarú*, fr. P, fr. Tupi *kumarú*] (1830) (cumarin) A dual-ring aromatic ketone, $C_9H_6O_2$, the sweet-smelling constituent of white clover, also produced synthetically. It is sometimes copolymerized with styrene to increase the deflection temperature above that of polystyrene.

Coumarone *n* (2,3-benzofuran, cumarone) C_8H_6O. Bicyclic ring compound. Parent substance for the coumarone resins. Properties: colorless liquid, aromatic odor; sp gr of 1.078; mp of $-18°C$ ($-0.4°F$); p of $177°C$ ($351°F$); insoluble in water; soluble in alcohol and ether; derived from the coal tar naphtha fraction boiling between $150°C$ and $200°C$ ($302–392°F$), and having the structure in ▶ Coumarone.

Syn: ▶ Benzofuran and ▶ Cumarone. See ▶ Coumarone-Indene Resins.

Coumarone-Indene Resin *n* Any of a family of resins produced by polymerizing a coal-tar naphtha containing coumarone and indene. The naphtha is first washed with sulfuric acid to remove some impurities, then is polymerized in the presence of sulfuric acid or stannic chloride as a catalyst. Remaining impurities determine the quality of the resin, which can range from a clear, viscous liquid to a dark, brittle solid. Coumarone-indene resins have no commercial applications when used alone, but are used primarily as processing aids, extenders, and plasticizers with other resins and with rubbers.

Coumarone-Indene Resin *n* (Cumar) Coal tar resins; indene resins; polycoumarone resins; polyindene resins. Resins obtained by heating mixtures of coumarone and indene (such as those that occur in the light-oil fraction from coal-tar refining) with sulfuric acid, so as to cause polymerization to thermoplastic materials with softening points of up to about 150°C (302°F). Properties: these vary from fairly viscous liquids to hard resins; color – pale yellow to nearly black; sp gr of 1.05–1.10; soluble in hydrocarbon solvents, pyridine, acetone, carbon disulfide, and carbon tetrachloride; insoluble in water and alcohol. Used as components in aluminum paints, concrete curing compounds, pipe oils, rubber compounding, adhesives, chewing gum, printing inks, floor tile binding, and phonograph records.

Coumarone Resins *n* Any of the group of thermosetting resins derived by the polymerization of mixtures of coumarone and indene.

Count *n* (1) A numerical designation of yarn size indicating the relationship of length to weight. Also see ▶ Yarn Number. (2) The number of warp yarns (ends) and filling yarns (picks) per inch in a woven fabric, or the number of wales and courses per inch in a knit fabric. For example, a fabric count of 68 × 52 indicates 68 ends per inch in the warp and 52 picks per inch in the filling.

Counter-Current *n* Process used in many industries in which material to be dried or extracted is caused to flow against the stream of drying agent or extracting liquid. A typical example is the tunnel drying of pigments. Here the wet filter press cakes and the stream of hot air enters the tunnel from opposite ends. The driest air meets the practically dry pigment so that all moisture is removed. By the time the air has reached the other end, contact with the wet pigment insures that the maximum amount of moisture has been removed by the air stream.

Counterion \❙kaun-tər-❙ī-ən, -❙än\ *n* (1940) An ion with a charge which is opposite in sign to that of some ion under consideration.

Counterions *n* The oppositely charged component of n electron pair; each ion at Na^+ has a counter ion Cl^- when AnCl dissolves in water; the oppositely charged ion accompanying an ionic initiated polymerization reaction.

Country Tar See ▶ Pine Tar *n*, Kiln Burned.

Couple \❙kə-pəl; "couple of" if often ❙kə-plə\ *n* [ME, pair, bond, fr. OF *cople*, fr. L *copula* bond, fr. co- + *apere* to fasten] (13c) Two equal and oppositely directed parallel but not collinear forces acting upon a body to form a couple. The moment of the couple or torque is given by the product of one of the forces by the perpendicular distance between them. Dimension, $[m\ l^2 t^{-2}]$.

Couple Acting on a Magnet *n* Magnetic moment *ml* in a field of strength *H*. If the magnet is perpendicular to the direction of the field

$$C = Hml = HM$$

If the angle between the magnet and the field is θ

$$C = Hml\sin\theta$$

The couple will be in dyne-cm for cgs electromagnetic units of *H*, *m* and *l*.

Coupling *n* The linking of a side effect to a principal effect. For composites an anisotropic laminate couples the shear to the normal components, while an unsymmetric one couples curvature with extension. Poisson coupling links lateral contraction to axial extension. The joining together of two or more polymer molecules, which contain terminal chemically reactive groups, by reaction with a third, usually small, molecule capable of reaction with the polymer functional groups which are normally the same.

Coupling Agent *n* A chemical capable of reacting with both the reinforcement and the resin matrix of a composite material to form or promote a stronger bond at the interface. The agent may be applied from the gas phase or a solution to the reinforcing fiber, or added to the resin, or both. *Also called Mutual Solvent or Cosolvent*. See also ▶ Silane Coupling Agent, ▶ Titanate Coupler, and ▶ Adhesion Promoter (Harper CA (ed) (2002) Handbook of plastics, elastomers and composites, 4th edn. McGraw-Hill, New York).

Coupon *n* A representative specimen of a material or sheet product, cut from the product and set aside for testing.

Courlene *n* Poly(ethylene) (fiber), manufactured by Courtaulds, Great Britain.

Courlene PY *n* Poly(propylene) (fiber), manufactured by Courtaulds, Great Britain.

Course *n* The row of loops or stitches running across a knit fabric, corresponding to the filling in woven fabrics.

Courtelle *n* Poly(acrylonitrile). Manufactured by Courtaulds, Great Britain.

Covalent Bond *n* A bond that results from sharing electrons (e.g., H_2O).

Covalent Solid *n* A solid in which atoms are bonded covalently to form a giant extended network.

Cove Ceiling *n* A ceiling which is rounded where it meets the wall.

Covellite *n* \kō-ˈve-ˌlīt\ *n* [F *covelline*, fr. Niccolò *Covelli* † 1829 Italian chemist] (1850) See ▶ Copper Blue.

Cover *n* (1) The degree of evenness of thread spacing. (2) The degree to which underlying structure is concealed by the surface material, as in carpets, the degree to which pile covers backing. (3) The ability of a dye to conceal defects in fabric.

Coverage *n* (1) Spreading rate generally expressed in ft^3/gal or $metres^2$/litre. In pigmented coatings, it is related to hiding power. In clear coatings, it refers to the area coated at a desired film thickness. (2) Description of the amount of paint per unit area which must be applied to achieve a specified contrast ratio. (3) The surface area covered by a given quantity of ink or coating material. See ▶ Mileage and ▶ Hiding Power (Paint / coatings dictionary (1978) Compiled by Definitions Committee of the Federation of Societies for Coatings Technology).

Cover Factor *n* The fraction of the surface area that is covered by yarns assuming round yarn shape.

Covering Power *n* (1) Term used occasionally in the paint industry as synonymous with hiding power but which actually has no precise meaning. (2) The ability of an ink to hide the material beneath (substrate) and to produce a uniform, opaque surface. Also see ▶ Opacity.

Coverstock *n* A lightweight nonwoven material used to contain and conceal an underlying core material. Examples are the facing materials that cover the absorbent cores of diapers, sanitary napkins, and adult incontinence products.

Covert *n* A medium weight to heavyweight wool or wool blend cloth woven with a steep twill from two or more shades of yarn-dyed fibers to produce a mottled or mélange effect.

Covulcanization *n* Simultneous vulcanization of a blend of two or more different rubbers to enhance their individual properties such as ozone resistance. Rubbers are often modified to improve covulcanization.

Cowoven Fabric *n* In aerospace textiles, a fabric in which a reinforcing fiber and a matrix fiber are adjacent to each other as one end in the warp and/or filling direction.

CP *n* (1) Abbreviation for ▶ Cellulose Propionate. See ▶ Cellulose Acetate Propionate. (2) Abbreviation for chemically pure, a designation for laboratory chemicals now largely superseded by *analytical reagent*.

Cp *n* Abbreviation for the deprecated but still widely used viscosity unit, CENTIPOISE.

CPE *n* Abbreviation for ▶ Chlorinated Polyethylene.

CPET *n* Abbreviation for Crystalline Polyethylene Tere-Phthalate.

CPVC *n* (1) Abbreviations for Chlorinated Poly(Vinyl Chloride). (2) In the paint industry, abbreviations for Critical Pigment Concentration, a source of confusion often encountered in the paint and color literature.

CR *n* Abbreviation for Chloroprene Rubber (British Standards Institution). See ▶ Neoprene.

Cr *n* Chemical symbol for the element chromium.

CR *n* Poly(chloroprene).

CR-39 *n* Abbreviation for "Carbonate Resin 39." See ▶ Diethylene Glycol Bis-(Allyl Carbonate).

Crab *n* A hand device used to stretch carpets in a small area.

Crabbing *n* The process of heating wool or hair fabrics, under tension, in a hot or boiling liquid, then cooling under tension, to provide the fabric with dimensional stability for further wet processing.

Crack *n* A defect in a woven fabric consisting of an open fillingwise streak extending partly or entirely across the fabric.

Cracking *n* (1) Generally, the splitting of a dry paint or varnish film, usually as a result of aging. The following terms are used to denote the nature and extend of this defect: *Hair-cracking*. Fine cracks which do not penetrate the top coat; they occur erratically and at random, *Checking*, Fine cracks which do not penetrate the top coat and are distributed over the surface, giving the semibalance of a small pattern. *Cracking*. Specifically, a breakdown in which the cracks penetrate at least one coat and which may be expected to result ultimately in complete failure. *Crazing*. Resembles checking, but the cracks are deeper and broader. *Crocodilian* or *ligaturing*. A drastic type of crazing, producing a pattern resembling the hide of a crocodile. (2) The process of breaking down certain hydrocarbons into simpler ones of lower boiling points, by means of excess heat, distillation under pressure, etc., in order to give a greater yield of low boiling products than could be obtained by simple distillation. (3) Cracking is also the treatment of rubber, uncured and cured, by passing it through moving corrugated rolls, as in preparing tires and other vulcanized rubber for reclaiming (Hare CH (2001) Paint film degradation – mechanisms and control. Steel Structures Paint Council; Koleske JV (1995) Paint and coating testing manual. American Society for Testing and Materials; Hess M (1965) Paint film defects. Wiley, New York).

Cracking Resistance *n* The ability of a coating to resist breaks of the film where the breaks extend through to the surface painted and the underlying surface is visible. The use of a minimum magnification of 10 diameters is recommended in cases where it is difficult to differentiate between cracking and checking. See ▶ Cracking and ▶ Checking Resistance.

Crackle Finish *n* Finish resulting from applying a top coat designed to shrink and crack and expose a more flexible undercoat, usually of a different color.

Crackle Varnish Clear protective top coat applied over a crackle finish.

Crack Mark *n* A sharp break or crease in the surface of a coated or laminated fabric.

Crack Stopper A method or material used or applied to delay the propagation of a potential or existing crack. Techniques include drilling of holes, installing a load-spreading doubler, or including in the design an interruption in part continuity.

Crammer-Feeder *n* (force feeder) A device fitted to the inlet port of an extruder that precompacts a low-density feedstock and propels it into the feed section of the extruder screw. As originally produced by Prodex Corp (now a division of HPM), the crammer-feeder consisted of a conical shell within which turned a decreasing-diameter screw driven independently of the extruder.

Crash *n* A course fabric with a rough, irregular surface made from thick, uneven yarns.

Crater *n* A small, shallow surface imperfection.

Cratering *n* Formation of small bowl-shaped depressions in a paint or varnish film that may or may not expose the underlying surface.

Crawl Shrinkage of milled and calendered stock after removal from rolls.

Crawling *n* (1) Defect in which a wet paint or varnish film recedes from small areas of the surface, leaving them apparently uncoated. British synonym is ▶ Cissing. (2) The contraction of an ink film into drops after printing on a surface which the ink does not wet completely.

Crawl Space *n* A shallow, unfinished space beneath the first floor of a house which has no basement, used for visual inspection and access to pipes and ducts. Also, a shallow space in the attic, immediately under the roof.

Crayon *n* A small stick of pigment in oil or wax. Usually covered with paper for ease of handling. Certain crayons contain water-soluble dyes and are prepared in an aqueous medium.

Crazing *n* An undesirable defect in plastics articles characterized by distinct surface cracks or minute frost-like internal cracks, resulting from stresses within the article that exceed the tensile strength of the plastic. Such stresses may result from molding shrinkage, or machining, flexing, impact shocks, temperature changes, or the action of chemicals and solvents. See also ▶ Stress Cracking (Hare CH (2001) Paint film degradation – mechanisms and control. Steel Structures Paint Council; Koleske JV (1995) Paint and coating testing manual. American Society for Testing and Materials).

Creaming of Emulsion *n* Separation of an emulsion into two layers of different concentration. The more concentrated top layer, i.e., the layer containing the greatest number of dispersed droplets per volume, has a creamy appearance. Gentle agitation of the two layers often effects uniformity of the emulsion.

Crease *n* A break or line in a fabric generally caused by a sharp fold. Creases may be either desirable or undesirable, depending upon the situation. A crease may be intentionally pressed into a fabric by application of pressure and heat and sometimes moisture.

Crease Recovery See ▶ Wrinkle Recovery.

Crease Resistance *n* Term used to indicate the capacity of a fabric to resist and/or to recover from, creases incidental to its usage.

Crease Retention *n* The ability of a fabric to maintain an inserted crease. Crease retention can be measured subjectively or by the relation of a crease in a subsequent state to the crease in the initial state. Crease retention may be strongly dependent on the conditions of use, e.g., normal wear, washing or tumble-drying.

Creel \ˈkrē(ə)l\ *n* [ME *creille*, *crele*] (14c) The spool and its supporting structure on which continuous strands or rovings of reinforcing material are would for use in the filament-winding process.

Creeling *n* The mounting of supply packages in a creel to feed fiber to a process, i.e., beaming or warping.

Creep \ˈkrēp\ *vt* [ME *crepen*, fr. OE *crēopan*; akin to ON *krjūpa* to creep] (before 12c) Due to its viscoelastic nature, a plastic subjected to a load for a period of time tends to deform more than it would from the same load released immediately after application. The degree of this deformation increases with the duration of the load and with rising temperature. Creep is the permanent deformation resulting from the prolonged application of a stress below the elastic limit. This deformation, after any time under stress is partly recoverable (*primary creep*) upon the release of the load and partly unrecoverable *secondary creep*). Creep at room temperature is sometimes called *cold flow*. See also ▶ Andrade Creep, ▶ Delated Deformation, ▶ Drift, ▶ Cold Flow, ▶ Compression Set, and ▶ Strain Relaxation (Elias Hans-Georg (2003) An introduction to plastics. Wiley, New York; Rosato DV (ed) (1992) Rosato's plastics encyclopedia and dictionary. Hanser-Gardner Publications, New York).

Creeping *n* Spontaneous spreading of a liquid on a surface. In the case of an applied film of paint, varnish or lacquer, it refers to the spread of the wet film beyond the area to which it was applied.

Creep Modulus *n* The total deformation measured at constant load over a period of time divided into the applied stress. Creep modulus, if known, simplifies some plastic part designs by allowing the designer to use standard formulas for deformation in which the creep modulus, for the expected service life under load, replaces the conventional, short-time modulus.

Creep Rubber *n* Natural rubber of a pale- to dark amber color prepared by coagulating natural-rubber latex with acid, then milling this coagulum into sheets. The other basic form of solid natural rubber (i.e., ribbed sheet) is prepared by drying the latex on rolls in the presence of smoke.

Creep Rupture *n* The rupture of a plastic under a continuously applied stress that is less than the short-time strength. This phenomenon is caused by the viscoelastic nature of plastics. Creep-rupture tests are generally conducted over a series of loads ranging from those causing rupture within a few minutes to those requiring several years or more.

Creep Strength *n* The initial stress at which failure occurs after a measured time under load. Thus, creep strength (at any temperature) must be labeled with the time to failure. Like ▶ Creep Modulus, creep strength is useful to designers for applications in which plastic articles or members will carry sustained loads.

Cremnitz White See ▶ Carbonate White Lead.

Crenular Cross Section See ▶ Cross Section.

Crepe \ ˈkrāp\ *n* [F *crêpe*] (1797) A lightweight fabric characterized by a crinkling surface obtained by the use of: (1) hard-twist filling yarns, (2) chemical treatment, (3) crepe weaves, and (4) embossing.

Crepe Rubber *n* A type of crude or sometimes synthetic rubber pressed into crinkled sheets

Cresol \ ˈkrē-ˌsól\ *n* [ISV, irreg. fr. *cresote*] (ca. 1869) (hydroxytoluene, methylphenol) $H_3CC_6H_4OH$. An important family of coal-tar derivatives, occurring in ortho, meta, and para isomers, and used in the production of phenol-formaldehyde resins and tricresyl phosphate, an important plasticizer for PVC. Three cresols are possible, namely: (a) *o*-cresol, mp 30°C; bp, 191°C; (b) *m*-cresol, mp 4°C; bp, 205°C; (c) *p*-cresol, mp 36°C; bp, 201°C, and these are found together in the crude cresylic acid from coal tar. The main use of the cresols is in the manufacture of cresol-formaldehyde resins, and cresylic acid, rich in the meta isomer, is usually chosen for this purpose. *Known also as Cresylic Acid*.

Cresol Resin *n* A phenolic-type resin obtained by condensing a cresol with an aldehyde.

Cresote *n* Heavy, high boiling oil obtained from coal and wood tars. A major use for creosote is as a wood preservative, or as a base for the same. Syn: ▶ Dead Oil and ▶ Pitch Oil.

Cresyl Diphenyl Phosphate *n* (CDP) $(H_3CC_6H_4O)PO(C_6H_5O)_2$. A plasticizer for cellulosics, vinyl chloride polymers and copolymers, with a high degree of flame resistance and good low-temperature properties. It is also acceptable for use in food-packaging films. It is most often used, in low percentages of the total plasticizer, as a flame retardant.

Cresylic Acid *n* A term sometimes applied to mixture of *o*-, *m*-, and *p*-cresol, which are mildly acidic, but also including wider fractions of phenolic compounds derived from coal tar or petroleum that contain xylenols and other higher-boiling phenols in addition to the cresols. It is used in the production of phenolic resins and tricresyl phosphate. See ▶ Cresol.

Cretonne \ ˈkrē-ˌtän\ *n* [F, fr. *Creton*, Normandy] (1870) See ▶ Chintz.

Crevice Corrosion *n* Corrosion which occurs within an adjacent to a crevice formed by contact with another piece of the same or another metal or with a nonmetallic material. When this occurs, the intensity of attack is usually more severe than on surrounding areas of the same surface.

Crimp *n* The waviness of a fiber. It determines the capacity of fibers to cohere under light pressure. Crimp is measured by either the number of crimps or waves per unit length or by the decrease in length upon crimping directed by the uncrimped length, expressed as a percent. (2) The difference in distance between two points on an unstretched fiber and the same two points when the fiber is straightened under specified tension. Crimp is expressed as a percentage of the unstretched length. (3) The difference in distance between two points on a yarn as it lies in a fabric and the same two points when the yarn has been removed from the fabric and straightened under specified tension, expressed as a percentage of the distance between the two points as the yarn lies in the fabric.

Crimp Amplitude *n* The height of displacement of the fiber from its uncrimped condition.

Crimp Deregistering *n* The process of opening a tow band by causing the peaks and valleys of the crimp to lay randomly rather that uniformly.

Crimped Yarn See ▶ Textured Yarns, (4).

Crimp Energy The amount of work required to uncrimp a fiber.

Crimp Frequency The crimp level, or number of crimps per inch in yarn or tow.

Crimping The process of imparting crimp to tow or filament yarn.

Crimp Setting *n* An after treatment to set the crimp in yarn or fiber. Usually heat and steam are used, although the treatment may be chemical in nature.

Crimson Antimony Syn: ▶ Antimony Vermillion.

Crimson Lake *n* Pigment derived from cochineal by precipitating the extract with aluminum and tin salts. It has a deep blue-red color. Alternative name is ▶ Florentine Lake.

Crimson Toner *n* Azo pigment produced by coupling 4-aminotoluene-3-sulfonic acid with β-oxynaphthoic acid, followed by conversion to the calcium lake.

Crinkle *n* (1) A wrinkled or puckered effect in fabric. It may be obtained either in the construction or in the finishing of the fabric. (2) The term is sometimes incorrectly used to describe the crimp of staple fiber.

Crinkle Finish Syn: ▶ Ripple Finish.

Crinkle Yarn See ▶ Textured Yarns.

Crinkling See ▶ Wrinkling.

Crinoline \ˈkri-nᵊl-ən\ *n* [F, fr. I *crinoline*, fr. *crino* horsehair (fr. L *crinis* hair) + *lino* flax, linen, fr. L *linum*] (1830) A stiff, heavily sized fabric used as an interlining or to support areas such as the edge of a hem.

Critical Angle (C) *n* (1873) The angle at which total reflection of a light ray passing from one medium to another occurs. The angle of incidence must be large (55°–60°).

Critical Conversion *n* The degree of reaction at which the molecular weight distribution curve first extends into the region of infinite molecular weight; gelation occurs after *critical conversion* and before the *upper limit of conversion*.

Critical Damping *n* In a damped vibrating system, damping so strong that the system, when displaced from rest, returns to rest in one-half cycle. Compare ▶ Logarithmic Decrement.

Critical Length See ▶ Breaking Length.

Critical Mass *n* (1964) The minimum mass the fissile material must have in order to maintain a spontaneous fission chain reaction. For pure U^{235} it is computed to be about 20 lb.

Critical Micelle Concentration *n* (CMC) The point beyond which the concentration of single molecules of surfactants remains relatively constant. Much more surfactant may be dissolved to produce clear solutions, but the added increments form micelles in the solution instead of appearing as individual molecules.

Critical Miscibility Temperature See ▶ Flory Temperature.

Critical Pigment Volume Concentration *n* (CPVC) That level of pigmentation, PVC, value in the dry paint, where just sufficient binder is present to fill the voids between the pigment particles. At this level, a sharp break occurs in film properties such as scrub resistance, hiding, corrosion resistance, ease of stain removal, etc. Different requirements for each product would dictate different PVC or CPVC ratios. Ceiling paints, for instance, are not required to be very washable and can be formulated at or above CPVC, whereas gloss paints and many exterior formulations are designed well below their CPVC, where CPVC has no significance. CPVC has significance only in flat paints.

Critical Point *n* (ca. 1889) The temperature and pressure above which the liquid and gaseous states become indistinguishable.

Critical Relative Humidity *n* The relative humidity at which the phase transition from the crystalline particle to the liquid droplet occurs.

Critical Shear Stress n In extrusion, the shear stress in the melt at the die wall that signals the onset of ▶ Melt Fracture. The stress is on the order of 0.1–0.4 MPa.

Critical Solution Temperature *n* The liquid-liquid critical point of a solution (solvent-polymer or other) denotes the limit of the two –phase region of the phase diagram (temperature vs. composition). This is point at which an infinitesimal change such as temperature or pressure will lead to separation of the mixture into two distinct separate phases. Two types of liquid-liquid critical points are the upper critical solution temperature (UCST) which denotes the warmest point at which cooling will induce phase separation and, and the lower critical solution temperature (LCST), which denotes the coolest point at which heating will induce phase separation

Critical Strain *n* In a strength test, the strain at the yield point.

Critical Surface Tension *n* The value of the surface tension of a liquid, γ_c, below which a drop of the liquid will wet and spread, forming a zero contact angle on a substrate whose surface energy is characterized by γ_c. This property is positively correlated with a polymer's solubility parameter. See also ▶ Surface Tension and ▶ Solubility Parameter.

Critical Temperature *n* (1) That temperature above which a gas cannot be liquefied by pressure alone. The pressure under which a substance may exist as a gas in equilibrium with the liquid at the critical temperature is the *critical pressure*. (2) The temperature at the critical point is called the *critical temperature T_c*. It is *the highest temperature at which gas can be liquefied by means of compression*. The pressure needed to liquefy a gas at its critical temperature is called the *critical pressure, P_e*. Above the critical temperature, no amount of compression will liquefy a gas, because molecular motion is so violent that no matter how closely together the molecules are crammed, intermolecular forces are not strong enough to hold them together as a liquid. Critical temperatures are high when intermolecular attractive forces are strong. In the Table below, the critical temperatures and critical temperature of all substances are tabulated. Helium has the lowest critical temperature of all substances. In helium the intermolecular forces are so weak that above 5K molecular motion prevents liquefaction, no matter how high the pressure is raised.

The liquid-gas equilibrium curve starts at the triple point. At the triple-point temperature the vapor pressure of the solid is the same as that of the liquid. Thus at this temperature and pressure *all three phases can coexist stably, in equilibrium*. This is the definition of a *triple point*.

The liquid-gas equilibrium line ends (at the high-temperature end) at a point called the *critical point*. This is *the temperature and pressure above which the distinction between gas and liquid vanishes*. How is this possible? As the temperature and pressure of a liquid-gas pair system are both increased, the increasing temperature tends to make the molecules in the liquid fly apart to form a gas, while the increasing pressure tends to force the molecules of the gas together into a liquid. In other words, the liquid becomes more like the gas and the gas more like the liquid. As the system moves up the liquid-gas equilibrium line toward the critical point, temperature and pressure increasing simultaneously so as to stay on the line, all the properties of the gas and liquid approach each other. At the critical point, temperature and pressure increasing simultaneously so as to stay on the line, all the properties of the gas and liquid approach each other. At the critical point itself, properties such as density, refractive index, thermal conductivity, viscosity, etc., are barely different in the two phases (and the difference *vanishes* at a slightly higher temperature and/or pressure).

The figure below shows the appearance of a cell containing a liquid and gas in equilibrium under conditions *A,B,C,* and *D* close to the critical point. These conditions are indicated in Fig. _____. At *A* and *B* the liquid and gas phases can be distinguished, but at the critical point *C* the *interface*, the meniscus at the top of the liquid, is fuzzy and indistinct. (The two phases have almost the same refractive indexes, and so the interface is almost invisible and, indeed, is almost nonexistent.) At *D*, beyond the critical point, only one phase is present.

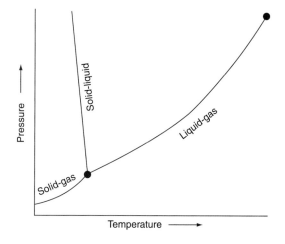

Conditions for phase equilibrium in water

Critical Constants

Substance	Critical temperature, K	Critical pressure, atm
He	5	2.3
H_2	33	12.8
N_2	126	33.5
O_2	154	49.7
CO_2	304	72.8
HCl	325	81.5
NH_3	406	111.3
Cl_2	417	76.1
H_2O	647	218.3
Hg	1,735	1,036.0

Crocheting *n* The interlocking of loops from a single thread with a hooked needle. Crocheting can be done either by hand or by machine.

Crocidolite \krō-'si-d°l-ˌīt\ *n* [Gr *Krokydolith*, fr. Gk *krokyd-, krokys* nap on cloth (akin to Gk *krekein* to weave) + Gr *–lith –lite*] (1835) Blue asbestos (Mineral Riebeckite). See ▶ Asbestos.

Crocking *n* (1) See ▶ Bleeding. (2) Physical transfer of color from one material to another. In ASTM D 1593, a test for resistance to crocking of flexible films subject the test film to rubbing with 5 cm² of white cotton cloth.

Crocking *n* (1) Removal of color on abrasion or rubbing. (2) Smudging, or rubbing off, of ink. (3) Color pigment in rubber which may not appear on the surface as a bloom but which will rub off and discolor an adjacent surface. (4) Staining of a white cloth by rubbing lightly over a colored surface.

Crocodiling See ▶ Cracking.

Crocoisite, Crocolite, Crocoite Native lead chromate mineral of red color.

Crocus Abrasive Either synthetic or natural iron oxide, crocus is the basis of the rouge used in many fine polishing and buffing operations. It is very soft, approximately 6 on the Mohs scale, bright red, and contains a small amount of silicon dioxide.

Crocus Cloth *n* An iron-oxide coated abrasive cloth, used as a polishing agent after most of the work has been done with emery or aluminum oxide, capable of giving a mirror-like finish to metal.

Crofon *n* Optical fibers from polymethyl methacrylate and polyethylene, manufactured by DuPont, U.S.

Crooked Cloth See ▶ Baggy Cloth.

Cross Coating *n* Application of a coat of paint by a series of strokes or spray passes, each at right angles to the previous series.

Cross Direction *n* The width dimension, within the plane of the fabric, that is perpendicular to the direction in which the fabric is being produced by the machine.

Cross Dyeing see ▶ Dyeing.

Cross-Flow Quency *n* In cooling extruded polymer filaments, refers to cooling air directed from one side cross the path of the filaments. There may be some type of suction on the opposite side to remove the heated air.

Cross Grain A pattern in which the fibers and other longitudinal elements deviate from a line parallel to the sides of the piece. Applies to either diagonal or spiral grain or a combination of the two.

Crosshead *n* (1) A device that receives a molten stream of plastic emerging from an extruder, diverts the flow to a direction usually 45° or 90° from the axis of the extruder, and forms the extrudate to a shape such as a parison for blow molding or a jacket around a wire. An essential element of the crosshead is the mandrel, a tubular core with grooves of various shapes and held in place by a perforated plate, web, or spider legs. Material emerging from the space between the mandrel and the crosshead housing is given its final shape by means of a die mounted on the end of the crosshead. (2) The moving member of a testing machine.

Crosshead Die See ▶ Crosshead.

Crossing *n* Method of obtaining even distribution of paint by means of a brush whereby the direction of brushing each series of strokes lies at right angles to that of the previous series.

Cross Laminate *n* A laminate in which the direction of greatest strength in some layers is perpendicular to that direction in other layers. A simple example is a laminate in which alternate layers contain unidirectional reinforcing fibers laid in perpendicular directions. See ▶ Laminated, Cross.

Crosslinked Polymerization *n* A bond, atom, or group linking the chains of atoms in a polymer, protein, or other complex organic molecule.

Crosslinked Polymers Polymers in which linear polymer chains are joined together by covalent chemical bonds.

Cross-linked Structures *n* (Crosslinked and Cross Linked) Theoretically, branched molecules are soluble in some solvent, and are thus distinguishable from cross-linked networks. Conversely, however, not all insoluble polymers are cross-linked networks. Irregular cross-linked networks are the result of either certain uncontrolled, nonstereospecific reactions, or they are produced by the subsequent cross-linking of linear or branched molecules. For network formation to occur, it is essential that each macromolecule be cross-linked at two or more sites to two other polymer chains (See Schematic diagram of a normal and crosslinked polymer). The cross-link can be of the same or of different atoms as occur in the main chain. An example of the first case is the product that results from the radical polymerization of butadiene to high conversions in the absence of transfer agent: A polymer with a large number of 1,4-*trans* links and a relatively high proportion of 1,2 double bonds is formed. Both the pendant double bonds and those in the main chain are less reactive than monomeric double bonds. The probability that the double bonds of the polymer will participate in the polymerization reaction is initially low and only becomes appreciable at higher conversions, when the concentration of the monomer is low and that of the polymer is large. An example of different atoms forming the cross-link is to be found in the vulcanization of poly(butadiene) with sulfur, in which sulfur bridges are formed between polymer chains. Another example is the cross-linking of unsaturated polyesters with styrene. True cross-link networks are only so designated when the molecule has so many cross-link points per primary chain that it is insoluble in all solvents. If the cross-linked molecule extends over the total volume of the reaction container, but the cross-linked product is still swollen by solvent, then this is known as a gel. Gels of very small size (i.e., between 300 and 1,000 nm) are known as microgels. Such microgels behave as tightly packed spheres because of their high branch densities. Because of spherical shape microgels are suspendable. Gel formation occurs after a certain conversion, the gel point. Just past the gel point, some of the initial monomer charge is present as part of the cross-linked network which encompasses the whole of the reaction area, whereas the rest of the original monomer is in the form of branched, although still soluble, molecules. Such a product is considered to be partly cross-linked. Partly cross-linked structures are thus the complete reaction product. Cross-linked networks are considered to be single molecules. Three-dimensional cross-linked networks are, according to definition, considered to be infinite in size. It is therefore pointless to consider their molecular weights. Such cross-link networks are classified according to the network chain lengths, branch type, and branch density. The number of chain links between two branch points is the network chain length. A branch point is defined as the point from which more than two chains, not necessarily of identical structure,

radiate. The formula number-average molecular weight of such a network chain is defined as

$$= (\bar{M}_c)_n = \bar{M}_u/x_c$$

where $\bar{M}_u$ is the average formula molecular weight of the monomeric unit and x_c is the degree of branching. x_c is also called the branch density and is given as the ratio of the moles of cross-linked monomeric units to the total moles of monomeric units present ($0 < x_c < 1$). Thus, it is the mole fraction of cross-linked monomeric units:

$$x_c = \frac{\text{moles of cross - linked monomeric units}}{\text{total moles of monomeric units present}}$$

The cross-link density or the cross-link index γ can be used to characterize network cross-link density. γ is the number of cross-linked monomeric units per primary chain, and is given by ($\bar{X}_n$ is the number-average degree of polymerization)

$$\gamma = \frac{(\bar{M}_n)_0}{(\bar{M}_c)_n} = \frac{(\bar{M}_n)_0}{(\bar{M}_u)} = x_c = \bar{X}_n x_c$$

Here, $(\bar{M}_n)_0$ is the number-average molecular weight of the primary chain. A primary chain is the linear molecule before cross-linking.

All the quantities so far defined relate to ideal networks, i.e., continuous branched structures without free chain ends. In reality, the number of such free chain ends increases with decreasing primary chain molecular weight. The molar concentration $[M_c]_{eff}$ in mol/g of effective network chains can, according to P.J. Flory, be calculated from the molar concentration $[M_c]$ of all the chains present for $(M_n)_0 > (\bar{M}_c)_n$,

$$[M_c]_{eff} = [M_c]\left[1 - 2\frac{(\bar{M}_c)_n}{(\bar{M}_n)_0}\right]$$

$(\bar{M}_c)_n$ is the number-average network chain length. With very extensive cross-linking this formula cannot be used, because in such a case the number of free ends is to too high. Under certain reaction conditions so-called macroreticular or macroporous networks are produced. In such cross-linked networks, the cross-linked chains of the polymeric substance are not completely randomly distributed over the whole of the volume occupied by the substance. They give a structure that is more or less porous. With equal branch densities, macroreticular networks are much more permeable to solute and solvent molecules, which leads to their use as ion-exchange and gel permeation chromatography supports. Because of their more rigid structures, these materials swell less than the normal irregular cross-linked networks. References: Elias H -G (1977) Macromolecules. Plenum Press, New York; Bhattacharva A, Ray P, Rawings JW (eds) (2008) Polymer grafting and crosslinking. Wiley, New York.

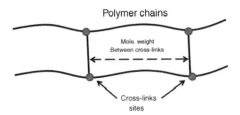

Crosslinking *n* Applied to polymer molecules, the setting up of chemical links among the molecular chains. When extensive, as in most cured thermosetting resins, crosslinking creates one infusible super-molecule from all the chains. Crosslinking can also occur between polymer molecules and other substances. For example, polyethylene can be crosslinked with carbon-black particles, which have sites to which polyethylene chains can link in the presence of a catalyst. The mixture of resin, filler, and catalyst can be molded as a thermoplastic, then transformed to a thermoset by crosslinking in the curing cycle. Crosslinking can be achieved by irradiation with high-energy electron beams, or by means of chemical crosslinking agents such as organic peroxides. The amount of crosslinking may be expressed in x-links per molecular weight or mass between x-links, weight between x-links sometimes referred to as crosslink density (Elias H -G (2003) An introduction to plastics. Wiley, New York; Elias H -G (1977) Macromolecules, vol 1–2. Plenum Press, New York).

Crosslinking Agent *n* A substance that promotes or regulates intermolecular covalent bonding between polymer chains.

Crosslinking Agents *n* (curing agents) An additive used with a polymer in order to bring about crosslinking. In rubber technology, the crosslinking agents used are vulcanization ingredients (sulfur, peroxide, etc.,). For thermosetting plastics, the crosslinking agents (hardeners) are amines or anhydrides for epoxy resins, or peroxides for unsaturated polyesters.

Crosslinking Index *n* The number of crosslinked units per primary polymer molecule, averaged over the whole specimen.

Cross Linking *n* The stabilization of cellulosic or manufactured fibers through chemical reaction with certain compounds in such a way that the cellulose or manufactured polymer chains are bridged across or "crosslinked." Cross-linking improves such mechanical factors as wrinkle resistance. Random cross-linking in manufactured polymers is undesirable and leads to brittleness and loss of tensile strength.

Crossply *n* A layer containing reinforcing fibers mainly running in one direction and perpendicular to the main fiber direction in adjacent layers. See also ▶ Cross Laminate.

Cross Section *n* (Nuclear cross section) A measure of the probability of a particular process. The nuclear cross section is expressed by a/bc, where a is the number of processes occurring, b the number of incident particles, and c the number of target nuclei per cm^3. There are nuclear cross sections for fussion, for slow neutron capture, for Compton collision, and for ionization by electron impact.

Cross Stitch See ▶ Pinhole.

Cross Termination *n* In free radical copolymerization, termination by reaction of two radicals terminated by monomer units of the opposite type, i.e., $\sim A^* + \sim B^* \rightarrow$ termination, by combination or disproportionation with rate constant k_{AB}. Cross-termination is often favored over termination by reaction between two like radicals due to polar effects.

Crotonaldehyde A colorless liquid synthesized by the aldol condensation of acetaldehyde, accompanied or followed by dehydration. It can be polymerized by triethylamine to a resin with film-forming properties, or copolymerized with many compounds. Other uses include solvent for PVC, short-stopped in the polymerization of vinyl chloride, and plasticizer synthesis.

Crotonic Acid *n* (*trans*-2-butenoic acid, *trans*- β-methacrylic acid) $CH_3-CH=CHCOOH$. A white crystalline solid prepared by the oxidation of crotonaldehyde. It forms copolymers with vinyl acetate, used as hot-melt adhesives. Esters of crotonic acid are used as plasticizers for acrylic and cellulosic plastics.

Crotonyl Peroxide *n* A catalyst for the polymerization of vinyl and vinylidene halides.

Crown *n* (1) Of a calender roll, a gradual, small increase in the diameter of a roll toward the center to compensate for the slight deflection due to bending of the roll under pressure. (2) The much ore convex surface of a transmission-belt pulley designed to keep the belt running in the center of the pulley.

Crowsfeet *n* A fabric defect consisting of breaks or wrinkles of varying degrees of intensity and size, resembling bird's footprints in shape, and occurring during wet processing of fabrics.

Crowsfooting *n* (1) Type of film defect where small wrinkles occur in a pattern resembling that of a crow's foot. See ▶ Wrinkling. (2) Type of crystallization on the surface of a varnish or paint film after exposure to the gas test. See ▶ Gas Checking.

Crude Resinous Liquid See ▶ Tall Oil.

Cryogenic \ˌkrī-ə-ˈje-nik\ *adj* (1896) Pertaining to very low temperatures, usually temperatures below about −150°C (123 K). Evaluations of plastics at cryogenic temperatures are conducted for potential space applications.

Cryogenic Finishing *n* Process by which a material is cryogenically tempered (deep freezing below −300°F). Cryogenic finishing relieves stress in the substrate, thus creating a more uniform micro-structure which extends the life of the material.

Cryogenic Grinding *n* (freeze grinding) Thermoplastics are difficult to grind to small particle sizes at ambient temperatures because they soften, cohere in lumpy masses, and clog screens. When chilled by dry ice, liquid carbon dioxide, or liquid nitrogen, thermoplastics can be finely ground to powders suitable for electrostatic spraying and other powder processes.

Cryohydrate *n* The solid which separates when a saturated solution freezes. It contains the solvent and the solute in the same proportions as they were in the saturated solution.

Cryoscopy *n* Measurement of polymer molecular weight based on freezing point depression.

Cryptometer *n* An instrument for measuring the hiding power or opacity of pigmented composition. It functions on the principle that a film or wedge of the paint is obtained with a uniformly varying thickness. By adjusting the instrumentation, the thickness of film necessary to obliterate can be determined.

Crystal *n* A homogeneous solid having an orderly and repetitive three-dimensional arrangement of its atoms. Crystalline polymers are never wholly crystalline but contain some amorphous material and many Crystallites. The "ideal crystal" is a homogeneous portion of crystalline matter, whether bounded by faces or not. Crystalline matter is matter that possesses a triperiodic structure on the atomic scale. It is characterized by discontinuous vectorial properties that give rise to "crystal planes" (1) crystal growth (faces); (2) cohesion (cleavage planes); (3) twinning (twin planes); (4) gliding (gliding planes); and (5) x-ray electron, or neutron diffraction ("reflecting" planes); all of which are parallel to the lattice planes (Hibbard MJ (2001) Mineralogy. McGraw-Hill, New York).

Crystal Lattice *n* The regular, repeating arrangement of particles (atoms, ions, molecules) in a crystal.

Crystalline \ˈkris-tə-lən\ *adj* [ME *cristallin*, fr. MF & L; MF, fr. L *crystallines*, fr. Gk *krystallinos*, fr. *krystallos*] (15c) A substance (usually solid but can be liquid) in which the atoms or molecules are arranged in a definite pattern that is repeated regularly in three dimensions. Crystals tend to develop forms bounded by definitely oriented plane surfaces that are harmonious with their internal structure. They may belong to any of six crystal systems: cubic, hexagonal, tetragonal, orthorhombic, monoclinic, or triclinic.

Crystalline-Amorphous Structures *n* The typical form of crystalline polymers. Crystalline polymers contain a large number of crystallites, as well as devoids of crystallinity. Most plastics are amorphous at processing temperatures, many retaining this state under all normal conditions.

Crystalline Growth *n* (1) The expansion and development of a crystal. The process involves diffusion of the crystallizing material to special sites on the surface of the crystal, incorporation of the molecules into the surface at these sites, and diffusion of heat away from the surface of the crystal. (2) The transformation of disoriented molecules, usually of the same substance, to a higher state of order. This process generally occurs rapidly for small molecules; however, the process is slow for polymer molecules and is arrested at temperatures below the glass transition temperature (Hibbard MJ (2001) Mineralogy. McGraw-Hill, New York).

Crystalline Melting *n* The heating of an ordered, crystalline structure polymer above its melts, it behaves like an amorphous polymer in that the molecular configurations become random.

Crystalline Polymers *n* Polymers containing both crystalline and amorphous material.

Crystalline Silica See ▶ Silica, ▶ Crystalline.

Crystalline Solid A true solid, one with a regular internal structure.

Crystallinity *n* (1) A state of molecular structure in some resins attributed to the existence of solid crystals with a definite geometric form. Such structures are characterized by uniformity and compactness. See ▶ Liquid-Crystal Polymer. (2) The percentage of a polymer sample that has formed crystals. Crystallinity in polymers occurs in two stages: nucleation and growth of nuclei. In most materials, crystallinity forms from growing nuclei to produce reoccurring/repeating structures that form lattices throughout the material; polymers can be crystalline, semicrystalline or non-crystalline (amorphous); crystallinity in a polymer exhibits a distinct endothermic energy transition or melting event (destruction of lattice structure) whereas a noncrystalline polymer shows a glass transition event or just softening.

Crystallite \ˈkris-tə-ˌlīt\ *n* [Gr Kristallit, fr. Gk krystallos] (1805) A perfect portion of an ordinary crystal; that is, a portion with its atoms and molecules arranged in a lattice free of defects. Ordinary crystals are composed of large numbers of crystallites, which may or may not be perfectly aligned with one another.

Crystallization *n* (For Ink) (1) A condition in which a dried ink film repels a second ink which must be printed on top of it. (2) Lacquers containing materials which crystallize out from the medium as the solvent evaporates. (3) Materials in which advantage is taken of the tendency of certain drying oils, notably tung oil, to "frost" or "crystallize" when dried under certain conditions (Leach RH, Pierce RJ, Hickman EP, Mackenzie MJ, Smith HG (eds) (1993) Printing ink manual, 5th edn. Blueprint, New York).

Crystallized Polyethylene Terephthalate *n* (CPET) PET resin to which a fractional percentage of nucleating agent has been added to encourage the development of crystallinity in extruded or molded products. The percent crystalline material is typically between 15 and 35, Modulus and strength increase with crystalline content. See also ▶ Polyethylene Terephthalate.

Crystal Polystyrene *n* Styrene homopolymer, which, through actually 100% amorphous, was so called because of its excellent clarity and the glitter of the early cube-cut pellets. See ▶ Polystyrene.

Crystal Structures *n* Homogeneous solids having an orderly and repetitive three-dimensional arrangement of its atoms.

CS *n* Abbreviation for CASEIN PLASTIC.

CSIRO *n* The Commonwealth Scientific and Industrial Research Organisation (CSIRO) is the national government body for scientific research in Australia. It was founded in 1926 originally as the Advisory Council of Science and Industry. Research highlights include the invention of atomic absorption spectroscopy, development of the first polymer banknote, invention of the insect repellent in Aerogard and the introduction of a series of biological controls into Australia, such as the introduction of Myxomatosis and Rabbit calicivirus which causes rabbit haemorrhagic disease for the control of rabbit populations. CSIRO's research into ICT technologies has resulted in advances such as the Pan-optic Search Engine (now known as Funnelback) and Annodex.

CSMA Abbreviation for Chemical Specialties Manufacturer Association.

CSR Chlorosulfonated poly(ethylene).

C-Stage *n* The final stage in the reaction of certain thermosetting resins in which the material is relatively insoluble and infusible. Certain thermosetting resins in a fully cured adhesive layer are in this stage. Sometimes referred to as *Resite*. See also ▶ A-Stage.

CTA See ▶ Cellulose Triacetate.

CTFE Resin See ▶ Polychlorotrifluoroethylene.

Cu Chemical symbol for the element copper (Latin: cuprum).

Cube Root Color Difference Equation *n* Specific color difference equation. The following equations are those recommended by the CIE in 1967 for study:

$$L = 25.29 G^{1/3} - 18.38$$

where $G = 0.0010X + 1.05Y + 0.0004Z$

$$a = K_a(R^{1/3} - G^{1/3})$$

where $R = 1.1084X + 0.0852Y - 0.1454Z$ and $K_a = 105.0$ for $R < G$; $Ka = R > G$

$$b = K_b(G^{1/3} - B^{1/3})$$

where $B = 0.0062X + 0.0394Y + 0.8192Z$ and $K_b = 30.5$ for $B < G$; $K_b = 53.6$ for $B > A$

$$\Delta E = [(\Delta L)^2 + (\Delta a)^2 + (\Delta b)^2]^{1/2}$$

The differences (Δ's) are calculated as sample minus standard (McDonald, Roderick, (1997) Colour physics for industry, 2nd edn. Society of Dyers and Colourists, West Yorkshire). See ▶ Color Difference Equations.

Cubic \ˈkyü-bik\ *adj* (1) Having the appearance of a cube. (2) Of algebraic equations, containing the unknown to the power 3 and not higher.

(3) Characterizing volume measure, as *cubic meter*. (4) The simplest of the six crystal systems, in which the three principal axes are mutually perpendicular and the atomic spacing is the same along all three.

Cull \ˈkəl\ *n* [ME, fr. MF *cuillir*, fr. L *colligere* to bind together] (13c) (1) A rejected material or product. (2) In transfer molding, the material remaining in the transfer pot after the mold has been filled. A certain amount of cull is usually necessary for the operator to be confident that the cavity has been properly filled.

Culture *n* (1) The process of securing the growth of fungi or other microorganisms upon artificial media. (2) The organisms resulting from the culturing process.

Cultured Stone *n* A term applied to decorative embedments of natural stones such as marble, granite, terrazzo, and slate in thermosetting resins. They are made by casting the resin, usually a polyester, in molds containing the stones. The embedments are used for counter tops, window sills, wall facings, flooring, giftware, etc.

Cumar Gum *n* A synthetic resin, used in varnishes to provide alkali-resistance properties.

Cumarone See ▶ Coumarone.

Cumene *n* (isopropylbenzene, isopropylbenzol, cumol) $C_6H_5CH(CH_3)_2$. A volatile liquid in the alkyl-aromatic family of hydrocarbons. It is used as a solvent and intermediate for the production of phenol, acetone, and α-methyl styrene; and as a catalyst for acrylic and polyester resins. Properties: bp, 153°C; sp gr, 0.862/20°C; refractive index, 1.506. *Known also as Isopropyl Benzene.*

Cumene Hydroperoxide *n* $C_6H_5C(CH_3)_2OOH$. A colorless liquid derived from an oxidize solution or emulsion of cumene, used as a polymerization catalyst.

Cumylphenol Derivative *n* One of a group of polymer intermediates based on cumylphenol that offer higher performance at lower cost than nonylphenol competitors. The free phenol is an accelerator for amine hardeners of epoxy resins. Cumylphenyl acetate and the glycidyl ether are reactive in epoxy systems, giving enhanced strength. The benzoate of cumylphenol aids extrusion of PVC compounds.

Cup Efflux cup. See ▶ Ford Cup.

Cup-Flow Test *n* A British Standard Test (B S 771) for measuring the flow properties of phenolic resins. A standard mold is charged with the specimen material and then closed under preset pressure. The time in seconds for the mold to close completely is the cup-flow index.

Cupioni A type of specialty or novelty yarn having slubs or enlarged sections of varying length.

Cuprammonium Rayon *n* A regenerated cellulose formed by dissolving cotton or wood-pulp linters in a solution of ammonia and copper oxide (from sulfate), then extruding the solution through spinnerets into warm water, where the filaments harden. The finest filaments (lowest denier) are made this way.

Cupric Sulfide \ˈkyü-prink\ *adj* (1799) See ▶ Copper Blue.

Cuprous (Copper) Oxide *n* Cu_2O. Occurs in nature as the mineral, cuprite. Prepared commercially by furnace-reduction of a mixture of cupric oxide and copper. Used as a toxin in anti-fouling compositions.

$$O^{--} \quad Cu^+$$
$$Cu^+ \quad Cu^{++}$$
$$O^{--}$$

Curing (1) In finishing fabrics, the process by which resins or plastics are set in or on textile materials, usually by heating. (2) In rubber processing, vulcanization. It is accomplished either by heat treatment or by treatment in cold sulfuryl chloride solution.

Cup Temperature *n* (1) In injection molding or extrusion, the measured temperature of a glob of melt collected at the injection nozzle or extrusion die in an insulated vessel and presumed to be equal to, or slightly less than, the average melt temperature leaving the machine. (2) Cup-mixing temperature, flow-average temperature In a flowing stream of fluid, the local product of velocity times temperature integrated over the stream cross section, said integral divided by the integral of the local velocity over the cross section (Strong AB (2000) Plastics materials and processing. Prentice Hall, Columbus).

Cup, Weight Per Gallon *n* Brass cup with a volume of exactly 83.3 ml used for quickly determining the wt per gal (or density) of a finished liquid product (Koleske JV (1995) Paint and coating testing manual. American Society for Testing and Materials; Wicks ZN, Jones FN,

Pappas SP (1999) Organic coatings science and technology, 2nd edn. Wiley, New York).

Curcas Oil *n* A vegetable oil obtained from the seeds of *Jatropha curcas*, grown in Central America, and in the Cape Verde Isles, the Comores Isles and in some of the former Portuguese colonies, Siam and the East Indies. The percentage fatty acids as triglycerides are oleic, 62%; linoleic, 19%, and saturated fatty acid, the remainder.

Curdle \ˈkər-dᵊl\ *v* [frequentative of ²*curd*] (1590) Syn: is ► Coagulation. See also ► Coagulation and ► Congeal.

Cure *n* (1) To change the properties of a polymeric system into a final, more stable, usable condition by the use of heat, radiation, or reaction with chemical additives (ASTM). (2) Synonymous with vulcanize. It includes time and temperature of vulcanization

Cure, Curing, also Vulcanization *n* To change the physical properties of a material by chemical reaction, which may be condensation, polymerization, or vulcanization.

Cure Cycle *n* The schedule of time periods at specified conditions to which a reacting thermosetting plastic or rubber composition is subjected to reach a specified property level.

Cure or Curing *n* Conversion of a wet coating or printing ink film to a solid film.

Cure Stress *n* Internal stress in cast or molded thermosetting parts, caused by unequal shrinkage in different sections of the parts. Depending on the directions of applied stress in service relative to the principal cure stress, parts may be considerably weaker, than their designers expected them to be.

Cure Time *n* The period of time that a reacting thermosetting plastic is exposed to specific conditions to reach a specified property level.

Curie \ˈkyúr-(ˌ)ē\ *n* [Marie & Pierre *Curie*] (1910) Unit for measuring radioactivity. One Curie = that quantity of any radioactive isotope undergoing 3.7×10^{10} disintegrations per second.

Curie's Law The intensity of magnetization,

$$I = \frac{AH}{T}$$

where H is the magnetic field strength, T the absolute temperature and A Currie's constant. Used for paramagnetic substances.

Curie Point *n* All ferro-magnetic substances have a definite temperature of transition at which the phenomena of ferro-magnetism disappear and the substances become merely paramagnetic. This temperature is called the "Curie Point" and is usually lower than the melting point (Serway RA, Faugh JS, Bennett CV (2005) College physics. Thomas, New York).

Curie-Weiss Law *n* the Curie law was modified by Weiss to state that the susceptibility of a paramagnetic substance above the Curie point varies inversely as the excess of the temperature above that point. This law is not valid at or below the Curie point (Serway RA, Faugh JS, Bennett CV (2005) College physics. Thomas, New York).

Curl See ► Kink.

Curing Agent *n* (hardener) A substance or mixture of substances added to a plastic or rubber composition to promote or control the curing reaction. It is also an additive which promotes the curing of a film. An agent that does not enter into the reaction is known as a *catalytic hardener* or *catalyst*. A *reactive curing agent* or *hardener* is generally used in much greater proportions than a catalyst, and is actually converted in the reaction. See also ► Accelerator, ► Hardener, ► Catalyst and ► Crosslinking. Syn: ► Hardener.

Curing Agent Blush *n* A blushing, blooming or sweating caused by applying coatings such as amine cured epoxies under conditions of high humidity.

Curing Agents for Epoxy Resins *n* A catalytic or reactive agent that brings about polymerization causing crosslinking.

Curing Temperature *n* The temperature to which a thermosetting or elastomeric material is brought in order to commence and complete its final stage of cure. See Temperature, Curing.

Curing Time (molding time) In the molding of thermosets, the time elapsing between the moment relative movement and between the mold parts creases, and the instant that pressure is released. Time necessary for curing.

Curl *n* In paper, distortion of the unrestrained sheet due to differences in structure or coatings from one side to the other. The curl side is the concave side of the sheet.

Curling *n* (1) See ► Coiling Soup. (2) Curling also refers to excessive warping of sheet goods, or distortion by uneven shrinkage. See also ► Crawl.

Current \ˈkər-ənt\ *adj* [ME *curraunt*, fr. OF *currant*, pp of *courre* to run, fr. L *currere*] (Electric) The rate of transfer of electricity. The transfer at the rate of one electrostatic unit of electricity in one second is the electrostatic unit of current. The electromagnetic unit of current is a current of such strength that one centimeter of the wire in which it flows is pushed sideways with a force of one dyne when the wire is at right angles to a magnetic field of unit intensity. The practical unit of current is the ampere a transfer of one coulomb per second, which is one tenth the electromagnetic unit. The International ampere is the unvarying electric current which, when passed through a solution of silver nitrate in accordance with certain specifications, deposits silver at the rate of 0.00111800 g/s. The international ampere is equivalent to 0.999835 absolute ampere. The ampere-turn is the magnetic potential produced between the two faces of a coil of one turn carrying one ampere. Dimensions,

$$\left[\varepsilon^{\frac{1}{2}} m^{\frac{1}{2}} l^{\frac{3}{2}} t^{-2}\right]; \left[\mu^{-\frac{1}{2}} m^{\frac{1}{2}} l^{\frac{1}{2}} t^{-1}\right]$$

(Serway RA, Faugh JS, Bennett CV (2005) College physics. Thomas, New York; Lide DR (ed) (2004 Version) CRC handbook of chemistry and physics. CRC Press, Boca Raton.)

Current Density Magnitude of current per unit area of a metal surface, usually expressed in milliamper per sq ft (mA/ft^2 or, A/m^2).

Current in a Simple Circuit *n* The current in a circuit including an external resistance R and a cell of electromotive force E and internal resistance r,

$$I = \frac{E}{R + r}$$

If E is in volts and r and R in ohms the current will be in amperes. For two cells in parallel,

$$I = \frac{E}{R + \frac{r}{2}}$$

For two cells in series,

$$I = \frac{2E}{R + 2r}$$

(Giambattista A, Richardson R, Richardson RC, Richardson B (2003) College physics. McGraw Hill, New York).

Curtain Coating *n* A method of applying paint to an object by moving it through a falling curtain of paint which may be used with low viscosity plastics or solutions, suspensions, or emulsions of plastics in which the substrate to be coated is passed through and perpendicular to a free flowing liquid "curtain" or "waterfall."

Curtaining Syn: ▶ Sagging.

Curvature of Field *n* The image plane formed by a lens is naturally curved. While one part of the field will be in good focus, the rest will need refocusing to be sharp. While the eye may partially correct for this, a camera lens will not, and the final image as photographed will not be in perfect focus over the entire image plane.

Cushion Back Carpet *n* (1) A unit of yarn number. The number of 100-yard lengths per pound avoirdupois of asbestos yarn or glass yarn, or the number of 300-yards lengths per pound avoirdupois of woolen yarn. (2) A length of woven cloth. (3) The number of needles per inch on a circular-knitting machine. A machine with 34 needles per inch is a 34-cut machine, and a fabric produced thereon is called a 34-cut fabric.

Cushion Carpet A carpet with padding made as an integral part of the backing.

Custom Color *n* Special colors made by adding colorant to paint or by intermixing colors, which permits the retailer to match a color selected by the consumer.

Custom Molder *n* (Brit: trade moulder) A firm specializing in the molding of items or components to the specifications o another firm that handles the sale and distribution of the item, or incorporates the custom-molded component into one of its own products.

Cut *n* (1) An expression commonly used to designate a typographic printing plate. (2) To dilute with ink, lacquer or varnish with solvents or with clear base; to thin. (3) The proportion of shellac gum in alcohol. A shellac varnish is referred to as a four pound or five pound cut, or weight, which means 4 lbs or 5 lbs of shellac gum to 1 gal of alcohol. (4) In the fiber industry, including glass and asbestos, the number of 100-yd lengths of fiber per pound. A now deprecated unit, 1 cut corresponds to a lineal density of 0.0049604 kg/m or 4,960.4 texes.

Cut-Layers As applied to laminated plastics, a condition of the surface of machined or ground rods and tubes and of sanded sheets in which cut edges of the surface layer or lower laminations are revealed (ASTM D 883).

Cut-off *n* (flash groove, pinch-off) In compression molding, the line where the two halves of a mold come together, often along a sharp mating ridge and groove.

Cut-off Saw (traveling cut-off) In extrusion of pipe, rod, and profiles, a circular saw that periodically swings forward, while moving downline at the same rate as the extruded product, to cut it into desired lengths. When the saw has completed its cut, it swings back at the same time quickly reversing travel to return to its starting point, poised for the next cut.

Cut Pile A pile surface obtained by cutting the loops of yarn in a tufted or woven carpet.

Cut Selvage *n* A cut or break occurring only in the selvage. A cut selvage is caused by incorrect loom adjustment during weaving or improper edge construction. The term also refers to loose edges cut during shearing of the fabric.

Cut Staple *v*(1) An inferior cotton fiber that was accidentally cut because it was too damp during ginning. (2) A term sometimes used to denote staple of manufactured fibers.

Cut Tape See ▶ Slit Tape.

Cutter (1) A mechanical device used to cut tow into staple. (2) A firm engaged in making up garments from finished fabrics. (3) A person employed in the wholesale garment industry whose specific work is to cut layers of fabric to be formed into garments. Cut Velvet: See ▶ Beaded Velvet.

Cutting In *v* Painting of a surface adjacent to another surface which must not be painted, for example, painting the frames of a window and avoiding painting the glass, or the painting of an area on a previously painted surface.

Cut Yarn A defective yarn, i.e., cut partially or completely through, resulting from malprocessing

Cyanoacrylate Adhesives *n* alkyl -2-cyanocrylates polymerize rapidly via anionic initiation in the presence of weak bases (water, alcohol) at ambient temperatures; highly exothermic reaction yielding brittle polymers; very useful for quickly setting anaerobic adhesives; also useful for suture less topical tissue adhesives (e.g., Dermabond®); and industrial grades are commonly known as "super glue."

Cyanoguanidine See ▶ Dicyandiamide.

Cyanuric Acid \ ˌsī-ə-ˈnúr-ik\ *n* [*cyan-* + *urea*] (1838) (1,3,4-triazine-2,4,6-triol, tricyanic acid, tricarbimide) An acid evolved from the blowing agent, azodicarbonamide, when it decomposes. The acid is corrosive, and is the chief cause of plate-out on components of extruders in the structural-foam process when azodicarbonamide is used as the blowing agent.

Cycle *n* The series of sequential operations entering into a repeating batch process or part of the process. In a molding operation, cycle time is the average time elapsing, over several normal cycles, between a particular occurrence in one cycle and the same occurrence in the next cycle.

Cyclic *adj* All cyclic compounds are recognized by their ring structure. Benzene is the simplest and most well-known cyclic compound, consisting of a single ring. Its monocyclic. When two or more rings are involved as with naphthalene, for example, the compound is bicyclic, tricycle, etc. Cyclic compounds may contain from three to six or more carbon atoms in the ring, e.g., Cyclopropane...3 atoms; Cyclobutane...4 atoms; Cyclopentane...5 atoms; Cyclohexane...6 atoms.

Cyclic Diolefins Resins *n* Polymerized products obtained from the cracked distillates of petroleum. They are unsaturated products, with excellent solubility, acid, alkali, and water resistance. A special application is in metallic paint media.

Cyclic Ketone Resins *n* Cyclohexanone is the primary constituent of these resins, and it is condensed either by the action of alkalis or acids. Mixed condensation products of cyclohexanone and aldehydes are also resinous. The cyclic ketone resins which are available commercially are notable chiefly by reason of their extremely pale colors and characteristic smell. They are usually very readily soluble in vegetable oils.

Cyclic Stress Strain *n* Repeated loading of a yarn on a tensile testing machine and the determination of the physical properties of the yarn during these cycles.

Cyclic Trimmer *n* Strictly, a polymer, in cyclic form, that contains three repeating groups. Cyclic trimmer is a by-product found in all commercial polyester and results in deposit buildup in package-dyeing equipment.

Cyclized Rubber *n* A thermoplastic resin produced by reacting natural rubber with stannic chloride or chlorostannic acid. This causes a reduction of the unsaturation and formation of condensed ring structures, typically with two or three rings being fused together. It is claimed (U.S. Patent 3,205,093) that a solution of cyclized rubber in toluene is one of the few lacquers known to adhere to polyolefins without their being pretreated or the lacquered surface being post-treated with radiation. Other uses include films and hot-melt coatings.

Cycloalkane A cyclic hydrocarbon with the general formula C_nH_{2n}.

Cyclohexane *n* \ ˌsī-klō-ˈhek-ˌsān\ *n* [ISV] (ca. 1909) (hexamethylene, hexahydrobenzene) C_6H_{12}. A saturated hydrocarbon with a six-membered ring, cyclohexane is derived from the catalytic hydrogenation of benzene, and is used as a solvent for cellulosics and as an intermediate in the production of nylon.

Cyclohexanol Acetate *n* CH₃COOC₆H₁₁. A nonflammable solvent for cellulosics and many other resins.

Cyclohexanone \- ▎hek-sə- ▎nōn\ *n* (ca. 1909) (pimelic ketone, ketohexamethylene) CH₂(CH₂)₄C=O. A colorless liquid produced by the oxidation of cyclohexane or cyclohexanol. Its most important use is for the manufacturer of adipic acid for nylon 6/6, and caprolactam for nylon 6. It is also an excellent high-boiling, slowly evaporating solvent for many resins including cellulosics, acrylics and vinyls. It is one of the most powerful solvents for PVC, and is often used in lacquers to improve their adhesion to PVC.

Cyclohexene Oxide *n* C₆H₁₀O. This epoxide is a highly reactive, colorless liquid that resembles ethylene oxide in most of its reactions. It is useful as an intermediate in the production of many organic chemicals used in plastics. Its epoxide structure is especially useful in applications where an HCI scavenger is required.

Cyclohexnol *n* (hexahydrophenol) CH₂(CH₂)₄CHOH. A colorless, viscous liquid prepared by the oxidation of cyclohexane or by the hydrogenation of phenol. It is used as an intermediate in the production of nylon 6/6, it is a solvent for cellulosic resins, and it is an intermediate in the manufacture of phthalate-ester plasticizers.

Cyclohexyl Methacrylate *n* H₂C=C(CH₃)COOC₆H₁₁. A colorless monomer, polymerizable to resins for optical lenses and dental parts, and useful for potting electrical components.

Cyclohexyl Stearate *n* C₆H₁₁OOCC₁₇H₃₅. A plasticizer for polystyrene, ethyl cellulose, and cellulose nitrate.

Cycloparaffins \- ▎par-ə-fən\ *n* (1900) Ring compounds of saturated hydrocarbon type based on groupings of methylene radicals (CH₂). Typical cycloparaffins are cyclopropane, cyclobutane, cycloheptane, etc. The cycloparaffins have very good solvent properties, and are constituents of crude petroleum's. *Known also as Naphthenes.*

Cyclopentane *n* (pentamethylene) C₅H₁₀. A solvent for cellulose ethers.

Cyclopolymerization *n* Polymerization in which ring structures are formed within the polymer chain.

Cyclotron \ ▎sī-klə- ▎trän\ *n* [*cycl-* + *-tron*; fr. the circular movement of the particles] (1935) The magnetic resonance accelerator for imparting very great velocities to heavier nuclear particles without the use of excessive voltages.

Cycolac *n* ABS, manufactured by Marbon, U.S.

Cycolon ABS, manufactured by Marbon, U.S.

Cylinder *n* (1) In carding, a large cast iron shell, with an outer diameter of 40–45 in., completely covered with card clothing on the surface. The shell is mounted rigidly on a shaft which projects at each end to rest in bearings. The cylinder must be accurately balanced since it rotates at speeds of 160 revolutions per minute and higher. (2) The main roll, or pressure bowl, on roller printing machines. The engraved rolls that apply color are arranged around the cylinder. (Also see ▶ Printing, ▶ Roller Printing.) (3) A slotted cylindrical housing for the needles in a circular-knitting machine. The number of slots per inch in the cylinder determines the cut of the machine. (4) See ▶ Drying Cylinders.

Cylinder Loading *n* Fibers imbedded so deeply in the wire clothing on a card cylinder that they resist transfer to the doffer cylinder according to the normal fiber path through the card. Causes include improper finish, excess moisture, or static on the fiber. The fiber builds up to such an extent that the carding operation is adversely affected. In extreme cases, the card will be slowed or stopped.

Cylindrical Weave *n* A type of woven, knitted, or braided sleeve generated as reinforcement in tubular reinforced-plastic structures such as pipe. Typically, the reinforcement will constitute about half the volume of the finished product; but it can be much less, as in knit-reinforced garden hose.

Cymatic Printing *n* This proprietary process owned by KBC is a method in which the oscillations of a musical chord are "caught" on a quartz plate and the vibration patterns photographed. The patterns thus obtained are used in making unique print fabrics of unusual variety and originality (Vincenti R (ed) (1994) Elsevier's textile dictionary. Elsevier Science and Technology Books, New York).

Cynoper See ▶ Mercuric Sulfide.